Biomathematics

Volume 1

Edited by

K. Krickeberg · R. C. Lewontin · J. Neyman
M. Schreiber

Mathematical Topics in Population Genetics

Edited by

Ken-ichi Kojima

With 55 Figures

Springer-Verlag Berlin·Heidelberg·New York 1970

ISBN-13:978-3-642-46246-7   e-ISBN-13:978-3-642-46244-3
DOI: 10.1007/978-3-642-46244-3

Title No. 7880.

Preface

A basic method of analyzing particulate gene systems is the probabilistic and statistical analyses. Mendel himself could not escape from an application of elementary probability analysis although he might have been unaware of this fact. Even Galtonian geneticists in the late 1800's and the early 1900's pursued problems of heredity by means of mathematics and mathematical statistics. They failed to find the principles of heredity, but succeeded to establish an interdisciplinary area between mathematics and biology, which we call now Biometrics, Biometry, or Applied Statistics.

A monumental work in the field of population genetics was published by the late R. A. Fisher, who analyzed "the correlation among relatives" based on Mendelian gene theory (1918). This theoretical analysis overcame "so-called blending inheritance" theory, and the orientation of Galtonian explanations for correlations among relatives for quantitative traits rapidly changed. We must not forget the experimental works of Johanson (1909) and Nilsson-Ehle (1909) which supported Mendelian gene theory. However, a large scale experiment for a test of segregation and linkage of Mendelian genes affecting quantitative traits was, probably for the first time, conducted by K. Mather and his associates and Panse in the 1940's.

By 1930 or thereabout, R. A. Fisher and J. B. S. Haldane in England, and Sewall Wright in the United States of America were identified as the founders of population genetics. A series of papers was published by Haldane on mathematical analyses of selection, mutation, etc., in populations of Mendelian genes, and the most of these analyses were assembled in his *The Causes of Evolution* (1932). Fisher's book called *The Genetical Theory of Natural Selection* was published in 1930. The inspiration one gets from reading these books is unmeasurable.

Sewall Wright has been contributing extensively to the development of population genetics ever since the publication dealing with the inheritance of coat color characteristics in guinea pigs (1916). In 1921, he published five papers in a row concerning "Systems of Mating". Then came a historical paper by Wright *Evolution in Mendelian Populations* in 1931. In this paper he introduced the concept of "Effective Population Size", and "Genetic Random Drift" in finite size populations. Undoubtedly, a three-volume series by Sewall Wright, of which the first volume appeared in 1968, will make another cornerstone in population genetics.

I am extremely grateful to have Professor Sewall Wright as one of the contributors in this volume.

There are dozens of very capable investigators whose backgrounds vary. Just mentioning several, Samuel Karlin and his group at Stanford University with *bona fide* mathematics backgrounds. There is Walter Bodmer, also at Stanford University, who is a brilliant investigator in both theoretical and experimental genetics. There are Oscar Kempthorne, Howard Levene, R. E. Comstock, Everett Dempster and many others with statistical backgrounds. However, there is a limit for this kind of undertaking, assembling authors, topics and matching them. Consequently, I had to reluctantly drop several investigators from a list of potential contributors, letting a higher priority go to those with biological backgrounds.

My original aim of this book was to document major issues and ideas that appeared during the development of population genetics in the past half century. However, some papers in this book are the results of original researches, and others are reviews of some aspects of population genetics.

While I have been editing this book, I have begun to feel a new trend and "revival" (according to I. M. Lerner) in population genetics. The new generation of population geneticists recognizes various factors such as population density, non-constant fitness values of a given genotype including frequency-dependent selection, biochemical aspects of population genetics, and ecological conditions for a population, and so forth. Thus, I am convinced that the studies of genetic populations have entered into a new phase, and this is a good reason to document various mathematical studies made up-to-date by geneticists in one volume.

Several authors in this book dedicated their papers to Professor Th. Dobzhansky for his seventieth birthday.

I am personally grateful for the encouragement and leniency for the time schedule for publication of this book given by Dr. Klaus Peters of Springer-Verlag, and the National Institutes of Health of the United States for their summer supports through the project program grant GM 15769. My personal thanks go to Mrs. JoBeth Porterfield who typed up almost all of the manuscripts in the final form and assisted with editorial work. I also acknowledge Dr. Yoshiko Tobari and Mrs. Myrtle Wing who supervised my laboratory work while I was heavily involved in editorial work.

Austin, September 1970   Ken-ichi Kojima

Table of Contents

List of Contributors

C. CLARK COCKERHAM, Department of Experimental Statistics, North Carolina State University, Raleigh, North Carolina/USA.

JAMES F. CROW, Laboratory of Genetics, Genetics Building, Madison, Wisconsin/USA.

WILLIAM G. HILL, Institute of Animal Genetics, Edinburgh/Scotland.

MOTOO KIMURA, National Institute of Genetics, Mishima/Japan.

KEN-ICHI KOJIMA, Department of Zoology, University of Texas, Austin, Texas/USA.

RICHARD LEVINS, Committee on Mathematical Biology, University of Chicago, Chicago, Illinois/USA.

R. C. LEWONTIN, Committee on Mathematical Biology, University of Chicago, Chicago, Illinois/USA.

C. C. LI, Graduate School of Public Health, University of Pittsburgh, Pittsburgh, Pennsylvania/USA.

OLIVER MAYO, Department of Genetics, University of Adelaide, Adelaide/South Australia.

R. H. RICHARDSON, Department of Zoology, University of Texas, Austin, Texas/USA.

ALAN ROBERTSON, Institute of Animal Genetics, Edinburgh/Scotland.

HENRY E. SCHAFFER, Department of Genetics, North Carolina State University, Raleigh, North Carolina/USA.

JOHN A. SVED, School of Biological Sciences, University of Sydney, Sydney/Australia.

JOHN R. TURNER, Department of Biology, University of York, Heslington/England.

SEWALL WRIGHT, Department of Genetics, University of Wisconsin, Madison, Wisconsin/USA.

Random Drift and the Shifting Balance Theory of Evolution[*][**]

S. WRIGHT

Introduction

The word "drift" has been used rather frequently in population genetics, but not always in the same sense. The author is probably responsible for its introduction by using it in papers in 1929, though with no intention of giving it a technical meaning. It has, however, come to be so used and thus is in need of careful definition.

My first use of "drift" in connection with population genetics was in a paper on R. A. Fisher's (1928) theory of the evolution of dominance of type alleles of deleterious mutations.

"In consequence of these facts there will be a gradual drift of the heterozygote toward wild type" (Wright, 1929 b).

"Drift" referred here to the results of a directed process, selection. It was used in a second note on the same subject in a different sense.

"Being random, such variations largely neutralize each other but there is a second order drift which cannot be ignored" (Wright, 1929 c).

In this case "drift" referred to the cumulative effects of a random process, accidents of sampling. I continued to use the term occasionally, in such expressions as "drifting at random" and it was taken up by others as a technical term in this second sense. I adopted it in a later paper (Wright, 1956 b) but pointed out the importance of always distinguishing "steady drift" and "random drift" and in the latter case of specifying whether from accidents of sampling or otherwise (as from fluctuations in selection pressure). Cain and Currey (1963) have suggested the convenient term "sampling drift" for the component due to accidents of sampling.

The Shifting Balance Theory

Most of the use of "random drift" has been in connection with the shifting balance theory of evolution proposed in 1929a (more fully in 1931 and 1932) and much of the confusion in its use seems to have been

* Paper No. 1203 from the Department of Genetics, University of Wisconsin.

** Aid is acknowledged from grant no. G-4920 from the National Science Foundation.

due to misunderstanding of the latter. In this theory, random processes (not necessarily sampling drift) play an essential role but only as an adjunct of selection. Many of those who have referred to the theory have insisted that random drift (specifically sampling drift) had been proposed as an alternative to selection, instead of merely as a source of raw material for selection at a higher level than that provided by mutation. In order to clarify this matter, it is desirable to begin with the major premises of the theory since in fairly recent papers some of these have been asserted to have been the opposite of what they actually were.

Premises

A. Multiple Factor Theory of Quantitative Variability. The aspect of evolution that is considered primarily is that of adaptive transformation based on the cumulative effects of changes in the frequencies of multiple minor factors, apparently responsible at any given time for merely quantitative variability. Species are assumed to carry numerous isoalleles at most loci, at fairly high frequencies, instead of being almost homallelic for a single type allele at each. The occasional utilization of major mutations is by no means excluded, but it is assumed that the establishment of one in the species depends on an adaptive process in the array of modifiers.

B. Universal Pleiotropy. The ramifications from the primary effect of any gene replacement through the network of metabolic reactions and of intercellular developmental processes to the characters which have selective differences, insure that each such replacement have effects on numerous characters (cf. Wright, 1963a).

C. Multiple Selective Peaks. The occurrence of joint reactions in the above networks implies extensive interaction effects among the effects of genes on the various characters. Moreover, all characters contribute to a single one, "selective value". Because of pleiotropy and interaction, there is always the potentiality for a vast number of different more or less harmonious combinations. These correspond to different "selective peaks" in the "surface" of selective values relative to the multidimensional field of gene frequencies. This contrasts with the *single* best type which would necessarily be present if each gene were favorable or unfavorable in itself. Finally, even where there is apparently additivity with respect to a quantitatively varying character, the usual intermediacy of the optimum grade insures the potentiality for a great many selective peaks, and pleiotropy insures that these be at diverse values.

D. Multiple Partially Isolated Demes. Most species contain many small, random breeding local populations (demes) that are sufficiently isolated, if only by distance, to permit differentiation of their sets of gene frequencies, but that are not so isolated as to prevent the gradual spreading of favorable gene complexes throughout the species from their centers of origin. This differentiation need not be associated with conspicuous phenotypic differences.

Phases of the Evolutionary Process

Change of gene frequency is treated as the elementary evolutionary process since it permits reduction of the effects of all factors to a common basis.

According to the theory built on the above premises, each significant step in the evolutionary process involves, in general, the conjunction of three phases. In principle, these would permit a continuing process as long as there is any ecological opportunity for this, with very little mutation and without necessarily involving any environmental differences among localities or any systematic environmental changes in time. In the first statement of these three phases, no such differences or changes are assumed.

1. Phase of Random Drift. In each deme the set of gene frequencies drifts at random in a multidimensional probability distribution about the equilibrium characteristic of a particular selective peak. The set of equilibrium values for each gene frequency is the resultant of three sorts of "pressures", those due to recurrent mutation, to recurrent immigration from other demes, and to selection. The fluctuations responsible for the random drift are in this special case the cumulative effects of accidents of sampling.

2. Phase of Mass Selection. From time to time, the set of gene frequencies drifts across one of the many two-factor saddles in the probability distribution in a deme. There ensues a period of relatively rapid change, dominated by selection among individuals (or families), until the set approaches the equilibrium associated with the new selective peak about which it now drifts at random, and thus returns to the first phase, but in general at a higher level.

3. Phase of Interdemic Selection. A deme that comes under control of a selective peak, superior to those controlling the neighboring demes, produces a greater surplus population and by excess dispersion systematically shifts the positions of equilibrium of these toward its own position, until the same saddle is crossed in them and they move autonomously to the same peak. This process spreads in concentric circles. Two such circles spreading from different centers may overlap and give rise to a new center which combines the two different favorable interaction systems and becomes a still more active population source. The virtually infinite field of interaction systems may be explored in this way with only a small number of novel mutations as alleles which had been rare, largely displace the previously more abundant ones.

There is a still higher level of adaptation at which a high mean rate of reproduction becomes an inadequate criterion of success. Such a rate may lead to overpopulation which threatens the persistence of the group by overcrowding and exhaustion of resources. In the long run, the most successful portions of the species are those in which the genetic system favors reproductive rates in balance with resources.

Effects of Fluctuating Environmental Change

The preceding model is unrealistic in postulating that there are no environmental changes of sufficient duration to be significant. Long period fluctuations in conditions, occurring more or less independently in different demes, simulate the effects of accidents of sampling but differ in being independent of population size. These are best treated mathematically as a component of the random drift within each deme, even though momentarily deterministic (Wright, 1931, 1935b, 1948). The same is true of fluctuating changes in immigration either in amount or quality (Wright, 1948).

Effects of Persistent Local Differences in Conditions

Conditions usually differ systematically in different regions in the range of a species. Thus the prevailing pattern of selective peaks in the species is subject to local modifications. Sufficiently great differences tend to lead to permanent differentiation and to splitting of the species. Even at a low level, most of the adaptive differences are only of local value and do not spread. Occasionally, however, a pattern of gene frequencies, arrived at by a succession of local adaptations, may have adaptive significance everywhere else, although one that could not have been reached except by the particular chain of events at its point of origin. It then spreads throughout the species as in phase 3 (Wright, 1940, 1965). In this case there is merely deterministic change within demes (from selection pressure in the deme of origin). Local differentiation functions in this case, however, as a random process from the standpoint of the species as a whole in supplying complex genetic material for intergroup selection.

Secular Change in Environmental Conditions

If there are secular changes in the conditions to which the species as a whole is exposed, there is a corresponding change in the whole surface of selective values and its peaks. Mass selection of the conventional sort (phase 2) at once takes over and tends to shift the set of gene frequencies toward the most available of the new peaks unless the change is so severe that it leads to extinction. In general, a rough adjustment to the new conditions is brought about. Fine adjustment may be brought about later by exploration of neighboring peaks under the three phase process.

If, however, such environmental changes succeed each other too rapidly for the slow process of fine adjustment, the course of evolution will be controlled wholly by mass selection but progress will be somewhat like that on a treadmill. The species merely holds its own as the acquirement of each new rough adaptation accompanies the undoing of an old one.

Ecologic Opportunity

In all cases, it is assumed that natural selection at one level or other, is the guiding principle, but this leaves it unexplained why some forms have changed only slightly in hundreds of millions of years, others have progressed steadily along rather narrowly restricted lines, while still others have developed with relatively explosive speed and often have branched in many directions to give rise to new higher categories (bradytely, horotely, and tachytely of Simpson, 1945).

The interpretation that fits best the rest of the theory is summed up in the words "ecologic opportunity". Most species are restricted to progress, if any, along a line of increasing perfection of adaptation in the niche which they occupy, because of the occupation of all slightly different niches by other species. Occasionally, however, an opportunity for evolution along one or more different lines may be presented: (1) by arrival of a colony in territory in which niches possible for it are unoccupied or (2) by some degree of accidental preadaptation to a drastically altered environment in which rival forms have become extinct or (3) attainment in the course of specialization for one way of life, of adaptations that happen to open up an extensive new way of life. A new higher category may evolve rapidly by branching out of the species in different directions in different localities, in the exploitation of any one of these opportunities (Wright, 1941 a, b, 1942, 1949).

As the shifting balance theory pertains directly only to the transformation of species as units, the equally important splitting of species will not be considered in detail here (cf. Wright, 1949).

Mathematical Framework

For more precise definition of the concept of "random drift" and its role in evolution, it is necessary to review briefly the mathematical theory of the operation of phases 1 and 2 within demes. The role of unpredictable chances in the operation of phase 3 in the evolution of the species as a whole precludes comparable mathematical treatment of it.

As noted earlier, the systematic pressures on gene frequency are conveniently grouped in three categories: those due to recurrent change in the genetic material itself (mutation pressure), those due to introduction from without (immigration pressure), and a wastebasket category, change in gene frequency without either of these (selection pressure). Selection pressure includes effects of differential viability at any stage, whether with respect to physiology or to adaptation to the external environment, differential emigration, differential mating, differential fecundity, and asymmetrical segregation (meiotic drive, Sandler and Novitski, 1957).

These are measured by the changes, $\bar{\Delta}q$, in gene frequency, q, which they tend to bring about per generation. The total pressure, where all are so slight that order is unimportant, is the sum of the partial pressures. In the simplest cases, with mutation rates v and u to and from the gene in question, replacement of the portion m of the population by immigrants with mean gene frequency Q, and with a momentary net selective advantage, s, of the gene over its alleles, collectively, the total pressure is as follows (Wright, 1931)

$$\bar{\Delta}q = [v(1-q) - uq] - m(q-Q) + sq(1-q) \qquad (1)$$

The various components have been elaborated to deal with situations more realistically. Thus mutation pressure in a system of multiple alleles can be represented by a set of equations of the type $\bar{\Delta}q_x = \sum[(u_{ix}q_i)] - (\sum u_{xi})q_x$ where u_{ix} refers to the rate of mutation of an allele to the gene, A_x, and u_{xi} is the reverse (Wright, 1949). The degree of dominance may be taken care of by expressing selection pressure in the form $(s + tq)q(1-q)$ (Wright, 1937). Much more general forms will be taken up later.

In most cases, some of the pressures are positive, others negative. Selection pressure by itself may involve opposed components. These give balanced polymorphism in one way or other. The net effect is to push gene frequencies toward an equilibrium value, $\hat{q}$, calculated by putting $\bar{\Delta}q = 0$.

Some random processes occur too rarely in the history of the population in question for statistical treatment and are best treated as unique events. There may be a unique introduction of a gene by migrant individuals or by an unusual hybridization. There may be a unique selective incident. The initial crossing of a particular saddle to initiate phase 2 is a unique event. Many random intrademic processes can, however, be represented as contributions to the variance of fluctuations about the equilibrium point per generation (Wright, 1931, 1948, 1956b). Letting N be the effective population size of the deme, σ_s^2 the variance of selective values of the gene in question, σ_m^2 the variance of amounts of replacement by immigrants, and σ_Q^2 the variance of fluctuations in the gene frequency of immigrants.

$$\sigma_{\Delta q}^2 = (1/2N)\, q(1-q) + \sigma_s^2 q^2 (1-q)^2 + \sigma_m^2 (q-Q)^2 + m^2 \sigma_Q^2 \qquad (2)$$

The first term is the contribution from accidents of sampling, the second is that from local fluctuations in the selection coefficient (simplest case), and the third and fourth are those from fluctuations in the amount and quality of immigration respectively. *These fluctuations do not constitute random drift* since they refer to contributions to variance in a single generation while *random drift implies a cumulative process.*

The directed and random processes operate jointly to determine a probability distribution which describes the extent of the random drift of the gene in question about its equilibrium value. Whatever the formulae for Δq and $\sigma_{\Delta q}^2$, this distribution is given by the following (Wright, 1938a, 1945b, 1952) in which C is a constant such that the total probability is one.

$$\phi(q) = (C/\sigma_{\Delta q}^2) \exp\left[2\int (\bar{\Delta} q/\sigma_{\Delta q}^2)\, dq\right], \qquad (3)$$

where

$$\int_0^1 \phi(q)\, dq = 1 .$$

This defines the amount of random drift under the given conditions. While the total pressure per generation on gene frequency, $\bar{\Delta} q$, and the variance of fluctuations, $\sigma_{\Delta q}^2$, also per generation, are sums which can be analyzed into contributions from the various factors; this is not true of the random drift, $\phi(q)$. This makes it impossible to define precisely the term sampling drift as a component if $\sigma_{\Delta q}^2$ is composite, but it may be defined as $\phi(q)$ in the absence of other sorts.

This one dimensional probability distribution does not, however, include the aspect of random drift which makes it of more than trivial significance in evolution: the possibility of crossing a saddle between the currently occupied selective peak and another. It is of significance only as cross-section of the total multidimensional distribution, within which the gene in question is drifting at random, defined by specified gene frequencies at all other interacting loci.

General Formulae for Selection Pressure

The symbol W has been defined as the absolute selective value of a genotype in an interaction system that is under consideration, in the sense that $\bar{W}$ $(= \sum f_i W_i$ where f_i is the frequency of the i-th genotype) is the ratio of effective population size, N, in one generation to that in the preceding generation under the specified conditions with respect to the rest of the genome and the environment (Wright, 1949). Random mating and Hardy-Weinberg frequencies are usually assumed within the deme, and also such weak selective differences within each locus that recombination keeps pace with changes in gene frequency sufficiently well that deviations from random combination among interacting loci may be ignored (Wright, 1942). If it is assumed that the genotypic frequencies in the population do not depart appreciably from Hardy-Weinberg ratios, then the allele frequency change is given by

$$\bar{\Delta}q_x = \frac{q_x(1 - q_x)}{2\bar{W}}\left[\sum W \frac{\partial f}{\partial q_x}\right]. \tag{4}$$

Where there are multiple alleles at a locus, evaluation depends on taking

$$\frac{\partial q_i}{\partial q_x} = \frac{-q_i}{1 - q_x}.$$

The appearance of this formula is deceptively simple. Its use in conjunction with other components of $\bar{\Delta}q$, and those of $\sigma^2_{\Delta q}$ in $\phi(q)$ is not such a gross oversimplification in principle as has sometimes been alleged. Thus in the case of four alleles at each of 100 loci, a rather simple system, the summation involves 10^{100} terms (a larger number than the number of elementary particles in the known universe). It does not, indeed, deal adequately with some types of chromosome aberration, or with such strong selective differences that the assumption of random combination is seriously invalidated, and does not take account of unique events or of cytoplasmic heredity. It is also, of course, intended to apply only to intrademic selection.

Obviously calculations can be made only from rather simple models, involving only a few loci, or simple patterns of interaction among many similarly behaving loci. In such calculations it is usually most convenient to deal with selective values that are relative to some standard, W_{st}, usually assigned to a particular genotype, $w = W/W_{st}$ (Wright, 1950). Since $w/\bar{w} = W/\bar{W}$

$$\bar{\Delta}q_x = \frac{q_x(1 - q_x)}{2\bar{w}}\left[\sum w \frac{\partial f}{\partial q_x}\right]. \tag{5}$$

Apart from application to simple systems, the greatest significance of this general formula is that its form brings out properties of systems that would not be apparent otherwise. This is especially the case where it can be assumed that constant values can be attributed to all genotypes as wholes (no frequency dependence). While probably never strictly true, this should often be a good approximation with respect to competition of individuals in relation to the external environment, though not where social interactions are involved. With constant w's, Eq. (5) reduces to the following, noting that the standard may often be chosen so that $\bar{w}$ in the denominator may be treated as 1 (Wright, 1935a, 1937).

$$\bar{\Delta} q_x = \frac{q_x(1-q_x)}{2\bar{w}} \frac{\partial \bar{w}}{\partial q} = (1/2)q_x(1-q_x)\frac{\partial \log \bar{w}}{\partial q_x}. \tag{6}$$

Where this formula holds, the point in the multidimensional "surface" of mean selective values, relative to the set of gene frequencies occupied by the deme, tends to move up the gradient, $\dfrac{\partial \bar{w}}{\partial q_x}$ or more accurately $\dfrac{\partial \log \bar{w}}{\partial q_x}$ except as qualified by the term $q_x(1-q_x)$ and by other pressures, not introduced into it.

This surface may have multiple selective peaks at each of which all Δq's are zero either because at a maximum, $\dfrac{\partial \bar{w}}{\partial q_x} = 0$, or because the gene is lost ($q_x = 0$) or fixed ($q_x = 1$).

A population that deviates from random combination within and among loci and so is not on this surface, for example F_1 of a cross between inbred lines, rapidly moves toward it, if the conditions assumed here are met, and on reaching it, moves along it until it comes to rest at a selective peak, except as diverted by other pressures. A "determinative" peak with respect to all pressures (mutation and immigration in addition to selection) has the property that Lerner (1954) has called genetic homeostasis. The population tends to return to it after any small displacement. After a large displacement, however, it may come to rest at a different peak.

The Multidimensional Distribution of Gene Frequencies

The formula for the total multidimensional probability distribution has not been obtained in terms of values of $\bar{\Delta} q$ and $\sigma^2_{\Delta q}$ of the various genes but some fairly general formulae have been given (Wright, 1937, 1949). For example, with given effective N and m for the population, Q for each gene and constant W for each genotype

$$\phi(q_1, q_2, \ldots, q_n) = C\overline{W}^{2N} \Pi_i[q_i^{4NmQ_i-1}]. \tag{7}$$

The product term has a factor for each allele at each locus.

This formula brings out at a glance how the multiple peaks in the surface of mean selective values, $\overline{W}$, tend to be reflected in greatly exaggerated form in multiple peaks in the probability distribution. It is obvious that there is only a very small chance that the set of gene frequencies will drift across any but an exceedingly shallow saddle from one peak to another unless the population passes through bottlenecks of extremely small size. Moreover, there cannot be much random drift unless $4Nm$ is small. The conditions under which effective N and effective m are much smaller than their apparent values will be considered later.

In general, the distribution for an interacting set of loci is restricted over a long period of time (if conditions are constant) within a shell of extremely low probabilities to the neighborhood of a single selective peak. This portion of the total multidimensional distribution is what constitutes the random drift.

Such a definition may seem unnecessarily complicated. It should be emphasized again, however, that no appreciable evolutionary significance has been attributed to the random drifting of a single gene frequency about its equilibrium point but only to such drifting of a set of many gene frequencies under conditions that permit occasional escape across a two-factor saddle.

A Simple Illustration

The interplay of random drift and intrademic selection may perhaps be grasped most easily from a graphical representation (Wright, 1963b). Figs. 1 to 4 give a highly simplified illustration of these processes in a single deme. It is assumed (Fig. 1) that the grade of the character in question is determined additively, without dominance, by four equivalent pairs of alleles, but that there is an optimum at the mid-value, converting this into a case of extreme factor interaction with respect to selective value. There are only six selective peaks here instead of the indefinitely large number expected if all characters are involved.

It is further assumed that there are small pleiotropic effects of two of the plus factors (A, B) which cause the peaks to be at three levels, a lowest (homallelic "CD" or, in full, $aabbCCDD$), four intermediate, and a highest (homallelic $AABBccdd$). Fig. 2 shows the relative selective values of the 16 possible homallelic populations. The system of gene frequencies is four dimensional.

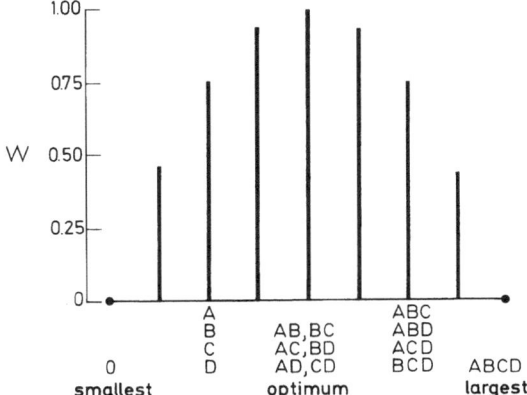

Fig. 1. Selective values (*W*) of genotypes with equivalent effects on a quantitative character, under the assumption that *W* falls off according to the square of the deviation from the midpoint. The letters refer to homozygous genotypes

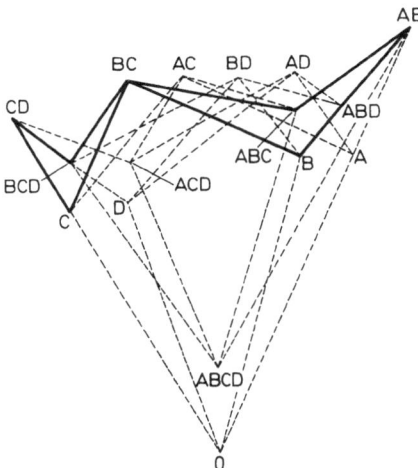

Fig. 2. Selective values (vertically) of the 16 homallelic populations in the system of Fig. 1 on the same scale, with the additional assumption that genes *A* and *B* have additive pleiotropic effects (0.0625 for each replacement of a or b). The three peaks dealt with in Figs. 3 and 4 are connected by solid lines

Fig. 3 shows trajectories of population change on two of the faces of this four dimensional system: that connecting the lowest selective peak with one of the intermediate ones, and that connecting the latter with the highest one. Calculations are based on formulae 16 and 19 in Wright

S. Wright

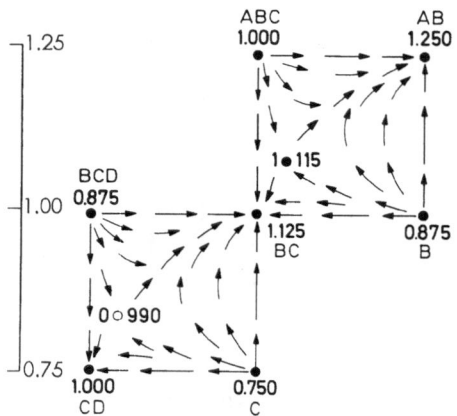

Fig. 3. Trajectories of gene frequency systems on surfaces of mean selective values on two faces of the 4-dimensional field

(1935a). It should be noted that for purposes of illustration, much greater differences are shown in selective values than would be probable in a real case. It must be assumed, moreover, that there is enough mutation, or immigration, from other demes to prevent complete fixation at any peak.

Assume that for historical reasons the deme starts in the neighborhood of the lowest peak (*CD* or, in full, *aabbCCDD*). Selection tends to hold it there in a four dimensional distribution, its random drift, of which only two dimensions are indicated. In the course of a very long time, an extreme joint deviation occurs with respect to the frequencies of genes *B* and *d*, that carries the set across the saddle at $q_B = 1/3$, $q_d = 1/6$, at which $\bar{W}$ is 99% of the peak value. The set now moves rapidly under direct selection toward the intermediate peak (homallelic *BC* or, in full, *aaBBCCdd*) at which $\bar{W}$ is 12.5% greater than at the low peak. After much random drifting about this peak, another extreme joint deviation carries the set across the shallow saddle at $q_A = 1/3$, $q_c = 1/6$, at which $\bar{W}$ is 1% below the intermediate peak and again mass selection takes over and rapidly carries the set to the neighborhood of the highest peak at which $\bar{W}$ is 25% above its value at the lowest peak.

The values of $\bar{W}$ along the easiest path are shown in profile in Fig. 4, which thus brings out the alternation between the very slow phase of random drift and the rapid one of intrademic selection. In the case illustrated, the process in the deme ceases once the peak at *AB* (in full *AABBccdd*) has been attained. It could continue indefinitely if the number of peaks is always large and new ones arise from time to time as novel mutations replace alleles that are lost.

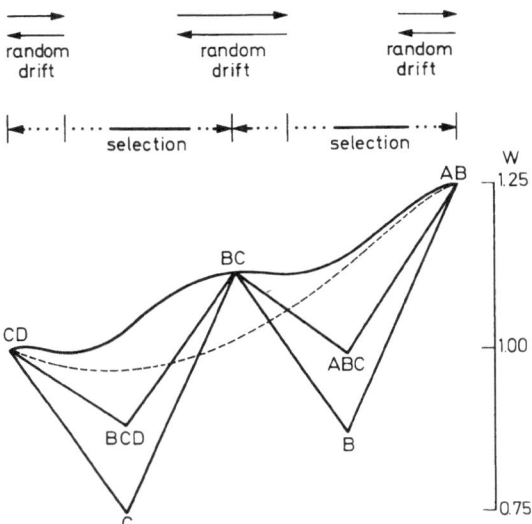

Fig. 4. Profile of mean selective values along path from lowest to highest peak through one of the four intermediate peaks (curved solid line) on same scale as Fig. 3. There is more depression along a path (broken line) avoiding all of the intermediate peaks

As the figures represent only one deme, they cannot represent the third phase of the evolutionary process, interdemic selection. This is represented in Fig. 5. Assume that there are many demes, all of which start from the lowest peak (CD). The species exhibits a pattern of trivial local differences due to independent random drifting in the demes.

Three are indicated as overcoming the low probability of crossing a saddle leading to establishment of control by one of the intermediate peaks. Excess growth of the population of these successful demes, followed by excess dispersion tends to carry the neighboring demes across the same saddle. All of these demes tend to shift rapidly, at first merely under immigration from the successful deme, but later under direct internal selection.

Again one may succeed in crossing the second saddle to come under control of the highest peak (AB). The influence of this will in turn spread in a concentric circle, but this highest combination may also be attained where two different ones such as BC and AD overlap.

At a certain stage, the species may show a pattern of strong differentiation of large areas: those still characterized by near fixation of $aabbCCDD$, those characterized by near fixation of $aaBBCCdd$ and other intermediate peaks, and a few that have attained near fixation of

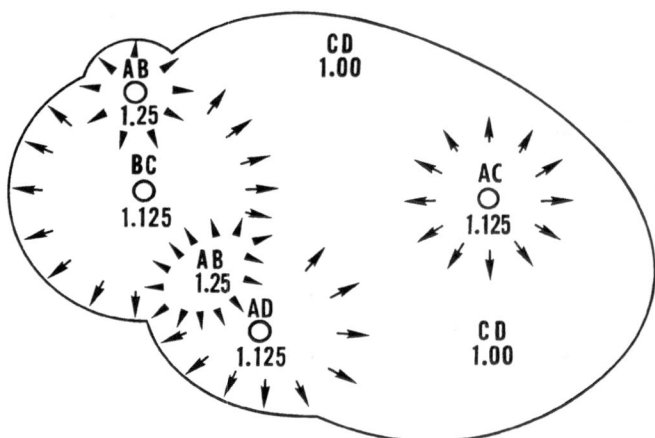

Fig. 5. Hypothetical area occupied by species to which Fig. 1 to 4 apply. Starting with lowest selective peak, CD (in full, homallelic $aabbCCDD$), intermediate peaks BC, AD and AC have been attained by random drift in three demes and have spread more or less. The highest peak, AB, has been attained by random drift from BC, and has also been attained by overlap of BC and AD

the most adaptive interaction system, $AABBccdd$, which is ultimately to become characteristic of the whole species.

There need be no environmental differences whatever among these large areas. An investigator could be excused for supposing that random drift could not be responsible for the large scale pattern of differentiation and that there must be some unknown environmental differences among the areas responsible for selective differentiation.

He would, of course, be right in one sense in attributing the result practically wholly to selection since only a few among hundreds of demes may drift far enough in the course of hundreds of generations to have had any significant effect on the course of evolution.

On the other hand, the nature of the pattern was largely determined by the random drift of these particular demes, and the selection was in relation to different genomes, not to different environments.

Population Structure

The formula for immigration pressure, $\Delta q = -m(q-Q)$, applies directly to a nearly isolated deme in which a small portion, m, is replaced in each generation by a random sample from the whole species, with equilibrium gene frequency, Q. The variance of gene frequencies established in a group of such "islands" of size N, has been shown to be about $\sigma_q^2 = Q(1-Q)/(4Nm+1)$ (Wright, 1931). Since Nm is the number of immigrants per generation, presumably rather large, this variance may appear to be too small to be significant. This, however, does not allow for the likelihood that effective N and effective m may be much smaller than actual N and m. With respect to m, immigrants would usually come largely from neighboring demes which differ much less in gene frequency from the one in question than does the average for the species as a whole, so that effective m, relative to the latter, is much less than the actual replacement by immigrants. There are various ways which may cause effective N (here called N_e) to be much smaller than actual N. N_ϵ refers only to sexually mature individuals. Again, with an unequal sex ratio, $N_e = 4N_m N_f/(N_m + N_f)$ where N_m and N_f are the numbers of males and females (Wright, 1931). With many more females than males this approaches $4N_m$. Another cause of reduction of N_e is inequalities in fecundity beyond that due to chance, $N_e \approx 4N/(2 + \sigma_k^2)$ in a static population in which each individual has on the average two offspring ($\bar{k} = 2$) that reach maturity with variance σ_k^2. With random (Poisson) variation, $\sigma_k^2 = 2$, giving $N_e = N$ but N_e is smaller if σ_k^2 is greater (Wright, 1939a). The most important reason for small N_e, however, is undoubtedly passage of the population through bottlenecks of small size. The effective number is given approximately by the harmonic mean of the numbers over a series of n generations that is too short for significant change in the gene frequencies by selection. $N_e = 1/\left[(1/n) \sum^{n} (1/N)\right]$ (Wright, 1939). The most favorable condition for a large amount of sampling drift has been described (Wright, 1938b, 1939, 1940) as that in which there are many localities in each of which the resident deme is likely to become extinct but a new one is soon founded by stray individuals from another locality. Under such conditions a large part of the species may trace its ancestry through a long succession of such very small bottlenecks.

At the opposite extreme from the ideal island model of species structure is that in which there is uniform density over a large area but such restricted dispersion that the effective size of "neighborhoods" (areas from which the parents of individuals may be treated as if drawn at random) is very small. The opposition between local inbreeding and dispersion in such cases tends to build up not only great differentiation among the neighborhoods themselves, but also among much larger areas up to a point at which this is prevented by recurrent mutation (or a supplementary means of long-range dispersion). The differentiation of areas consisting of tens of thousands of neighborhoods may theoretically be great if the rate of recurrent mutation is of the order of 10^{-5} or less and effective N of neighborhoods is of the order of 10. The differentiation of areas may be appreciable if of the order of 100, but negligible if of the order of 1000 (Wright, 1940, 1943, 1946, 1951b).

This is in a continuous area. In a linear continuum; the amount of differentiation at all levels is enormously greater (Wright, 1943).

These are ideal situations which in nature would nearly always be complicated by selective differences within and among neighborhoods. The process of the building up of differences among large areas by isolation by distance is so slow that very small selective differences take precedence. It is, however, important to recognize that if extensive differentiation is brought about by temporary selective differences, a large degree of everchanging differentiation tends to be maintained indefinitely under isolation by distance.

An intermediate structure, in which there are numerous large clusters with very restricted dispersion between them, is probably more frequent than either the ideal island pattern or a uniform continuum. The effective neighborhood size here is not the population number of the cluster but the increment per generation in the size of ancestral populations based on the amount of dispersion between clusters. Sampling drift can easily play an important role with this structure (Wright, 1951 b).

Deviations from Random Combination

It was noted that formulae 4 to 6 for selection pressure in the presence of interaction, depended on the assumption of random combination. It has been shown that there are no serious departures from their validity if the interactive selection is of lower order than the amount of recombination as is usually the case for genes in different chromosomes or with loose linkage in the same chromosome (Wright, 1945a, 1952, 1965). Even where there is considerable departure from random combination, the pattern of selective values, with numerous selective peaks, holds qualitatively so that the theory is not seriously affected in qualitative terms. There is a "surface of mean selective values" but the populations on it deviate somewhat from random combination (Wright, 1967).

Mean Selective Value and Fitness

The differences in the implications of Eq. (6) for constant genotypic selective values and (4) or (5), where they do not reduce to (6) because of frequency dependent selective values, are important in the theory. The meaning in the latter case is less intuitively obvious than in the case of (6). The nature of the difference may be seen by writing the term $\sum \left(W_i \dfrac{\partial f_i}{\partial q_x} \right)$ in the form $\left[\dfrac{\partial \overline{W}}{\partial q_x} - \left(\dfrac{\overline{\partial W}}{\partial q_x} \right) \right]$ in which the difference from (6) is wholly in the second component. It is apparent that with frequency dependence, the course of change is not controlled by the gradient of the surface $\overline{W}$ to the extent that it is under constancy of the genotypic selective values. The population does not in general tend to come to rest at the selective peaks of $\overline{W}$ in so far as controlled by selection pressure.

In some cases the expression $\sum \left[(W/\overline{W}) \dfrac{\partial f}{\partial q} \right]$ can be integrated giving an expression $F(W/\overline{W})$ which defines a surface relative to the sets of gene frequencies in the same way that $\overline{W}$ (or more precisely $\log \overline{W}$) does if the W's are constant, but it may be a very different surface.

$$\overline{\Delta q_x} = (1/2)\, q_x (1 - q_x) \frac{\partial F(W/\overline{W})}{\partial q_x} . \tag{8}$$

It is convenient to distinguish $\overline{W}$ and $F(W/\overline{W})$ where the latter exists at all, as surfaces of "mean selective value" and of "fitness" respectively. The latter term is appropriate because it is its slope that controls the course of mass selection according to Eq. (8).

This fitness function can always be found in the case of a mere pair of alleles. It can also be found in cases in which the *relative* selective values (w's) are constant but the *absolute* ones (W's) are not, either because the competition among individuals is of a sort that has no effect on population size ($W = k(w/\overline{w})$, $\overline{W} = k$, a constant, while $F(W/\overline{W}) = \log \overline{w}$), or because the absolute values all involve the same function, ψ, of one or more of the gene frequencies ($W = \psi w$, $\overline{W} = \psi \overline{w}$ but $F(W/\overline{W}) = \log \overline{w}$ because of cancellation of ψ in $W/\overline{W}$). Finally, if the relative selective value of each genotype is a function of its own frequency that can be expressed in powers of the latter and $\overline{w}$ in the denominator can be treated as 1, integration is possible and gives an expression for $F(w/\overline{w})$ (Wright, 1949, 1960a, 1964).

In most cases, however, there is no such function. Eq. (4) and (5) hold for all gene frequencies in the interaction system but $F(W/\overline{W})$ of Eq. (8) does not exist. Nevertheless, the population tends to move along a determinable course, often of a strongly spiral sort, in the field of gene frequencies, toward a position at which all $\overline{\Delta q}$'s are zero. There may be many such positions which may be called "selective goals" even though not peak values of any definable "surface". The situation is analogous to movement in a vector field in which there is no potential function because of a curl (Wright, 1956b).

Whether $F(W/\overline{W})$ exists or not, the points toward which selection may carry a panmictic population and the selective peaks, with respect to mean selective value may differ greatly if the genotypic selective values are frequency dependent. A genotype that parasitizes its own species may move toward fixation even though its presence is injurious to all individuals in proportion to a function, ψ, of its frequency. As it approaches fixation, the absolute mean selective value, $\overline{W}$, declines and may become less than 1, leading to extinction. Conversely a genotype that benefits the population in proportion to its frequency but at its own expense and thus of low "fitness", tends toward extinction, again with decline in absolute mean selective value (unless there is sufficient benefit to close relatives).

In R. A. Fisher's (1930) fundamental theorem of natural selection, "the rate of increase of fitness of any organism at any time is equal to its genetic variance in fitness at that time", the word "fitness" is used in two senses. In the second sense it is a property of individual genotypes W which have a "genetic" (additive) "variance". In the first sense, it is a property of the population that has a "rate of increase". This, it turns out, is $F(W/\overline{W})$ if the latter exists, and if it does not, the quantity $\left[\dfrac{\partial \overline{W}}{\partial t} - \left(\dfrac{\overline{\partial W}}{\partial t} \right) \right] / \overline{W}$ plays the role of rate of increase of fitness in the theorem (assuming random combination, otherwise there is a residual term) (Wright, 1956 b). Letting W_A be the sum of additive effects of genes,

$$\frac{\partial F(W/\overline{W})}{\partial t} = \sigma^2_{W_A}/\overline{W} \quad \text{if} \quad F(W/\overline{W}) \text{ exists.} \tag{9}$$

In any case

$$\left[\frac{\partial \overline{W}}{\partial t} - \left(\frac{\overline{\partial W}}{\partial t} \right) \right] / \overline{W} = \sigma^2_{W_A}/\overline{W}. \tag{10}$$

It has been generally recognized that "fittest" in the expression "the survival of the fittest" has only a tantologous meaning. In this sense it is appropriate to designate the property of the population that automatically increases under selection as the "fitness function." It seems desirable, however, to use a different term for $\overline{W}$, "mean selective value", a property of the population that may decrease under selection, in order to avoid ambiguity. This implies use of "selective value" rather than "fitness" for the property of individuals of which $\overline{W}$ is the population mean. This involves the paradox, however, that increase in the fitness function of a population may imply decrease in population size ($\partial N/\partial t = (\overline{W} - 1)$ under given conditions) and may even lead to extinction.

Complications in Phase 3

Phase 3 depends on the shifting balance between local selection and immigration as far as each deme is concerned. In the simplest case $\overline{\Delta}q = sq(1 - q) - m(q - Q)$ with equilibrium frequency of the gene, $\hat{q}$. This is related to the values of $s(=\bar{W}-1)$, m and Q approximately as below (Wright, 1931):

	$\lvert s \rvert \gg m$	$\lvert s \rvert = m$	$\lvert s \rvert \ll m$
Favorable selection $(s > 0)$	$1 - \dfrac{m}{s}(1-Q)$	$\sqrt{Q}$	$Q\left[1 + \dfrac{s}{m}(1-Q)\right]$
Unfavorable selection $(s < 0)$	$\dfrac{mQ}{(-s)}$	$1 - \sqrt{1-Q}$	$Q\left[1 - \dfrac{(-s)}{m}(1-Q)\right]$

If a particular deme reaches a high selective peak and becomes a much greater source of emigration than before, gene frequencies in neighboring demes tend to shift from left to right to come under control of the gene frequencies, Q's, in the source.

Much more material for evolutionary change is to be expected in an array of partially isolated demes than in an equally large panmictic population. Various types of balance occur, indeed, in panmictic populations: balancing of adverse selection by recurrent mutation (Haldane, 1927), selective advantage of heterozygotes over both homozygotes (Fisher, 1922), opposition between alleles that are superior in different aspects of selective value if superiority is associated with more than semidominance (Haldane, 1962), selective advantage of whichever allele falls below a critical level in frequency, likely to occur in a region with diverse niches (Wright and Dobzhansky, 1946) and even mutation pressure in the case of almost neutral alleles (Wright, 1931, 1966). These are all somewhat special cases, however, in comparison with the situation in a large array of partially isolated demes. In the latter, selection is practically certain to be occurring in somewhat different directions in different demes, because their environments are somewhat different or, if not, because they have come to occupy different selective peaks. This insures that the species as a whole have more than one allele at high frequencies, while dispersion insures that this be the case in each deme.

These considerations make it probable that population growth will occur unequally throughout the range. There should be an evershifting pattern of population sources and sinks that will insure continual change in composition of the population as a whole and of the selective peaks that predominate. There has, however, been no adequate mathematical theory for the operation of this third phase as a whole and none can be developed because of the role played by unique events. Even a rather small number of strongly heterallelic loci provide the basis for a virtually infinite number of combinations and a very large and changing array of selective peaks. The times and places at which there is a primary shift from control by one to control by another, and what this other will be, are wholly unpredictable matters.

There is a complication in the fact that the selective peaks that are of significance in phase 2 are those with a high fitness function $F(W/\bar{W})$ while those that are of significance in phase 3 are those characterized by rapid population growth and high dispersive and invasive capacities. This depends on more than high mean selective value $\bar{W}$, but the latter is a direct measure of rate of population growth and much more of an indicator of the whole set than is $F(W/\bar{W})$ where there is a difference.

There is little conflict unless selection is strongly frequency dependent. Where there is conflict the course of evolution is presumably a compromise of some sort. In phase 3 there is interference with the spreading through the species of types of individual "fitness" that are deleterious to the group and promotion of the spreading of types of behavior that are beneficial to the group at the expense of the individual, but not necessarily with full success (Wright, 1956b, 1964).

Major Mutations

The local shifts from control by one selective peak to control by another on which phase 3 depends, are likely to occur only across extremely shallow saddles and thus where selective differences are very slight. This accounts for the emphasis (in premise A) on multifactorial quantitative variability. This, however, by no means precludes an important role in evolution of mutations that by themselves present the opportunity for a major step.

It is unlikely that such a mutation will be free from side effects which put it at a net selective disadvantage. Its usefulness depends on a process of domestication in which these side effects are overcome by a suitable combination of modifiers. If this mutation is continually recurring in demes throughout the range of the species, it may encounter somewhere, at some time, a set of modifiers that puts it above the critical line of net selective advantage, whereupon the whole interaction system, major gene and modifiers, starts spreading slowly throughout the species (Wright, 1952, 1963b).

Selection Among Clones

There is an extreme form of the shifting balance theory which should be mentioned. This is the process that occurs in a species with prevailing uniparental reproduction but with an occasional crossing of clones. Crossing gives rise to an enormous variety of recombinant clones. Selection among these, necessarily according to genotypes as wholes, rapidly eliminates all but the superior ones. Crosses among these provide a new array of genc types, which may have some common basis of superiority (Wright, 1931, 1956a).

It is a violent process, likely to eliminate all but one clone in a locality and thus to destroy the balance between recombination and clonal selection on which continued progress depends, apart from rare favorable mutations. It lends itself better to a succession of readaptations to changes in local conditions, of the treadmill sort, than to orderly evolutionary advance of the species as a whole.

Other Theories Based on Chromosomal Heredity

A survey of what is known of the course of evolution in any favorable case indicates that there is no one invariable mode. The relation between theories is thus likely to be one of supplementation, leading to weighing of degrees of importance rather than of complete opposition. There is moreover a contrast between theories that are concerned almost wholly with the kinds of genetic material that play a role, to the exclusion of the dynamics of the process of incorporation, and those concerned primarily with dynamics. The former are, usually, concerned with evidence that certain types of major chromosomal change have actually been important while the latter are usually concerned, as in the case of the shifting balance theory, with minor factors and the possibility that the accumulation of these may lead to major change.

The shifting balance theory has least in common with the theories of abrupt origin of species (de Vries, 1905; Willis, 1940; Goldschmidt, 1940). Major chromosomal mutations seem to have more to do with the splitting of species by reproductive isolation than with adaptive transformation. The adaptiveness of species that arise by amphidiploidy is, for example, largely a reflection of processes that occurred in the parent species. Reciprocal translocations are balanced mutations which similarly are much more significant as steps toward splitting of species than of transformation of characters.

Edgar Anderson (1939) has shown that hybridization followed by "introgression" into one of the parent species by repeated backcrossing has been an important process. This, however, consists in introducing blocks of genetic material that have proved useful in another context, and is a supplement to mutation and not an alternative to theories concerned with the dynamics of incorporation of genetic materials into a harmonious system.

Dobzhansky (1956), and others, have shown that large alternative blocks of genes in one or more chromosomes, within which crossing over is prevented by inversions, are widespread in *Drosophila* species and permit segregation of different "coadaptive" complexes as wholes and thus permit local populations to produce individuals adapted to rather widely different conditions (Wright and Dobzhansky, 1946). It may be looked upon as a device by which a single deme may simultaneously occupy two or more peaks of selective value. The dynamics of the process by which those alternative patterns are built up remains to be accounted for. In other cases, a similar result seems to be brought about by polymorphic switch genes (Sheppard, 1961). This again has to do with the range of adaptability of the species, an end result of processes of adaptive transformation, not with the dynamics of the latter.

Bridges (1935) and Metz (1947), working with *Drosophila* and *Sciara*, respectively, were impressed with the importance of small chromosomal duplications in providing new material which can be worked over in the evolutionary process. The breaking up of what were at first considered single loci into extensive pseudoallelic system suggests that gene duplication by unequal crossing over followed by differentiation has been a general phenomenon of great importance in evolution. Because of this process and of inversions, each chromonema tends to become subdivided into regions of more or less similar material, within which crossing over is greatly reduced, bounded by unconformities (cf. Wright, 1959). All of this, however, has to do with the raw material for evolution, not with the dynamics of adaptive transformation.

Morgan (1930) summed up his views on evolution in the statement, "If we had the complete ancestry of any animal or plant, living today, we should expect to find a series of forms differing at each step by a single mutant change in one or another of the genes, and each a better adapted or differently adapted form from the preceding."

Morgan was clearly more concerned with the material utilized in evolution than the dynamics of the selection process that is implied. He evidently thought of the species as homallelic, or nearly so at all loci at any given time with progress based on the occasional occurrence and fixation by selection of a favorable mutation. This is in marked contrast with the shifting balance theory which required strong heterallelism at many or all loci (Wright, 1929a, 1931).

With respect to dynamics, Castle (1903) seems to have been the first to work out generation by generation the effects of selection on a gene frequency: that of a recessive lethal. Norton (in Punnett, 1915) worked out the general cases for favorable dominants or recessives. A discussion by Chetverikov (1929) of the implications of Norton's tables, stimulated extensive study of hidden variability from rare recessives by Russian geneticists.

The most comprehensive investigations of the dynamics of selection in genetic systems have been those of Haldane, beginning in 1924, summarized in 1932, and continued almost to his death. He described the progress toward fixation of favorable mutations under a great variety of complicating conditions. Factor interaction was among these but for the most part as an impediment to rapid progress rather than as something which enormously amplifies the field of significant variability provided that there is a way in which there may be selection among systems.

Haldane, however, recognized that evolution is a process which can occur in many different ways. While his studies were almost wholly on the course of selection in panmictic populations from a deterministic standpoint, he recognized the probable validity of the shifting balance theory for the early evolution of the human species (Haldane, 1949).

Fisher's theory (1930) centered in his "fundamental theorem" of natural selection, already quoted. According to this any reduction by factor interaction of the portion of the total variance attributable to the additive ("genetic") effects of the separate genes, cuts down by just that much the "rate of increase of fitness". This theorem seemed to be a refutation of the possibility of any selection among interaction systems. He did indeed recognize the possibility that the progress toward fixation of a favorable mutation may sometimes reverse the relative selective values at interacting loci and thus lead to chains of readjustment. This process is severely limited, however, by the fact that a mutation has a favorable net effect only if it fits well into the current interaction system from the first. Fisher assumed that the whole species is effectively panmictic (unless in process of fission). The most important difference from the shifting balance theory was in the last premise of the latter, (D) fine scaled subdivision of species, in at least parts of their ranges, into partially isolated demes. He also did not accept pleiotropy (premise B) as being nearly as universal as I did, as illustrated by his theory (1928) of the evolutionary origin of the prevailing dominance of type genes over recurrent *deleterious* mutations, which he showed gave a selection intensity for dominance modifiers at most of the order of the mutation rate of the type gene in question. a rate so low that it seemed to me (1929b) (and to Haldane (1930)) and others that some selective effect of the modifier in homozygous wild-type would almost always take precedence over the effect on heterozygotes which Fisher postulated. Along with this, he did not accept the concept of multiple selective peaks (premise C).

Criticisms and Misinterpretations

A number of authors have bracketed the theories proposed by Haldane, Fisher, and myself in the 1920's as essentially the same and as involving such oversimplification as to be of little value. The most extensive, fairly recent criticism of this sort included the following statement (Mayr, 1959; c.f. Wright, 1960b).

"In order to permit mathematical treatment, numerous simplifying assumptions had to be made such as that of the absolute selective values of any given gene. This period was one of gross oversimplification. Evolutionary change was essentially presented as an input or output of genes, as the adding of certain beans to a bag and the withdrawal of others."

The last sentence applies in a sense to Haldane's and Fisher's deductions for essentially homogeneous populations, not because they ignored factor interactions, which they did not, but because evolutionary change is necessarily of this "bean bag" sort in such a population (as demonstrated by Fisher's fundamental theorem, which is accurate except for some qualification due to linkage). Haldane (1964) has adequately presented the importance of thoroughgoing analysis under these conditions in his paper "a defense of bean bag genetics."

The statement is, of course, wholly incorrect with respect to the shifting balance theory, which was from the first designed to bring out the conditions under which there could be selection among interaction systems, in which genes are favorable or unfavorable according to the genes with which they are associated. Absolute selective values were thus not attributed to such genes and the steps in evolution were not input or output of genes but shifts in control by selective peaks. Mayr himself urged the importance of selection of interaction systems but rejected the only process that has been suggested (interdemic selection) by which this may occur under predominant biparental reproduction.

The most frequent criticisms have been by those who have maintained that random drift was being presented as an *alternative* to natural selection in the transformation of populations (Fisher and Ford, 1947, 1950; Wright, 1948, 1951a). This has taken its most extreme form where critics have supposed that I attributed significance to complete fixation of some random combination of genes such as occurs under very close inbreeding. My judgement in 1931 of the evolutionary significance of such fixation was as follows:

"In too small a population there is nearly complete fixation, little variation, little effect of selection and thus a static condition, modified occasionally by chance fixation of a new mutation, leading inevitably to degeneration and extinction."

More frequently, sampling drift has been discussed as a factor to be considered only in accounting for trivial differences in gene frequencies, observed among small neighboring localities. Thus Cain and Currey (1963) on finding striking differences among populations of snails *(Cepaea nemoralis)* of certain large areas for which there was no obvious explanation in terms of protective coloration (as described by Cain and Sheppard, 1954):

"These area effects are too homogeneous over large areas containing very large numbers of individuals to be due to sampling drift or any other random process acting at present."

Their conclusion that the cause is selection, related to subtle climatic differences, may well be correct, but it cannot be proved by elimination. If the color factors in this case are involved in interaction systems, it is possible that the area effect may be due to local selection (phase 2) and spreading (phase 3) related to different selective peaks, arrived at by random drift (phase 1) without there having been any significant differences in the external environment (cf. Fig. 5).

This sort of criticism ignores the point that no evolutionary significance has ever been attributed to random drift except as a trigger which occasionally causes selection to be directed toward a new and superior selective peak in the surface of selective values. To ascribe an area effect brought about in this way to either random drift by itself or selection by itself obscures the fact that both have played indispensable roles (Wright, 1965).

The point is also often overlooked that sampling drift has never been treated as the only random process which can act as a trigger in the above sense but merely as the one which happens to be the easiest to deal with mathematically.

The most serious criticisms of the theory have been that local populations are not in general small enough to give a basis for appreciable sampling drift or sufficiently isolated for enough differentiation of their genetic systems to give a basis for interdemic selection (Simpson, 1953; Mayr, 1963).

I was not unaware of these difficulties in my first major accounts (1931, 1935) but held that *effective* numbers and amount of immigration would often be sufficiently small, for reasons already discussed, to permit significant sampling drift and local differentiation and, moreover, that where numbers are not small enough other forms of random drift may be operating. It is not necessary that conditions be such that all three phases of the theory operate in all parts of all species all of the time.

A theory that applies best to a large population that includes a great many partially isolated demes and is heterallelic at many, if not all, loci, at many of which the local selective intensities (s), the effective amount of immigration (m), and the sampling variance (coefficient $1/2 N$) or other forms of random drift, are balanced against each other within one or two orders of magnitude, is not easy to test experimentally. Nevertheless, the theory has been supported by experiments by Dobzhansky and Pavlovsky (1957) and Dobzhansky and Spassky (1962) in which the amount of dif-

ferentiation of subpopulations of *Drosophila pseudoobscura*, derived
from a common source, was found to be significantly greater among ones
which have passed through one or more bottlenecks of small size, than
among ones which have not.

Probably the strongest support comes from studies of natural populations of many species in which persistent statistical differences in quantitative characters or in gene frequencies at segregating loci have been
found to be the rule among local populations even in the absence of
appreciable environmental differences. Again it is to be emphasized that
such differences are not interpreted under the theory as manifestations
of random drift by itself but as selective differences, associated with different selective peaks merely triggered by random drift of some sort. The
human species is not among the least mobile but from the universality of
fine scaled differentiation among relatively undisturbed rural populations and the historic record of expansion from some centers, contraction
at others, it would appear that the three phases of the theory have probably been operating throughout its history to bring about changes in the
genetic composition of the species as a whole even though in the last
0.1 % or so of this history most local populations have been so dense that
sampling drift is negligible (Wright, 1932).

A criticism that has been made, orally at least, is that the theory
cannot be considered scientific since it implies that the course of evolution is essentially unpredictable. This is a semantic question on the scope
of science. It seems to me that a chain of deductions from well understood
phenomena such as Mendelian heredity, and parameters which are potentially describable with accuracy, such as those of the various evolutionary pressures and population structures, is part of science, even
though the only positive long-term conclusion is that with a favorable
structure and a favorable ecologic opportunity, there will be an indefinitely continuing evolutionary process, and though it reaches the
negative conclusion that the detailed course of the process is wholly
unpredictable, apart from short-term extrapolation from the current
trend.

There is indeed more predictability, in a sense, if the population is
known to be essentially panmictic and it is assumed that conditions will
remain constant, since in this case the short-term projection of the trend
is all there will be (apart from the exceedingly slow incorporation of
novel favorable mutations). This latter aspect, in so far as significant, is
again wholly unpredictable in detail.

Artificial selection constitutes a violent change in conditions. It can
be predicted that in a panmictic population, the phenotype will move in
the desired direction. The rate and, to a less extent, the ultimate degree of
success may be predicted on the basis of the initial response and such

knowledge as it is possible to obtain of the genetics of the variations. It can also be predicted that progress in the desired direction will be associated with disharmonies and deterioration in other respects.

Artificial selection of somewhat varied sorts, in an array of lines, each of which is kept isolated as long as it is doing well, but is out-crossed to superior lines if not, simulates on a small scale all of the phases of the shifting balance theory. It gives a basis for continued progress over longer periods, than selection in a single panmictic population of the same total size. The breeds of livestock were developed in somewhat this way (McPhee and Wright, 1925).

The theory that long period evolution is the result of a long succession of changes in the environment gives more predictability than the shifting balance theory, provided that this succession of environmental changes is known. Actually the long-term course is equally unpredictable because of the unpredictability of the succession.

Evolution is, in short, a process for which it is unlikely that there will ever be long-term prediction. I exclude here the prediction that in the long run the fate of any particular species will be extinction because sooner or later conditions will arise with which it cannot cope.

Summary

The term "drift" was introduced into population genetics in metaphorical senses for both directed and random processes but has come to be restricted largely to the latter as a somewhat technical term. The meaning should, however, always be clearly indicated by prefixing "steady" or "random" as the case may be.

"Random drift" has been applied most frequently to the cumulative effects of accidents of sampling, but since all modes of local differentiation that are uncorrelated with the course of evolution of the species as a whole play a similar role and have been included under random drift, it is desirable to use Cain and Currey's term "sampling drift" if only the former is meant.

Since random drift is of most significance in the shifting balance theory of evolution, the more important premises of this theory (multifactorial heredity of quantitative variability, universal pleiotropy, prevalence of interaction systems, especially where there is an intermediate optimum, and occurrence of numerous partially isolated demes) and its three phases (local random drift, local mass selection, and interdemic selection) are briefly reviewed. The effects of fluctuating environmental conditions, of persistent local differences, of secular changes and the ultimate control by ecological opportunity are discussed.

In the mathematical formulation for the intrademic phases, random drift is defined as the portion of the multidimensional probability distribution of the set of gene frequencies in the neighborhood of the currently controlling selective peak.

Various complications in the three phases are discussed. The effective population size of demes and the effective amount of immigration are in general much smaller than the apparent values, with the consequence that sampling drift is more important than is immediately apparent. Where selective values are frequency dependent, change of gene frequency is controlled by the "surface" of a fitness function, where it exists, a surface that may differ radically from that of mean selective values. Where there is no such function there still may be multiple "selective goals" different from the peaks of mean selective value. The course of change in phase 3 is controlled largely by local rates of population growth and diffusion which depend as a rule on mean selective value. The overall control of evolution is thus based on a compromise between control by "fitness" and control by mean selective value.

The mode of utilization of major mutations by the shifting balance theory through adaptation of modifiers is discussed.

An extreme variant of the theory occurs where there is predominant but not exclusive uniparental reproduction.

Theories concerned with the materials for evolution and with its dynamics should not be confused. The shifting balance theory is contrasted with other Mendelian theories of evolutionary dynamics. The theories of Haldane and Fisher apply to phase 2, mass selection in an essentially panmictic population. In these theories, interaction is a hindrance to evolution. The shifting balance theory is the only one applicable to species with predominant biparental reproduction which permits selection among interaction systems.

Various misinterpretations and criticisms are discussed briefly.

References

Anderson, E.: Introgressive hybridization. New York: John Wiley & Sons 1949.

Bridges, C. B.: Salivary chromosome maps with a key to the banding of chromosomes of Drosophila melanogaster. J. Hered. **26**, 60 − 64 (1935).

Cain, A. J., and J. D. Currey: Area effects in Cepaea. Phil. Trans. Roy. Soc. London B **246**, 1 − 81 (1963).

− , and P. M. Sheppard: Natural selection in Cepaea. Genetics **39**, 89 − 116 (1954).

Castle, W. E.: The laws of Galton and Mendel and some laws governing race improvement by selection. Proc. Amer. Acad. Arts Sci. **39**, 233 − 242 (1903).

Chetverikov, S. S.: On certain aspects of the evolutionary process from the standpoint of modern genetics. Proc. Amer. Phil. Soc. **105**, 167 – 195 (1961); translated from the Russian by Malina Barker, edited by I. Michael Lerner, from the original in: Zh. Eksp. noi Biol. A **2**, 3 – 54 (1926).

Dobzhansky, Th.: A review of some fundamental concepts and problems of population genetics. Cold Spr. Harb. Symp. Quant. Biol. **20**, 1 – 15 (1956).

–, and O. Pavlovsky: An experimental study of interaction between genetic drift and natural selection. Evolution **11**, 311 – 319 (1957).

–, and N. P. Spassky: Genetic drift and natural selection in experimental populations of *Drosophila pseudoobscura*. Proc. Natl. Acad. Sci. (Wash.) **48**, 148 – 156 (1962).

Fisher, R. A.: On the dominance ratio. Proc. Roy. Soc. Edinburgh **42**, 321 – 341 (1922).

– The possible modification of the response of the wild type to recurrent mutation. Amer. Natur. **62**, 115 – 126 (1928).

– The genetical theory of natural selection. Oxford: Clarendon Press 1930.

–, and E. B. Ford: The spread of a gene in natural conditions in a colony of the moth, Panaxia dominula. Heredity **1**, 143 – 174 (1947).

– – The "Sewall Wright" effect. Heredity **4**, 117 – 119 (1950).

Goldschmidt, R.: The material basis of evolution. New Haven: Yale Univ. Press 1940.

Haldane, J. B. S.: A mathematical theory of natural and artificial selection I. Trans. Cambridge Phil. Soc. **23**, 19 – 41 (1924).

– A mathematical theory of natural and artificial selection V. Selection and mutation. Proc. Cambridge Phil. Soc. **23**, 838 – 844 (1927).

– A note on Fisher's theory of the origin of dominance and on a correlation between dominance and linkage. Amer. Natur. **64**, 87 – 90 (1930).

– The causes of evolution. London: Harper and Bros. 1932.

– Human evolution: past and future. In: Genetics, paleontology and evolution, pp. 405 to 418. Ed. by G. L. Jepsen, G. G. Simpson and E. Mayr. Princeton, N. J.: Princeton Univ. Press 1949.

– Conditions for stable polymorphism at an autosomal locus. Nature (Lond.) **193**, 1108 (1962).

– A defense of "bean bag" genetics. Perspect. Biol. Med. **7**, 343 (1964).

Lerner, I. M.: Genetic homeostasis. New York: John Wiley and Sons 1954.

Mayr, E.: Where are we? Genetics and twentieth century Darwinism. Cold Spr. Harb. Symp. Quant. Biol. **24**, 1 – 14 (1959).

– Animal species and evolution. Cambridge: Harvard Univ. Press 1963.

McPhee, H. C., and S. Wright: Mendelian analysis of the purebreeds of livestock. III The Shorthorns. J. Hered. **16**, 205 – 215 (1925).

Metz, C. W.: Duplication of chromosome parts as a factor in evolution. Amer. Natur. **81**, 81 – 103 (1947).

Morgan, T. H.: The scientific basis of evolution. New York: W. W. Norton & Co. 1930.

Punnett, R. C.: Mimicry in butterflies. Cambridge: Cambridge Univ. Press 1915.

Sandler, L., and E. Novitski: Meiotic drive as an evolutionary force. Amer. Natur. **91**, 105 – 110 (1957).

Sheppard, P. M.: Some contributions to population genetics resulting from the study of the *Lepidoptera*. Advan. Genet. **10**, 165 – 216 (1961).

Simpson, G. G.: Tempo and mode in evolution. New York: Columbia Univ. Press 1944.

— The major features of evolution. New York: Columbia Univ. Press 1953.

de Vries, H.: Species and varieties. Their origin by mutation. Chicago: Open Court Publ. Co. 1905.

Willis, J. C.: The course of evolution. Cambridge: Cambridge Univ. Press 1940.

Wright, S.: Evolution in a Mendelian population. Anat. Rec. **44**, 287 (1929 a).

— Fisher's theory of dominance. Amer. Natur. **63**, 274 – 279 (1929 b).

— The evolution of dominance. Amer. Natur. **63**, 556 – 561 (1929 c).

— Evolution in Mendelian populations. Genetics **16**, 97 – 159 (1931).

— The roles of mutation, inbreeding, crossbreeding, and selection in evolution. Proc. 6 th Int. Cong. Genet. **1**, 356 – 366 (1932).

— The analysis of variance and the correlations between relatives with respect to deviations from an optimum. J. Genet. **30**, 243 – 256 (1935 a).

— Evolution in populations in approximate equilibrium. J. Genet. **30**, 257 – 266 (1935 b).

— The distribution of gene frequencies in populations. Proc. Natl. Acad. Sci. (Wash.) **23**, 307 – 320 (1937).

— The distribution of gene frequencies under irreversible mutation. Proc. Natl. Acad. Sci. (Wash.) **24**, 223 – 239 (1938 a).

— Size of population and breeding structure in relation to evolution. Science **87**, 430 – 431 (1938 b).

— Statistical genetics in relation to evolution. Actualites scientifique et industrielles, Vol. 802. Paris: Hermann et Cie 1939 a.

— Breeding structure of populations in relation to speciation. Amer. Natur. **74**, 232 – 248 (1940).

— The material basis of evolution. Sci. Monthly **53**, 165 – 170 (1941 a).

— The "age and area" concept extended. Ecology **22**, 345 – 347 (1941 b).

— Statistical genetics and evolution. Bull. Am. Math. Soc. **48**, 223 – 246 (1942).

— Isolation by distance. Genetics **28**, 114 – 138 (1943).

— Tempo and mode in evolution: a critical review. Ecology **26**, 415 – 419 (1945 a).

— The differential equation of the distribution of gene frequencies. Proc. Natl. Acad. Sci. (Wash.) **31**, 382 – 389 (1945 b).

— Isolation by distance under diverse systems of mating. Genetics **31**, 39 – 59 (1946).

— On the roles of directed and random changes in gene frequencies in the genetics of populations. Evolution **2**, 279 – 294 (1948).

— Adaptation and selection. In: Genetics, paleontology and evolution, pp. 365 – 389. Ed. by G. L. Jepsen, J. J. Simpson, and E. Mayr. Princeton, N. J.: Princeton Univ. Press 1949.

— Discussion on population genetics and radiation. J. Cell Comp. Physiol. **35**, 187 – 210 (1950).

— Fisher and Ford on the "Sewall Wright" effect. Amer. Sci. **39**, 452 – 458 (1951 a).

— The genetical structure of populations. Ann. Eugen. **15**, 323 – 354 (1951 b).

— The theoretical variance within and among subdivisions of a population that is in a steady state. Genetics **27**, 312 – 321 (1952).

— Modes of selection. Amer. Natur. **90**, 5 – 24 (1956 a).

— Classification of the factors of evolution. Cold Spr. Harb. Symp. Quant. Biol. **20**, 16 – 24 D (1956 b).

— Genetics, the gene and the hierarchy of biological sciences. Proc. X Int. Cong. Genet. **1**, 475 – 489 (1959).

— Physiological genetics, ecology of populations and natural selection. In: The evolution of life, pp. 429 – 475. Ed. by Sol Tax. Chicago: University of Chicago Press 1960a.

— Genetics and twentieth century Darwinism: a review and discussion. Amer. J. Human Genet. **12**, 365 – 372 (1960 b).

— Genic interaction. In: Methodology in mammalian genetics, pp. 158 – 188. Ed. by W. J. Burdette. San Francisco: Holden-Day 1963 a.

— Discussion of plant and animal improvement in the presence of multiple selective peaks. In: Statical genetics and plant breeding. Ed. by W. D. Hanson and H. F. Robinson. Natl. Acad. Sci. — Natl. Res. Council, Publ. 982 (Washington) 1963 b.

— Stochastic processes in evolution. In: Symposium on stochastic models in medicine and biology. Ed. by J. Gurland. Madison: The University of Wisconsin Press 1964.

— Factor interaction and linkage in evolution. Proc. Roy. Soc. London B **162**, 30 – 104 (1965).

— Polyallelic random drift in relation to evolution. Proc. Natl. Acad. Sci. (Wash.) **55**, 1074 to 1081 (1966).

— "Surfaces" of selective value. Proc. Natl. Acad. Sci. (Wash.) **58**, 165 – 172 (1967).

— , and Th. Dobzhansky: Genetics of natural populations. XII Experimental reproduction of some of the changes caused by natural selection in certain populations of *Drosophila pseudoobscura*. Genetics **31**, 125 – 156 (1946).

Changes in Mean Fitness under Natural Selection

J. R. G. TURNER

The Survival of the Fittest

The Reformation in Europe changed Man's view of his relation with God; no longer dependent on a hierarchy of ordained vicars, the individual must seek God by his own efforts. There grew, particularly in England, the tradition of Natural Theology: not only was God approached through His Works of Nature, but those very Works were held to be evidence of His Divine Wisdom. This idea was applied especially to living organisms, whose adaptations, whose fitness to the ends they served, suggested powerfully that they had been designed by an intelligent, beneficent creator. This "Argument from Design", a very popular proof of the existence of a Christian God, is well summarized by Robert Boyle (1688):

> And I confess, that when I assist at a well-administered Anatomy, I do so wonder at the admirable Contrivance of a Humane Body, that I cannot but somewhat wonder, that there should be found among Philosophers, men that can ascribe it to blind Chance. The Stoick that in Cicero asked an Epicurean, why Chance did not make Palaces and other Buildings, seems not to me to have made an impertinent Question.

The great opposition to Darwin's theory of descent by natural selection arose, not merely because it conflicted with a literal interpretation of the Bible, but because natural selection, by providing a "materialistic", causal − one might say "atheistic", Epicurean − explanation of the fitness of organisms, cut away the Argument from Design, and with it much of Natural Theology (see Ellegård, 1958, for a detailed account). As Natural History and Natural Theology were intimately connected, Darwin's ideas were opposed both by scientists and by theologians.

This is perhaps one reason why evolution by natural selection is so often described as the "survival of the fittest"; biologists were at pains to emphasize that their causal explanations were adequate to account for the "fitness" of organisms, and that Divine intervention was not necessary. Another reason for the popularity of the phrase was its notorious application to capitalist social practice and *laissez-faire*. The expression was invented by Spencer and incorporated, at the suggestion of Wallace, in the *Origin* (see Peckham, 1959):

> *I have called this principle ... by the term Natural Selection ... But* [adds Darwin in the fifth edition] *the expression often used by Mr. Herbert Spencer of the Survival of the Fittest* [sic], *and is more accurate* [sic], *and is sometimes equally convenient.*

So, as a result of the religious basis of science, and the pithiness and erstwhile social acceptability of Spencer's phrase, we now tend to think of evolution by natural selection as the survival of the fittest. But even if we strip "fittest" of all its social and moral overtones of "best" and make it mean, objectively, the ability to survive and to have offspring, and then allow that the phrase although a tautology is a useful tautology; even then I believe that it is a misleading description of evolution by natural selection. In a sexual organism, during the course of a generation, some genotypes have a better chance of survival and reproduction than others; one may if one likes call this the survival of the fittest. At the end of the generation the genotypes of the diploid adults are split into haploid portions, and then reassorted by mating. The haploid portions are themselves shuffled by recombination. If the organism is haploid for most of its life cycle, the genes are shuffled by recombination during the diploid phase. Therefore although natural selection may be the survival of the fittest, evolution by selection and sexual reproduction is not the survival of the fittest as the fittest genotypes are repeatedly broken down. In the long term of course those genetic systems, species, and taxa which are "fittest" will survive, but the kind of "fitness" which promotes this survival is not necessarily that which is favoured by natural selection within individual populations.

The discovery of mathematical theorems in science may be said to involve finding equations with the same "shape" as the physical system we are describing. Two theorems in population genetics take the shape of the "survival of the fittest", as they equate change in the mean fitness of a population to the variance in fitness; as the variance is positive, the mean always increases. Li's theorem is that in the absence of mating the change in mean fitness is equal to the genotypic variance in fitness (Li, 1955b); this seems unassailable, and conforms with the description of natural selection as the survival of the fittest. The other proposed by Fisher (1930) as the "Fundamental Theorem of Natural Selection" is

> *The rate of increase in* [mean] *fitness of any organism at any time is equal to its* [additive] *genetic variance in fitness at that time.*

This theorem includes the round of mating, and just as evolution by selection is not the survival of the fittest, so does this theorem turn out not to be true in general; Fisher has proposed a theorem, not of natural selection, but of evolution. An analysis of its mathematical basis shows, as will be seen later, that Fisher, because he regarded two different parameters as identical, believed his theorem was much more general than it is; we will see that it is strictly true only with random mating, constant genotypic selection, and no dominance or recombination.

I have criticized Fisher's theorem from the above view, as the "survival of the fittest" is one of the obvious interpretations of the theorem. Fisher himself seems not to have used Spencer's phrase and to have taken a much more subtle and not readily understood view of the meaning of his theorem. As far as I understand the argument presented, without studying his other works or correspondence, it is as follows (Fisher, 1930, 1941, 1958).

Natural selection acts because individuals in a population have, as a result of their genetic makeup, different abilities to survive and reproduce, that is, because there is genetic variance in fitness, using "fitness" in the sense of ability to survive and reproduce. Natural selection acts by consuming this genetic variance. The rate of consumption was unknown to Fisher; he considered instead the effect of selection on the mean fitness of the whole population, showing erroneously by a certain lack of mathematical rigour, that this mean increased at a rate equal to the variance. From the definition of fitness, the mean fitness of the population is simply its rate of growth; Fisher (1941) was very critical of the view, which he thought he detected in the equations of Sewall Wright, that selection worked for the "good of the species" and that it was this kind of general "fitness" that was increased. However, it was obvious that the rate of growth of populations could not simply go on increasing, so Fisher allowed for decrease in fitness due to mutation and deterioration of the environment and due to the effects of overpopulation (what we would now call density-dependent factors). This left his mean-fitness-whose increase-was-equal-to-the-variance without any particular physical meaning, as it was no longer the actual rate of population growth, and he fell back on rather hazy concepts of "adaptation" and the statement that increase in mean fitness resulting from selection (after deduction of the effects of environmental change) would result not in an increase in the rate of growth, which because of ecological factors would remain about zero, but in an increase in the standing population (this is a *non sequitur*, see the end of this chapter). Fisher's view of the meaning of "fitness" is further obscured by his statement (1930) that "*although measured by a uniform method*, [it] *is qualitatively different for every different* [species of (?)] *organism*." This can hardly refer to population growth.

The main thing which Fisher seems to have liked about his theorem was that it resembled, with certain important differences, the physical law that entropy increases (the second law of thermodynamics). *"It is not a little instructive that so similar a law* [to the entropy law] *should hold the supreme position among the biological sciences"* (1930). I find this claim extravagant; the theorem is lacking in mathematical rigour and the concept of fitness is hazy. However this theorem, proposed in the pioneer days of population genetics with that rare courage to seek the deepest laws, has inspired many discussions, both mathematical and ecological, which I shall try to review.

Fisher was aware of some of the difficulties in his theorem; he introduces (1958) an oblique discussion of dominance, dealing with its effects on the mating system, and his idea (1930) that linkage is increased between interacting loci can be shown to depend on the fact that his fitness theorem could not be strictly true for two loci (Turner, 1967a).

To avoid misinterpretation, I had better say that I am criticizing Fisher's theorem as a statement of fact in 1968, not attempting a historian's judgement of Fisher's contribution to science in 1930.

Predictive Models, Genetic Loads, and Ecology

The changes in mean fitness of populations under selection are interesting not only because of their relation to Fisher's fundamental theorem, but because the maximization of mean fitness enables one to construct a comparatively simple pictorial model (an "adaptive topography") which can be used for theoretical explanations (Wright, 1932; Li, 1955a) and for investigating real populations (Lewontin and White, 1960); because the mean fitness relates directly to the "genetic load" (proportion of the population dying from genetic causes); and because the mean absolute fitness is the rate of population growth — an important ecological parameter.

We will therefore in the following pages consider the equations for the change in mean fitness, the construction of adaptive topographies, genetic loads, and population ecology, but not all in equal detail.

Basic Models and Equations

The notation used here is roughly the conventional one, with the subscripts 0 and 1 representing the alleles (a, A) at a particular locus. When generations are separate, Q_{ij} and q_i refer to a zygotic population in a particular generation; when they overlap the symbols refer to the whole population.

Changes in Gene Frequency

In a population in which generations do not overlap, as in a single-brooded insect or an annual plant in a seasonal climate, if the probabilities of survival and reproduction of the genotypes aa, aA, AA are W_{00}, W_{01}, W_{11}, their zygotic frequencies are Q_{00}, Q_{01}, Q_{11} and the gene frequencies are q_0, q_1, remembering that

$$q_0 = Q_{00} + \tfrac{1}{2}Q_{01}; \quad q_1 = \tfrac{1}{2}Q_{01} + Q_{11} \tag{1}$$

it is clear that after a generation of selection the new genotype frequencies are

$$Q'_{00} = Q_{00}W_{00}/\bar{W} \tag{2}$$

and so on, and the new gene frequency is

$$q'_0 = Q'_{00} + \tfrac{1}{2}Q'_{01} = (Q_{00}W_{00} + \tfrac{1}{2}Q_{01}W_{01})/\bar{W}$$
$$= q_0 W_0/\bar{W} \tag{3}$$

where

$$W_0 = (Q_{00}W_{00} + \tfrac{1}{2}Q_{01}W_{01})/q_0 \tag{4}$$

is the mean fitness of the allele a,

$$W_1 = (\tfrac{1}{2}Q_{01}W_{01} + Q_{11}W_{11})/q_1 \tag{5}$$

is the mean fitness of the allele A and

$$\bar{W} = Q_{00}W_{00} + Q_{01}W_{01} + Q_{11}W_{11} \tag{6}$$
$$= q_0 W_0 + q_1 W_1 \tag{7}$$

is the mean fitness of the whole population. $\bar{W}$ is the rate of population growth. From Eq. (3) the change of gene frequency is (Wright, 1949)

$$\Delta q_0 = q_0(W_0 - \bar{W})/\bar{W} . \tag{8}$$

Remembering Eq. (7) we find that

$$W_0 - \bar{W} = q_1(W_0 - W_1) \tag{9}$$

this being the average excess of the allele a, so that

$$\Delta q_0 = q_0 q_1 (W_0 - W_1)/\bar{W} . \tag{10}$$

Defining the average excess of the gene substitution as

$$a = (W_0 - W_1) \tag{11}$$

we finally have an equivalent of Fisher's equation for change in gene frequency (Fisher, 1930, 1941):

$$\Delta q_0 = q_0 q_1 a/\bar{W} . \tag{12}$$

If mating is random, so that

$$Q_{00} = q_0^2, \quad Q_{01} = 2q_0 q_1, \quad Q_{11} = q_1^2$$

we find that

$$W_0 = q_0 W_{00} + q_1 W_{01}, \qquad W_1 = q_0 W_{01} + q_1 W_{11}, \qquad (13)$$

$$W_0 - W_1 = (d\bar{W}/dq_0)/2, \qquad (14)$$

so that (Wright, 1949)

$$\Delta q_0 = \frac{q_0 q_1}{2\bar{W}} \frac{d\bar{W}}{dq_0}. \qquad (15)$$

A population containing three genotypes can be represented as a point in an equilateral triangle, each of the homogeneous axes representing the frequency of one of the genotypes (Li, 1955b, citing de Finetti). Populations resulting from random mating lie on a parabola, and the course followed by a population under selection and random mating is a series of zigzags (Fig. 1), selection pulling the population away from the

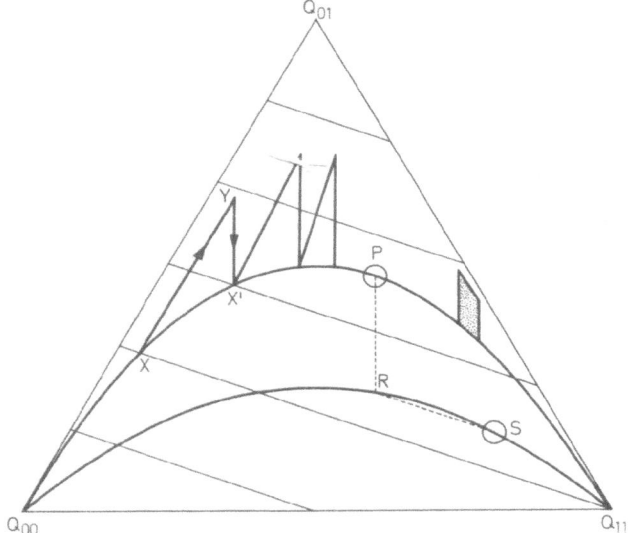

Fig. 1a–c. Homogeneous 3 co-ordinate graphs including all possible populations made up of the three genotypes AA, Aa, aa, the vertices representing populations with 100% of the genotype indicated, the side opposite representing populations from which that genotype is absent.

(a) Upper parabola is locus of populations in Hardy-Weinberg equilibrium, lower parabola of inbred populations with $F = \frac{1}{2}$; thin lines are contours of equal $\bar{W}$, if $W_{00} = \frac{1}{4}$, $W_{01} = 1$, $W_{11} = \frac{1}{2}$ (high $\bar{W}$ at upper vertex) and the zigzag line X, Y, X'... is the path followed by a random mated population under selection with these fitness values; the shaded block is a population with overlapping generations under the same selection, and the circles are stable equilibria; under random mating P is at the maximum value of $\bar{W}$ within the parabola; with inbreeding the equilibrium (not at maximum $\bar{W}$) can be found by the construction PRS.

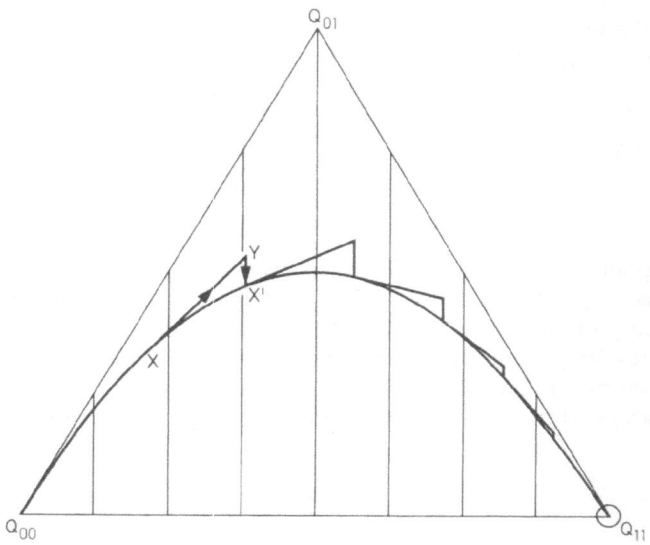

(b) As (a) except that fitness values are $W_{00} = \frac{1}{5}$, $W_{01} = \frac{3}{5}$, $W_{11} = 1$ (no dominance).

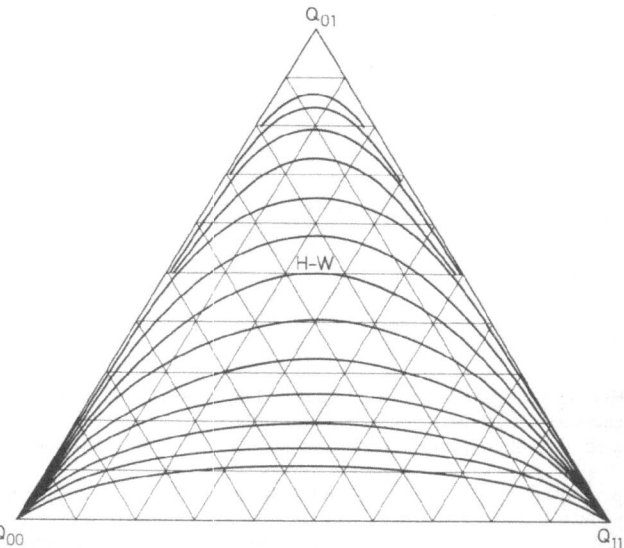

(c) Conics of constant λ, the values being (from top to bottom) 64, 32, 16, 8, 4, 2 (hyperbolas), 1 (Hardy-Weinberg parabola, labelled H-W), $\frac{1}{2}$, $\frac{1}{4}$, $\frac{1}{8}, \frac{1}{16}, \frac{1}{32}, \frac{1}{64}$ (ellipses and circle). The Y points in Figs 1 (a, b) all lie on conics of this type. After Cannings and Edwards (1968), Li (1955b, 1967b), Turner (1967b)

parabola, and random mating returning it; the path followed at mating is vertical because gene frequencies do not change, and gene frequency is constant along the vertical axis of the triangle (see Cannings and Edwards, 1968).

This has been discussed in detail, as the form of the equations is basic to what follows. Three similar sets of equations can be devised. We may use only relative values of fitness, dividing each value by that of the heterozygote to obtain

$$w_{00} = W_{00}/W_{01} \tag{16}$$

and so on. The equations are the same as Eq. (2) to (15) except that w replaces W; in this way we can ignore changes in reproductive potential resulting from changes in population density.

If the generations overlap, that is if all age-classes are always present in the population and *if the age-structure is constant*, we may write a similar set of equations in which time derivatives (dq/dt) replace finite differences (Δq), the Malthusian parameter, M (instantaneous rate of birth minus death) replaces W, and the mean fitness in the denominator is removed (it is needed in the finite-difference equations in order to keep frequencies adding to unity; infinitesimal difference equations do not need this). For derivations, see Kimura (1958).

The equations can also be generalized for any number of alleles. For illustration, here are the equations using relative Malthusian parameters, m (that is Malthusian parameters scaled in the same way as w parameters, see above):

$$m_i = \Sigma_i Q_{ij} m_{ij}/q_i, \tag{17}$$

$$\bar{m} = \Sigma_{ij} Q_{ij} m_{ij} = \Sigma_i q_i m_i, \tag{18}$$

$$dq_i/dt = q_i(m_i - \bar{m}), \tag{19}$$

$$\frac{dq_i}{dt} = \frac{q_i(1 - q_i)}{2} \frac{\partial \bar{m}}{\partial q_i}. \tag{20}$$

Eq. (20) assumes that the Q_{ij} are measured in zygotes, which is not strictly correct. The equivalence (a rough equivalence only) of the equations is seen from the fact that

$$m_{ij} = \log_e w_{ij} \tag{21}$$

so that

$$\Delta q_0 = \frac{q_0 q_1}{2 \bar{w}} \frac{d\bar{w}}{dq_0} = \frac{q_0 q_1}{2} \frac{d(\log \bar{w})}{dq_0} \tag{22}$$

is similar in form to

$$\frac{dq_0}{dt} = \frac{q_0 q_1}{2} \frac{d\bar{m}}{dq_0} = \frac{q_0 q_1}{2} \frac{d(\overline{\log w})}{dq_0}. \tag{23}$$

See Wright (1955) and Moran (1962) for the form of the derivatives in Eq. (20).

In the triangular graph, a population with overlapping generations is represented by an area stretching away from the parabola (Fig. 1); its exact shape depends on the way we classify the age-classes. Under selection it moves more or less uniformly across the triangle, changing shape somewhat as it does so.

Many organisms have separate generations; it is unlikely that many have overlapping generations, but no adequate theory exists for populations with partly overlapping generations (changing age-structure), which are probably common and as will be shown, virtually a mathematical necessity under natural selection.

Average Excess and Average Effect

These concepts were explained in some detail by Fisher (1941), and have been very clearly expounded by Edwards (1968), on whose account the following is based. Except where stated, the equations are general, and are not dependent on the assumption of Hardy-Weinberg equilibrium, which is the case where all cases become special. The equations use W; w, M (absolute) or m (relative Malthusian parameters), or indeed any quantitative character, would be equally appropriate.

We have already defined the *average excess* in fitness of the gene substitution as the difference between the mean fitness of the two alleles:

$$a = W_0 - W_1 = (Q_{00}W_{00} + \tfrac{1}{2}Q_{01}W_{01})/q_0 - (\tfrac{1}{2}Q_{01}W_{01} + Q_{11}W_{11})/q_1$$

$$= \frac{q_1(Q_{00}W_{00} + \tfrac{1}{2}Q_{01}W_{01}) - q_0(\tfrac{1}{2}Q_{01}W_{01} + Q_{11}W_{11})}{q_0 q_1} \tag{24}$$

$$= \frac{Q_{00}q_1(W_{00} - W_{01}) + Q_{11}q_0(W_{01} - W_{11})}{Q_{00}Q_{11} + \tfrac{1}{2}Q_{00}Q_{01} + \tfrac{1}{2}Q_{01}Q_{11} + \tfrac{1}{4}Q_{01}^2} \tag{25}$$

(remembering Eq. (1)).

The *average effect* of the gene substitution is defined as the regression, by least squares, of the fitness of a genotype on the number of a alleles in the genotype. Graphically, the fitnesses of the three genotypes can be plotted against the number of a genes (Fig. 2); we then require three points, B_{00}, B_{01}, B_{11} such that the straight line joining them is the best fit to the values of the genotypes. In other words we need three points such that

$$\begin{aligned} B_{00} &= \mu + \alpha, \\ B_{01} &= \mu, \\ B_{11} &= \mu - \alpha, \end{aligned} \tag{26}$$

these being known as the additive or breeding values of the genotypes. Fisher (1941) solved the least squares equation and found that the re-

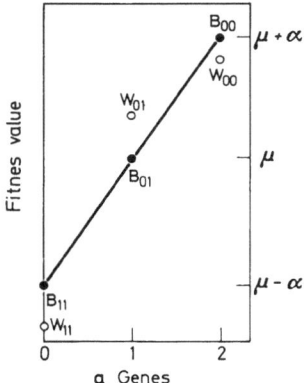

Fig. 2. Actual fitness values (W_{ij}) and breeding values (B_{ij}) of the three genotypes against the number of a genes in the genotype. Line is least squares regression fitted to the open circles. See text for other symbols. After Falconer (1961), Li (1955 b)

quired *average effect* is given by

$$\alpha = \frac{Q_{00}q_1(W_{00} - W_{01}) + Q_{11}q_0(W_{01} - W_{11})}{\frac{1}{2}Q_{00}Q_{01} + \frac{1}{2}Q_{01}Q_{11} + 2Q_{00}Q_{11}} \tag{27}$$

$$= \frac{Q_{00}q_1(W_{00} - W_{01}) + Q_{11}q_0(W_{01} - W_{11})}{\frac{1}{2}\{Q_{00} + Q_{11} - (Q_{00} - Q_{11})^2\}}. \tag{28}$$

Another important parameter is the derivative of $\bar{W}$ with respect to q_0:

$$\frac{d\bar{W}}{dq_0} = \frac{dQ_{00}}{dq_0} W_{00} + \frac{dQ_{01}}{dq_0} W_{01} + \frac{dQ_{11}}{dq_0} W_{11}. \tag{29}$$

In general the quantities a, α and $d\bar{W}/dq$ have no simple relation to each other (however, see Eq. (57)). Thus comparing Eq. (25) and (27) we see that $a = \alpha$ only if

$$\tfrac{1}{4}Q_{01}^2 = Q_{00}Q_{11}. \tag{30}$$

Likewise differentiation of $\bar{W}$ with respect to q_0 will yield $2a$ if, for example,

$$dQ_{00}/dq_0 = 2Q_{00}/q_0 \tag{31}$$

which is satisfied by the Hardy-Weinberg proportion

$$Q_{00} = q_0^2 \tag{32}$$

and similarly for the other terms (see Kojima, 1959). Edwards (1968) gives a full proof (based on Fisher, 1941) that $d\bar{W}/dq_0 = 2\alpha$ only if

$$\tfrac{1}{4}Q_{01}^2 = \lambda Q_{00}Q_{11} \tag{33}$$

where λ is a constant. Now Eq. (30) is the equation for populations in Hardy-Weinberg equilibrium (Li, 1955 b) and there appear to be no values of λ other than unity which give a genetically meaningful form to Eq. (33) (see later).

Thus finally

$$a = \alpha = \tfrac{1}{2} d\bar{W}/dq_0 \tag{34}$$

if and only if the population is in Hardy-Weinberg equilibrium. In general none of these quantities equals any other.

Note that the condition is Hardy-Weinberg equilibrium (binomial proportions), not random mating. The equation of random mating with binomial proportions is a most unfortunate error; a population is random-mated if the individuals mate without exercising choice and mate with their relatives at a frequency determined by chance. It is a perversion of the term to say that a population is random-mated if all the genetic material is distributed at random, both within and between loci, as does Blakley (1967, following Feller). Less obviously, a population is not in Hardy-Weinberg equilibrium even at a single locus under random mating if generations overlap, for the population then consists in part of adult individuals which have survived selection and whose genotypes are not, in general, in binomial proportions (Edwards, 1968; Turner, 1967 b). This fact is crucial to a correct understanding of the theory of overlapping generations.

Variance

Three measures of variance will be used. The *genotypic variance* in fitness is defined, for diploids, as the mean square deviation of the fitness of the diploid genotypes from the mean fitness:

$$V_G = \Sigma_{ij} Q_{ij}(W_{ij} - \bar{W})^2 . \tag{35}$$

The *haploid variance* in fitness is a similar deviation of the mean fitnesses of the haploid genotypes (Turner, 1967 b):

$$V_H = \Sigma_i q_i(W_i - \bar{W})^2 . \tag{36}$$

For multiple loci the i represents gametic types, which are treated as though they were alleles at a single locus; V_H is then V_m of Kojima and Kelleher (1961) and $\tfrac{1}{2} V_{TC}$ of Kimura (1965).

For two alleles we see, remembering (Eq. (9), (11)) that:

$$\begin{aligned} V_H &= q_0(W_0 - \bar{W})^2 + q_1(W_1 - \bar{W})^2 \\ &= q_0 q_1(W_0 - W_1)^2 = q_0 q_1 a^2 . \end{aligned} \tag{37}$$

The *additive* or *genic variance* in fitness is the mean square deviation of the additive (breeding) values of fitness of the genotypes about their own mean:

$$V_A = \Sigma_{ij} Q_{ij} (B_{ij} - \bar{B})^2 \tag{38}$$

where $\bar{B}$ is the mean breeding value. To compute this for a single locus, we substitute Eq. (26), but as we are interested only in variance we may subtract μ from all values. Then

$$\bar{B} = \alpha(Q_{00} - Q_{11}), \tag{39}$$

$$\begin{aligned} V_A &= Q_{11}(-\alpha)^2 + Q_{00}\alpha^2 - \alpha^2(Q_{00} - Q_{11})^2 \\ &= \alpha^2 \{Q_{00} + Q_{11} - (Q_{00} - Q_{11})^2\} \,. \end{aligned} \tag{40}$$

Another form of this is obtained by noting (from Eq. (25))

$$a = \frac{Q_{00}q_1(W_{00} - W_{01}) + Q_{11}q_0(W_{01} - W_{11})}{q_0 q_1} \tag{41}$$

and comparing this with Eq. (28), whence

$$\alpha = 2q_0 q_1 a / \{Q_{00} + Q_{11} - (Q_{00} - Q_{11})^2\} \tag{42}$$

so that

$$V_A = 2q_0 q_1 a\alpha \tag{43}$$

which is the form used by Fisher.

From Eq. (37) and (34), if the population is in Hardy-Weinberg equilibrium

$$V_A = 2q_0 q_1 a^2 = 2q_0 q_1 \alpha^2 = q_0 q_1 (d\bar{W}/dq_0)^2/2 = 2V_H \tag{44}$$

but in general this is not so (cf. Eq. (58)).

If may be helpful to think of V_H as the variance of Fisher's average excess (a) and of V_A as the variance of Fisher's average effect (α).

Dominance

The dominance of a character is the difference between the value of the heterozygote and the mean of the values of the associated homozygotes. Thus for two alleles the dominance in fitness is

$$d = W_{01} - \tfrac{1}{2}(W_{00} + W_{11}) \,. \tag{45}$$

We now specify the genotype frequencies more explicitly than hitherto; for any system of mating, and any population composition, the genotype frequencies can be described in terms of the inbreeding coefficient, F, which may be positive or negative (see any textbook):

$$\begin{aligned} Q_{00} &= (1 - F)q_0^2 + Fq_0, \\ Q_{01} &= 2(1 - F)q_0 q_1, \\ Q_{11} &= (1 - F)q_1^2 + Fq_1 \,. \end{aligned} \tag{46}$$

If $F = 0$ the population is in Hardy-Weinberg equilibrium. Some of the proofs given earlier using only the Q_{ij} can be carried out with slightly less difficulty using Eq. (46). It is easy to show that the second derivative of the mean fitness on gene frequency is

$$d^2 \bar{W}/dq_0^2 = (1 - F)\{2(W_{00} + W_{11}) - 4W_{01}\} = -4d(1 - F). \quad (47)$$

All higher derivatives are zero. If there is no dominance, $d^2 \bar{W}/dq_0^2$ is zero, and the graph of $\bar{W}$ in gene frequency is linear. Further, in the absence of dominance the graph of the three genotypic values against the number of a genes in the genotype is linear, and the additive (breeding) values of the genotypes are the same as the genotypic values (Fig. 2). Then, obviously, the genotypic variance is comprised entirely of the additive (genic) variance:

$$V_G = V_A. \quad (48)$$

If in addition the population is in Hardy-Weinberg equilibrium,

$$V_G = 2V_H, \quad (49)$$

but neither Eq. (48) nor (49) is true in general.

If a population becomes entirely inbred, so that it contains only homozygotes, then its mean fitness is

$$\bar{W}_I = \Sigma_i q_i W_{ii}. \quad (50)$$

Thus $\bar{W}_I$ is the fitness the population would have if it became completely inbred but did not change its gene frequency. It is easy to show that only when there is no dominance this is the same as the mean population fitness:

$$\bar{W} = \bar{W}_I. \quad (51)$$

Further, for any value of the inbreeding coefficient,

$$\bar{W}_I = \bar{W} - 2(1 - F)q_0 q_1 d. \quad (52)$$

When there is no dominance, the relations of a, α, and $d\bar{W}/dq_0$ become special. Thus

$$d\bar{W}/dq_0 = d\bar{W}_I/dq_0 = W_{00} - W_{11} \quad (53)$$

and, as the regression of breeding value on genotype is a straight line passing through W_{00} and W_{11}, it is obvious from the definition of α that

$$\alpha = \tfrac{1}{2}(W_{00} - W_{11}). \quad (54)$$

If we spell out a in terms of Eq. (46) we find

$$a = (1 - F)\{q_0 W_{00} + q_1 W_{01} - q_0 W_{01} - q_1 W_{11}\} + F(W_{00} - W_{11}) \quad (55)$$

and substituting $W_{01} = \frac{1}{2}(W_{00} + W_{11})$ for lack of dominance:

$$a = \frac{1}{2}(1 + F)(W_{00} - W_{11}).$$ (56)

Thus for any system of mating

$$d\bar{W}/dq_0 = 2\alpha = 2a/(1 + F),$$ (57)

$$V_A = 2q_0q_1a\alpha = 2q_0q_1\alpha^2(1 + F) = 2q_0q_1a^2/(1 + F)$$
$$= 2q_0q_1a(d\bar{W}/dq_0) = 2V_H/(1 + F) = V_G$$ (58)

if there is no dominance (cf. Eq. (44)). Kempthorne (1957) shows that $\alpha = a/(1 + F)$ and $V_A = 2V_H/(1 + F)$ and so on for any degree of dominance, but this need not concern us.

The standard textbooks of population and biometrical genetics (Falconer, 1961; Kempthorne, 1957; Li, 1955b) give much of the above mathematics, but frequently restricted to Hardy-Weinberg equilibrium, except in Kempthorne. Those who find the above algebra heavy-footed should consult the much more economical treatment by Moran (1962).

Changes in Mean Fitness under Selection Only

In an apomictic population, or in a sexual population between the beginning and end of a generation, that is excluding the rounds of mating, or in a haploid population with a negligible diploid phase (treating single loci only), the only process occurring is natural selection, which by itself can be called the survival of the fittest. Li (1955b) has expressed this mathematically, describing his theorem as a simplified version of Fisher's Fundamental Theorem (below); it is in fact a more rigorous, more general, less useful, and decidedly different theorem, as it excludes the round of mating, which is included in Fisher's theorem.

If a population consists of any number of genotypes, with survival rates W_{ij} (survival and reproduction rates in a clonal population), then from Eq. (2) the genotype frequencies after one generation are

$$Q'_{ij} = Q_{ij}W_{ij}/\bar{W}$$ (59)

so that the new mean fitness (growth rate) is

$$\bar{W}' = \Sigma_{ij}Q'_{ij}W_{ij} = \Sigma_{ij}Q_{ij}W_{ij}^2/\bar{W}$$ (60)

and the change in mean fitness is

$$\begin{aligned}
\Delta\bar{W} &= \bar{W}' - \bar{W} \\
&= (\Sigma_{ij}Q_{ij}W_{ij}^2)/\bar{W} - \bar{W} \\
&= (\Sigma_{ij}Q_{ij}W_{ij}^2 - \bar{W}^2)/\bar{W} \\
&= \Sigma_{ij}Q_{ij}(W_{ij} - \bar{W})^2/\bar{W} \\
&= V_G/\bar{W}
\end{aligned}$$ (61)

so that the gain in fitness is equal to the genotypic variance in fitness or the total variance in fitness at the beginning of the generation (strictly, the total coefficient of variation). As the variance is always positive, the mean fitness of a genetically mixed population increases; this can easily be seen, as the fitter genotypes tend to become more common (Edwards, 1968).

Although the above proof is given for diploids, it obviously applies to any kind of organism; if generations overlap, then

$$d\bar{M}/dt = \Sigma_{ij} Q_{ij}(M_{ij} - \bar{M})^2 = V_G \tag{62}$$

so that, as Feller (1967) observes, a genetically mixed population is not Malthusian but has a steadily increasing Malthusian parameter.

The theorems which follow include the process of sexual reproduction and are therefore more elaborate.

Changes in Mean Fitness at a Single Locus

Non-overlapping Generations According to Li

If the fitness values of the diploid genotypes are constant, and the population is in Hardy-Weinberg equilibrium, or is inbred with constant F, it is easy to show that $\bar{W}$ and q_0 are related by a quadratic equation. Hence to find the change in mean fitness caused by a change in gene frequency, we need the first two terms of Taylor's expansion, so following Li (1967a)

$$\Delta\bar{W} = (\Delta q_0)\frac{d\bar{W}}{dq_0} + \frac{(\Delta q_0)^2}{2!}\frac{d^2\bar{W}}{dq_0^2}, \tag{63}$$

and remembering Eq. (15) and (47) this becomes, for random mating

$$\Delta\bar{W} = \frac{q_0 q_1}{2\bar{W}}\left(\frac{d\bar{W}}{dq_0}\right)^2 + \frac{q_0^2 q_1^2}{8\bar{W}^2}\left(\frac{d\bar{W}}{dq_0}\right)^2(-4d), \tag{64}$$

and so from Eq. (44) and (52)

$$\begin{aligned}
\Delta\bar{W} &= V_H\left(2 - \frac{2q_0 q_1 d}{\bar{W}}\right)\Big/\bar{W} \\
&= V_H\left(\frac{\bar{W} + \bar{W} - 2q_0 q_1 d}{\bar{W}}\right)\Big/\bar{W} \tag{65} \\
&= \frac{V_H}{\bar{W}} \times \frac{\bar{W} + \bar{W}_I}{\bar{W}}
\end{aligned}$$

if mating is random, the derivatives and means being taken in the zygotic population.

If there is no dominance, from Eq. (51), Eq. (58)

$$\Delta \bar{W} = 2V_H/\bar{W} = V_A/\bar{W} \ . \tag{66}$$

If both homozygotes are lethal, $\bar{W}_I = 0$ and

$$\Delta \bar{W} = V_H/\bar{W} \ . \tag{67}$$

If a is a dominant lethal,

$$\Delta \bar{W} = V_H/\bar{W} \times 1 + 1/q_1 \tag{68}$$

which tends to infinity as q_1 tends to zero. Thus there is always a gain in fitness (V_H being positive), which is between one and infinity times the haploid variance, but which in normal circumstances will be about twice the haploid variance, or about the additive variance in fitness.

In the de Finetti triangle (Fig. 1), the zigzag path of the population crosses contours of equal $\bar{W}$, which can be shown to be linear (Cannings and Edwards, 1968). Under selection the change in fitness is $V_G/\bar{W}$. At mating some of the fitness gained is lost, and the net gain is given by Eq. (65). If there is no dominance, the initial gain ($V_G/\bar{W}$) equals the net gain ($V_A/\bar{W}$) as $V_G = V_A$ (Eq. (58)). No fitness is lost at mating as the population moves vertically and the $\bar{W}$ contours are vertical (obviously, as $\bar{W} = \bar{W}_I$) (Fig. 1). More simply, without dominance the gain under selection equals the net gain, as selection of genotypes is virtually the same as selection of genes.

The consequence of Li's theorem is thus that fitness maximizes and that the change per generation is roughly equal to the additive variance. The stable equilibrium is at the point of maximum fitness.

The form of Eq. (65) is not precisely that due to Li, but a contracted version, due to Turner (1967b), of the form derived, by another method, by Scheuer and Mandel (1959); this can be shown to be identical with the equation obtained by Li (Mandel, 1968).

Non-overlapping Generations with Inbreeding

Li's theorem is readily extended to populations with any value of F. First note that with inbreeding

$$W_0 - W_1 = \tfrac{1}{2}d(\bar{W} + F\bar{W}_I)/dq_0 \tag{69}$$

(Li, 1955a). So

$$\begin{aligned} d\bar{W}/dq_0 &= 2(W_0 - W_1) - F(d\bar{W}_I/dq_0) \\ &= 2a - F(W_{00} - W_{11}) \ . \end{aligned} \tag{70}$$

Then, remembering Eq. (45), (46), (47), and (52), $\bar{W} = q_0 W_0 + q_1 W_1$, and taking higher derivatives in

$$\bar{W} = (1 - F)(q_0^2 W_{00} + 2q_0 q_1 W_{01} + q_1^2 W_{11}) + F(q_0 W_{00} + q_1 W_{11}) \quad (71)$$

we find

$$\begin{aligned}
\Delta \bar{W} &= (\Delta q_0) \frac{d\bar{W}}{dq_0} + \frac{(\Delta q_0)^2}{2!} \frac{d^2 \bar{W}}{dq_0^2} + (\Delta F) \frac{d\bar{W}}{dF} \\
&\quad + \frac{2(\Delta F)(\Delta q_0)}{2!} \frac{d^2 \bar{W}}{dF dq_0} + \frac{3(\Delta F)(\Delta q_0)^2}{3!} \frac{d^3 \bar{W}}{dF dq_0^2} \\
&= (q_0 q_1 a^2 / \bar{W})\{2 - 2q_0 q_1 d(1 - F)/\bar{W}\} - F q_0 q_1 a(W_{00} - W_{11})/\bar{W} \\
&\quad - (\Delta F) 2q_0 q_1 d \{1 - a(q_0 - q_1)/\bar{W} - q_0 q_1 a^2 / \bar{W}^2\} \quad (72) \\
&= \frac{V_H}{\bar{W}} \times \frac{\bar{W} + \bar{W}_I}{\bar{W}} - F \frac{\text{COV}(W_H, W_G)}{\bar{W}} \\
&\quad - \frac{(\Delta F) 2q_0 q_1 d W_0 W_1}{\bar{W}}
\end{aligned}$$

where ΔF is the *change* (not increment) in inbreeding, and the covariance between genic and genotypic fitness is

$$\text{COV}(W_H, W_G) = q_0 q_1 (W_0 - W_1)(W_{00} - W_{11}). \quad (73)$$

The change in fitness given by Eq. (72) is not necessarily positive, even with a constant value of F. This is most easily seen from the fact that with no change in inbreeding

$$\Delta \bar{W}_I = \text{COV}(W_H, W_G)/\bar{W} \quad (74)$$

$$\Delta(\bar{W} + F\bar{W}_I) = \frac{V_H}{\bar{W}} \times \frac{\bar{W} + \bar{W}_I}{\bar{W}} . \quad (75)$$

so that with a stable equilibrium it is the function $(\bar{W} + F\bar{W}_I)$ which increases to a maximum. As the maximum of this does not, obviously, correspond with the maximum of $\bar{W}$, the mean fitness must decrease when a population travels between the maximum of $\bar{W}$ and the maximum of $(\bar{W} + F\bar{W}_I)$ (see also Fig. 1a).

In the absence of dominance, we apply the same geometrical properties of the de Finetti triangle as above, and see that the initial and net gain in fitness must be $V_G/\bar{W} = V_A/\bar{W}$. This can be confirmed, using Eq. (58) and (72) as follows:

$$\Delta \bar{W} = \frac{2V_H}{\bar{W}} \left(1 - \frac{F}{1 + F}\right) = 2V_H/(1 + F)\bar{W} = V_A/\bar{W} = V_G/\bar{W} . \quad (76)$$

This does not take into account the variance in gene frequency, nor does it say what will be the equilibrium value of F, which is discussed by Jain and Workman (1967).

Overlapping Generations According to Fisher

Fisher's treatment (1930, 1941, and 1958) is, to me, very difficult to follow, and perhaps other geneticists find this also; the very clear analysis by Edwards (1968) is a great help towards understanding, as is Kempthorne's (1957) detailed treatment of the statistics.

Having proved the formula for the variance (Eq. (43) of this paper) Fisher argues as follows (I have changed the notation and the argument slightly in the hope of making it clearer; Fisher's proof of 1958 is considerably revised, notably by the inclusion of the factor 2, which he omitted in 1930):

The change in gene frequency is (see Eq. (12))

$$dq_0/dt = q_0 q_1 a . \tag{77}$$

For a change in gene frequency the change in mean fitness is given by

$$d\bar{M}/dt = 2\alpha(dq_0/dt) . \tag{78}$$

Hence the change in mean fitness is (from Eq. (43))

$$d\bar{M}/dt = 2q_0 q_1 a\alpha = V_A . \tag{79}$$

Fisher gives no justification for Eq. (78); he states it as intuitively obvious. The obvious equation is

$$\frac{d\bar{M}}{dt} = \frac{d\bar{M}}{dq_0} \times \frac{dq_0}{dt} \tag{80}$$

and this is only the same as Eq. (78) when the population is in Hardy-Weinberg equilibrium and hence $d\bar{M}/dq_0 = 2\alpha$ (see Eq. (34)) or if there is no dominance (below). As Fisher did not restrict the conditions for his theorem, he seems not to have realized that the derivative was not in general equal to twice the average effect, and to have confused regression of additive value on genotype with the derivative of mean value on gene frequency.

I think this is well shown by his presentation of 1941. He defines average effect verbally as the effect on the mean of the population of substituting one allele for another. We might expect this to be, in mathematics, the derivative of mean on gene frequency, but he then asserts that it is the partial regression of mean on the number of alleles in the three possible genotypes (i.e. the definition of α given earlier in this paper). He then goes on to show that $a = \alpha$ only under "random mating" [sic],and follows with a very obscure paragraph which proves that $\alpha = \frac{1}{2}d\bar{M}/dq_0$ (i.e. that changes in mean can be predicted by using the regression coefficient) only if $\frac{1}{4}Q_{01} = \lambda Q_{00} Q_{11}$ (Edwards, 1968). He does not clearly explain the meaning of this last equation; as shown below, it implies Hardy-Weinberg equilibrium, but there may perhaps be other bizarre special cases which satisfy it. So, initially (Fisher, 1930) he seems to have been convinced that the derivative was the same as the regression, this being the reason he omitted the factor 2 in 1930. By 1941 he had introduced the factor 2 and a condition which restricted the circumstances in which the derivative was the same as (twice) the regression, but he had not realized the implications of the restriction.

Fisher's (1941) uncouth criticism of Wright's equation (Eq. (15) in this paper) confirms that Fisher's thinking on α and $d\bar{M}/dq_0$ was confused; I will use Eq. (20) to avoid discussing logarithms. The two-allele version of Eq. (20) can be written, using absolute fitness

$$(dq_i/dt)/q_0q_1 = \tfrac{1}{2}(d\bar{M}/dq_0). \tag{81}$$

Noting that

$$dq_i/dt = q_0q_1a \tag{82}$$

Fisher points out that the left hand side of Wright's Eq. (81) is a, and that the right hand side is α; therefore, he says, Wright is merely asserting that $a = \alpha$. Of course, the right hand side is only α under Hardy Weinberg equilibrium, and as this is in any event the condition for the derivation of Wright's equation, and as this makes $a = \alpha$, Fisher has not destroyed Wright's equation, but merely discovered its restrictions via the back door.

It follows from Eq. (57) (with $\bar{W}$ replaced by $\bar{M}$), that Fisher's statement of his theorem is valid also for any mating system if there is no dominance.

Kempthorne (1957) gives a different proof that Fisher's theorem is exactly true only for a population with a constant value of λ, this being generalized for any number of alleles.

Populations with Constant λ

In summary then, Fisher recognized that it was necessary that $\tfrac{1}{4}Q_{01}^2 = \lambda Q_{00}Q_{11}$, where λ = constant (Eq. 33), for α to be $\tfrac{1}{2}d\bar{M}/dq_0$, and hence for his derivation of the fundamental theorem to be valid, but he did not realize how restrictive this condition was. Up to now I have asserted that the only genetically meaningful interpretation is $\lambda = 1$ and hence Hardy Weinberg equilibrium; this condition is satisfied by the zygotic portion of a random-mating population, but not by the whole population, and hence there are no real populations with overlapping generations for which Fisher's theorem is true. This assertion is now defended.

Cannings and Edwards (1968) studied the family of curves given by Eq. (33) in the de Finetti diagram (Fig. 1). They are a series of conics, parts of hyperbolas for $\lambda > 1$, a parabola for $\lambda = 1$, and segments of ellipses for $\lambda < 1$. It seems most unlikely that any system of mating can give a sequence of populations lying on one of these conics, with the exception of random mating giving the parabola. Inbreeding with constant F gives a set of parabolas with the equation

$$\tfrac{1}{4}Q_{01}^2 = Q_{00}Q_{11} - 2q_0q_1F \tag{83}$$

(Haldane and Moshinsky, cited by Cannings and Edwards, 1968), which is obviously not the same as the ellipses with $\lambda < 1$. Any system of assortative mating giving one of the conics would involve frequency dependent selection and thus invalidate the fundamental theorem.

Initially it seems that a random mated population under selection has $\lambda = $ constant. Imagine a population with two age classes, zygotes and adults, in which zygotes are subjected to selection before becoming adults. Following Cannings and Edwards, the genotype frequencies in the adults are

$$Q_{00}' = q_0^2 W_{00}/\bar{W}, \qquad Q_{01}' = 2q_0q_1 W_{01}/\bar{W}, \qquad Q_{11}' = q_1^2 W_{11}/\bar{W} \tag{84}$$

and these satisfy the equation (dropping the primes)

$$\tfrac{1}{4} W_{00} W_{11} Q_{01}^2 = W_{01}^2 Q_{00}Q_{11} \tag{85}$$

and $\tfrac{1}{4}Q_{01}^2 = \lambda Q_{00}Q_{11}$ where $\lambda = W_{01}^2/W_{00}W_{11} = $ constant. The average value of λ for the whole population is therefore some fraction of the constant $W_{01}^2/W_{00}W_{11}$, the fraction depending on the age-structure. However, this fraction is not constant, as the higher the value of $\bar{W}$ the more individuals survive to adulthood, and the higher the proportion of adults in the population. Thus for the whole population λ is not constant, and although special cases may be discovered, it is unlikely that Eq. (33) is satisfied for any real population.

Introduction of changing age-structure may be said to destroy an essential assumption of the overlapping generation equations, and to stretch them to a limit for which they were not intended; Kimura (1958) is careful to point out that they are only an approximation. In this way, we can regard Fisher's theorem as valid for random mating, and it is fair to say that introducing age-structure in this'way hardly makes the equations realistic in ecological terms. The equations for separate generations, which deal with zygotes only, are not subject to these difficulties.

Note that although all the conics of constant λ give $\alpha = \tfrac{1}{2}d\bar{M}/dq_0$, only the Hardy-Weinberg parabola gives $a = \tfrac{1}{2}d\bar{M}/dq_0$; this will be used at the end of the chapter to show that $\bar{M}$ is not maximum at a non-trivial equilibrium.

Overlapping Generations According to Kimura

This brilliant treatment of the Fundamental Theorem (Kimura, 1958) eliminates the flaws in Fisher's theorem. Basically, Kimura finds the value of the discrepancy caused by the fact that 2α is not in general equal to $d\bar{M}/dq_0$. Readers should consult the original for details of the manipulations or Kimura (1960) for a two-allele version, but an outline may be helpful (see also Moran, 1962). If the M_{ij} are constant, then

$$
\begin{aligned}
d\bar{M}/dt &= \Sigma_{ij}(d\bar{M}/dQ_{ij})\,(dQ_{ij}/dt) \\
&= \Sigma_{ij}M_{ij}(dQ_{ij}/dt)\,.
\end{aligned}
\tag{86}
$$

If we let

$$
Q_{ij} = q_i q_j \theta_{ij}
\tag{87}
$$

where

$$
\theta_{ij} = Q_{ij}/q_i q_j
\tag{88}
$$

measures the departure of the genotype frequencies from Hardy-Weinberg proportions, the time derivative in Eq. (86) may be written

$$
dQ_{ij}/dt = q_i q_j(d\theta_{ij}/dt) + q_i \theta_{ij}(dq_j/dt) + q_j \theta_{ij}(dq_i/dt)\,.
\tag{89}
$$

Further

$$
M_{ij} = \bar{M} + (M_{ij} - \bar{M})
\tag{90}
$$

so that

$$
\begin{aligned}
d\bar{M}/dt &= \Sigma_{ij}M_{ij}(dQ_{ij}/dt) \\
&= \bar{M}\Sigma_{ij}(dQ_{ij}/dt) + \Sigma_{ij}(M_{ij} - \bar{M})\,(dQ_{ij}/dt) \\
&= \Sigma_{ij}(M_{ij} - \bar{M})\,(dQ_{ij}/dt)
\end{aligned}
\tag{91}
$$

as the sum of the changes in genotype frequencies must be zero. Remembering that $dq_i/dt = q_i(M_i - \bar{M})$ Kimura then substitutes Eq. (89) in (91), and, using the least squares equation for regression, extracts the term $\Sigma_i q_i(M_i - \bar{M})\alpha_i = \frac{1}{2}V_A$ from each of two of the terms (α_i being the partial regression of phenotypic value in the i-th allele). The remainder of these two terms, plus the first term from Eq. (89) sum to

$$
\begin{aligned}
\Sigma_{ij}(M_{ij} - \bar{M} - \alpha_i - \alpha_j)q_i q_j(d\theta_{ij}/dt) \\
= \Sigma_{ij}d_{ij}Q_{ij}(d\theta_{ij}/dt)/\theta_{ij} \\
= \Sigma_{ij}Q_{ij}d_{ij}(d\log\theta_{ij}/dt)\,,
\end{aligned}
\tag{92}
$$

d_{ij} being the dominance deviation of the ij-th genotype

$$
d_{ij} = M_{ij} - B_{ij},
\tag{93}
$$

if B_{ij} is the breeding value of fitness (as in Eq. (26)). Thus finally

$$d\bar{M}/dt = 2\Sigma_i q_i \alpha_i (M_i - \bar{M}) + \Sigma_{ij} Q_{ij} d_{ij}(d\log\theta_{ij}/dt)$$
$$= V_A + \overline{d_{ij}(d\log\theta_{ij}/dt)} \qquad (94)$$

the bar denoting the mean. The definition of V_A here is an extension of the two-allele definition (Eq. (43)).

So the rate of change of fitness is equal to the additive variance if, as we noted above, the population is in permanent Hardy-Weinberg equilibrium ($\theta_{ij} = 1$ for all ij) or if there is no dominance ($d_{ij} = 0$ for all ij); or, as will be argued, if λ is constant; otherwise there is an additional quantity in the rate of change, given by the average product of the dominance deviation of a genotype and the rate of change of the logarithm of its coefficient of departure from Hardy-Weinberg equilibrium.

There is nothing in this statement to make us think that fitness necessarily increases; while V_A must be positive, the second term might be negative and greater than V_A. With no dominance fitness must increase, and any system with constant θ_{ij} makes the second term zero; however, permanent Hardy-Weinberg equilibrium seems to be the only example (inbreeding with constant F does not give constant θ_{ij}), and with overlapping generations this is extremely difficult, if not impossible, to attain. It will be shown later that $\bar{M}$ is not maximum at nontrivial equilibrium, and therefore it follows that $\bar{M}$ can decrease under selection and random mating.

Multiple Alleles

Both Fisher (1958) and Kimura derive their theorems – Eq. (79) and (94) – for any number of alleles; Fisher's equation is, like his two-allele equation, approximate, simply stating that the rate of change of mean is the additive variance, summed over all alleles, and using the average excess of the individual alleles (Eq. (9)) instead of a, and partial regressions, α_i, instead of α. What is the multi-allele version of Li's Eq. (65)? I have attempted to prove by symmetry and induction that Eq. (65) is valid for any number of alleles where the variance and means are summed for all alleles (Turner, 1967b). Professor Li (private communication) has pointed out to me that this formula does not give the correct value for $\Delta\bar{W}$ for three alleles, as can be shown by computing the changes in the allele frequencies, then using these to obtain the new genotype frequencies after random mating, and hence calculating $\bar{W}'$, the new value of $\bar{W}$. The values of $\bar{W}' - \bar{W}$ and $V_H(\bar{W} + \bar{W}_I)/\bar{W}^2$ are compared, for various schemes of selection, in Table 1. The error in the proof by symmetry and induction

Table 1. *Values of $\Delta \bar{W}$ calculated by simulation and by various formulae, for four schemes of selection with three alleles (initial frequencies $q_0 = 0.5$, $q_1 = 0.1$, $q_2 = 0,4$) and one scheme for four alleles ($q_0 = 0.4$, $q_1 = 0.1$, $q_2 = 0.3$, $q_3 = 0.2$)*

Selection scheme	$\bar{W}' - \bar{W}$	$2V_H/\bar{W}$	$V_H(\bar{W} + \bar{W}_1)/\bar{W}^2$	$2V_H/\bar{W} + \text{COVAR}$ $(W_i, W_j, W_{ij})/\bar{W}^2$
I (overdominance)	0.000,770	0.000,779	0.000,508	0.000,770
II (underdominance)	0.459,035	0.394,709	0.471,703	0.459,035
III (mixed dominance)	0.130,620	0.145,532	0.134,694	0.130,620
IV (no dominance)	0.809,091	0.809,091	0.809,091	0.809,091
V (mixed dominance)	0.105,287	0.093,191	0.076,338	0.105,287

is the assumption that the formula for $\Delta \bar{W}$ must consist entirely of elements of the variance and of the two means, an assumption which is clearly not justified.

The correct formula for $\Delta \bar{W}$ for multiple alleles seems to be derived as follows:

$$\Delta \bar{W} = \Sigma_i(\Delta q_i)(\partial \bar{W}/\partial q_i) + \Sigma_{ij}(\Delta q_i)(\Delta q_j)(\partial^2 \bar{W}/\partial q_i q_j)/2, \quad (ij \neq ji) \quad (95)$$

higher derivatives being zero. If the first term is $2V_H/\bar{W}$ where V_H sums over all alleles and if

$$\partial^2 \bar{W}/\partial q_i q_j = 2W_{ij}, \quad (96)$$

then (95) yields

$$\Delta \bar{W} = 2V_H/\bar{W} + \text{COVAR}(W_i, W_j, W_{ij})/\bar{W}^2 \quad (97)$$

where

$$\text{COVAR}(W_i, W_j, W_{ij}) = \Sigma_{ij} q_i q_j (W_i - \bar{W})(W_j - \bar{W})W_{ij} \quad (98)$$

or, altering the restrictions to $i \neq j$, $ij = ji$, so that homozygotes are separated from heterozygotes

$$\text{COVAR}(W_i, W_j, W_{ij}) = \Sigma_i q_i^2 (W_i - \bar{W})^2 W_{ii}$$
$$+ 2\Sigma_{ij} q_i q_j (W_i - \bar{W})(W_j - \bar{W})W_{ij}. \quad (99)$$

I have not succeeded in proving with any degree of rigour that the assumptions about the first term of Eq. (95) (but see Kimura, 1958) and about the derivatives Eq. (96) are true; an intuitive indication is given in Turner (1969), where Eq. (97) is generalized for any metric character. But from the numerical examples (Table 1), Eq. (97) does seem to be the correct equation. The genotypic fitness values used in deriving Table 1 are as follows:

Three alleles:

(I) $W_{00} = 1,$ $W_{11} = W_{01} = 2,$ $W_{22} = 0,$ $W_{02} = 4,$ $W_{12} = 3,$

(II) $W_{00} = 4,$ $W_{11} = 3,$ $W_{22} = W_{02} = 2,$ $W_{01} = 0,$ $W_{12} = 1,$

$$(100)$$

(III) $W_{00} = W_{01} = W_{02} = 1,$ $W_{11} = 3,$ $W_{22} = 0,$ $W_{12} = 2,$

(IV) $W_{00} = 4,$ $W_{11} = W_{02} = 2,$ $W_{22} = 0,$ $W_{01} = 3,$ $W_{12} = 1$

Four alleles:

(V) $W_{00} = W_{03} = W_{23} = 1,$ $W_{11} = W_{01} = 2,$ $W_{22} = W_{13} = 0,$

$$(101)$$

$W_{33} = W_{12} = 3,$ $W_{02} = 4.$

It is noteworthy that in example (V) the factor $(\bar{W} + \bar{W}_I)/\bar{W}$ reduces the value of $\varDelta \bar{W}$ below $2V_H/\bar{W}$, whereas the actual change is greater than $2V_H/\bar{W}$.

It can be shown, by working from the change in fitness without mating (Eq. (61)), that the second term in Eq. (97) is zero when there is no dominance. What is not clear from Eq. (97) is whether $\varDelta \bar{W}$ is necessarily positive under all circumstances; but this is proved to be so by the *General Inequalities* discussed below. For two alleles, Eq. (97) reduces to (65).

Since this chapter went to press, Li (1969) has given an identical proof of Eq. (97), which he writes very concisely in matrix notation.

Equivalence of Kimura's and Li's Theorems

In the absence of dominance, for any system of mating, we have seen that both with overlapping and separate generations the change in mean fitness equals the additive variance in fitness (compare Eq. (76) and (94)). It might be thought that with dominance and random mating the theorems are not equivalent, as Kimura's appears to give $\theta_{ij} = 1$, $d\bar{M}/dt = V_A = 2V_H$ whereas Li's gives only approximately $2V_H$. But random mating does not give $\theta_{ij} = 1$, or even $\theta_{ij} = $ constant, in the same way that it does not give $\lambda = $ constant, and there is little discrepancy between the theorems. (Here ij includes ii.)

The following demonstration is not completely rigorous, but improves on the one given before (Turner, 1967 b):

If dominance is positive, in Eq. (65)

$$(\bar{W} + \bar{W}_I)/\bar{W} = 2 - 2q_0 q_1 d/\bar{W} < 2 \qquad (102)$$

so that $\varDelta \bar{W}$ is less than $2V_H$.

In a population with overlapping generations, if $\lambda = $ constant, as we have seen $d\bar{M}/dt = V_A$, and therefore the θ term of Kimura's equation must be zero. Using the dominance deviations (Falconer, 1961)

$$d_{00} = -2q_1^2 d\,,$$
$$d_{01} = 2q_0 q_1 d\,, \tag{103}$$
$$d_{11} = -2q_0^2 d$$

we can write the term in full

$$2q_0^2 q_1^2 d(-d\log\theta_{00}/dt + d\log\theta_{01}/dt - d\log\theta_{11}/dt) \tag{104}$$

assuming that Hardy-Weinberg proportions give a close enough approximation. It is easy enough to show that for constant λ, θ_{01} is constant, so that the middle term of Eq. (104) is zero, and for the whole expression to be zero the other terms must cancel each other. But if dominance is positive, then in the selected portion of the population heterozygotes are in excess of expectation ($\theta_{ij} > 1$) and homozygotes are deficient ($\theta_{ii} < 1$). If we allow that the age-structure can change, then under selection more of the zygotes survive to adulthood and the excess of heterozygotes over expectation becomes less, so that $d\log\theta_{01}/dt$ is negative. Similarly $d\log\theta_{00}/dt$ and $d\log\theta_{11}/dt$ tend a little more to the positive than they would if age structure were constant. Thus given the signs in Eq. (104), the total θ term is negative, and $d\bar{M}/dt$ is rather less than V_A, or less than $2V_H$, just as $\Delta\bar{W}$ is in Li's theorem. Similarly in both formulae the change in fitness will be greater than $2V_H$ if dominance is negative.

This argument clearly requires generalizing to the form of an integral over any number of age-classes.

Thus for a single locus Kimura's term in θ_{ij} describes several different phenomena: departure from Hardy-Weinberg equilibrium caused by selection, changes in nonrandom assortment caused by inbreeding with constant F, and changes in the value of F; these are given as separate terms in Eq. (72).

Non-overlapping Generations According to Kempthorne

By a fairly elaborate proof, Kempthorne (1957) derives a theorem for the change in the mean of any metric character under selection; reduced to the change in fitness this is

$$\Delta\bar{W} = V_A/\bar{W} + \tfrac{1}{2}\Sigma_{ij}Q_{ij}(\Delta\lambda_{ij})\,d_{ij}/\lambda_{ij} \quad (i \neq j) \tag{105}$$

where $\lambda_{ij} = Q_{ij}^2/Q_{ii}Q_{jj}$. With non-overlapping generations under random mating one may consider only zygotes, so that the λ_{ij} term is zero in this case, and we see by comparison with Eq. (65) that this theorem (Eq (105)) must be a first order approximation. If the difference terms are changed to time derivatives, the theorem seems to be the same as Kimura's and would give a change of less than V_A under random mating with positive dominance, but this needs further investigation.

Summary for Single Loci

With non-overlapping generations and random mating, mean fitness cannot decrease; its increase is equal roughly to the additive variance in fitness. With inbreeding, even though F is constant, mean fitness may decrease under selection. With overlapping generations and random mating the rate of change of fitness is equal roughly to the additive variance. With no dominance, the rate of change of fitness is equal to the additive variance, which in turn is equal to the total genotypic variance, for any system of generations or of mating. The formulae for overlapping and non-overlapping generations give very similar results.

General Inequalities

Several authors (Scheuer and Mandel, 1959; Mulholland and Smith, 1959; Atkinson, Watterson, and Moran, 1960; Kingman, 1961) have proved that for any number of alleles under random mating the mean fitness of a zygotic population is greater than or equal to the mean fitness of the previous zygotic generation:

$$\bar{W}' \geqq \bar{W}, \quad \Delta \bar{W} \geqq 0 . \tag{106}$$

From a genetical point of view this is important, as it shows that the change given by Eq.(97) cannot be negative. Also, if as Blakley (1967) states, these inequalities are of very general application, for example in learning theory, their further exploration and application may be interesting biologically as well as mathematically.

Changes in Mean Fitness at Two Loci

We introduce two new kinds of parameter – departure from random combination of alleles at the two loci (gametic excess, linkage disequilibrium) and departure from additive interaction in fitness between the loci (epistasis). If the gametes ab, aB, Ab, AB occur with frequencies q_0, q_1, q_2, q_3 and the genes with frequencies q_a, q_A, q_b, q_B, then gametic excess is measured by the determinant

$$D = q_0 q_3 - q_1 q_2 \tag{107}$$

so that the gamete frequencies are

$$q_0 = q_a q_b + D \,,$$
$$q_1 = q_a q_B - D \,,$$
$$q_2 = q_A q_b - D \,,$$
$$q_3 = q_A q_B + D \,.$$

$$(108)$$

With random combination of the two loci D is zero. Alternatively non-random association is measured by

$$X = q_0 q_3 / q_1 q_2 \tag{109}$$

which is unity with random combination.

The meaning of epistasis is most easily seen by writing the fitness in terms of selection coefficients (Table 2); the u_i are the parameters of epistasis. The parameters usually used (e.g. Cockerham, 1954) have the same values as the u_i, but different signs. To convert the u_i of Table 2, which are relative parameters, to absolute U_i, it is necessary to multiply all fitnesses by W_h, the fitness of the double heterozygote $AaBb$.

Table 2. *Fitness values* (w_{ij}) *at two loci, written in terms of selection coefficients*

	AA	Aa	aa
BB	$1 - s_1 - t_1 - u_3$	$1 - t_1$	$1 - s_0 - t_1 - u_1$
Bb	$1 - s_1$	1	$1 - s_0$
bb	$1 - s_1 - t_0 - u_2$	$1 - t_0$	$1 - s_0 - t_0 - u_0$

If we write the genotype frequencies under random mating in terms of the gene frequencies and D and multiply them by the fitnesses in Table 1 to obtain $\bar{W}$, we find (Bodmer and Felsenstein, 1967; Turner, 1967a, c)

$$\bar{W} = \bar{W}^* - 2D(q_a q_b U_0 - q_a q_B U_1 - q_A q_b U_2 + q_A q_B U_3)$$
$$- D^2(U_0 + U_1 + U_2 + U_3)$$

$$(110)$$

where $\bar{W}^*$ is the fitness the population would have if D was reduced to zero but the gene frequencies held constant. For brevity rewrite Eq. (110):

$$\bar{W} = \bar{W}^* - 2D\eta - D^2 \Sigma U_i \,. \tag{111}$$

Then it is possible to show (Turner, 1967c) that

$$W_0 - W_1 - W_2 + W_3 = -(\eta + D\Sigma U_i) = \tfrac{1}{2}(d\bar{W}/dD) \tag{112}$$

$$-\Sigma U_i = \tfrac{1}{2}(d^2 \bar{W}/dD^2) \tag{113}$$

where the W_i are defined as before Eq. (17).

The change in fitness is found as follows: If there were no recombination, the change in fitness would be the same as for four alleles:

$$\Delta \bar{W} = V_{H(2)}/\bar{W} + \text{COVAR}(W_i, W_j, W_{ij})/\bar{W}^2 \tag{114}$$

where the means and variances sum over all four gametes. $V_{H(2)}$ is thus half the total additive variance at the two loci jointly and includes the additive variances at both loci individually and part of the additive × additive interaction variance (Kojima and Kelleher, 1961).

Kimura (1966) gives a detailed breakdown of V_H ($\frac{1}{2} V_{TC}$ in his notation) for two loci and finds it to be comprised as follows, assuming random mating

$$V_{H(2)} = \frac{1}{2} \Sigma V_A + (\frac{1}{2} d\bar{W}/dD)^2/J \tag{115}$$

$$= \Sigma V_{H(1)} + f V_{AA} \tag{116}$$

where the summation is over both loci (but not orthogonally, see below), the number subscripts indicate how many loci are considered ($V_{H(1)}$ being V_H in previous sections), $f V_{AA}$ indicates a portion of the additive × additive epistatic variance, and

$$J = 1/q_0 + 1/q_1 + 1/q_2 + 1/q_3. \tag{117}$$

The exact size of f seems to be unknown, as a full understanding of the analysis of variance for two loci has not yet been achieved, not by the writer at least.

The effects of recombination are found by noting that the change in D under recombination is (Turner, 1967d)

$$\Delta D = - R D_z W_h/\bar{W} \tag{118}$$

where R is the recombination fraction and D_z the value of D in the zygotic population. Substitution of Eq. (112)–(113), (118) in Taylor's expansion

$$\Delta \bar{W} = (\Delta D)(d\bar{W}/dD) + \frac{1}{2}(\Delta D)^2 (d^2 \bar{W}/dD^2) \tag{119}$$

gives the loss in fitness due to recombination and thus the total change in fitness is

$$\Delta \bar{W} = \frac{2 V_{H(2)}}{\bar{W}} + \frac{\text{COVAR}(W_i, W_j, W_{ij})}{\bar{W}^2} - \frac{R D_z W_h}{\bar{W}} \left(-2\eta - 2 D_m \Sigma U_i + \frac{R D_z W_h \Sigma U_i}{\bar{W}} \right), \tag{120}$$

D_m being the value of D in the mature, selected population. If mutation is included, the last term becomes

$$4\{ -\mu D_m - (\tfrac{1}{2} - \mu) R D_z W_h/\bar{W} \} \\ \times [-\eta - \Sigma U_i \{ D_m - 2\mu D_m - (\tfrac{1}{2} - \mu) R D_z W_h/\bar{W} \}] \tag{121}$$

where all four mutations $(A \to a,\ a \to A,\ B \to b,\ b \to B)$ occur at the same rate μ. For the rate of change in D from mutation, see Turner (1967d); the version of Eq. (121) given in the appendix to that paper is incorrect, as a result of an error in the algebra; the sign of the last term within the bracket of the various forms of Eq. (120) given in Turner (1967b) is accidentally reversed, and the first part of the equations in that paper employ $V_{H(2)}(\bar{W} + \bar{W}_I)/\bar{W}^2$ which I then mistakenly thought was the change under selection.

If dominance is slight and ΣU_i is small, which it may be even if epistasis is strong, Eq. (120) can be rewritten

$$\Delta \bar{W} = 2 V_{H(2)}/\bar{W} + 2 R D_z W_h (\eta + D_m \Sigma U_i)/\bar{W}\ . \tag{122}$$

An equation of this form was first derived by Kojima and Kelleher (1961), using a different method, for overlapping generations. They found

$$d\bar{W}/dt = 2V_m - 2\alpha_{AB} W_h R D \tag{123}$$

where W's are measured by Malthusian parameters, V_m is equivalent to V_H, and α_{AB} is the additive $\times$ additive interaction or $(W_0 - W_1 - W_2 + W_3)$. Remembering Eq. (112) and replacing W by M, this becomes in the present notation:

$$\begin{aligned} d\bar{M}/dt &= 2V_{H(2)} - 2RD M_h(M_0 - M_1 - M_2 + M_3) \\ &= 2V_{H(2)} - RD M_h(d\bar{M}/dD) \\ &= 2V_{H(2)} + 2RD M_h(\eta + D\Sigma U_i) \end{aligned} \tag{124}$$

where all symbols are the same as before, except that they involve M and not W. This is obviously equivalent to Eq. (122).

These equations divide the change in fitness into two parts: the first term, always positive, is the change under selection and random mating; the second term is the change resulting from recombination and mutation; it may be positive or negative, although populations rapidly change into a state where D and $d\bar{W}/dD$ have the same sign, making the term negative. The total change in mean fitness may be positive or negative. The implications of these theorems are discussed in more detail by Kojima and Kelleher (1961) and Turner (1967b, c), and relevant theory will be found in the extensive literature on two loci (references in Bodmer and Felsenstein, 1967). Most of the above equations can be extended to include differences in fitness and for recombination between the coupling (AB/ab) and repulsion (Ab/aB) heterozygotes.

Kimura (1958) extends his theorem to multiple loci. Instead of the term in θ for a single locus, as in Eq. (94), the equation includes this term for each of the loci separately and a similar term for the loci jointly.

Thus for two loci we have

$$d\bar{M}/dt = V_A + \Sigma_{ij}Q_{ij(A)}d_{ij(A)}(d\log\theta_{ij(A)}/dt)$$

$$+ \Sigma_{kl}Q_{kl(B)}d_{kl(B)}(d\log\theta_{kl(B)}/dt) \qquad (125)$$

$$+ \Sigma_{ijkl}Q_{ijkl(AB)}e_{ijkl(AB)}(d\log\theta_{ijkl(AB)}/dt)$$

where the terms with double subscripts are parameters for single loci (as indicated in the brackets) as previously defined, and the terms with quadruple subscripts are for both loci jointly, Q_{ijkl} being the frequency of a two-locus genotype, e_{ijkl} its epistatic deviation — that is the difference between its fitness and the expected value found by summing the least squares additive and dominance effects of its component genes — and

$$\theta_{ijkl(AB)} = Q_{ijkl(AB)}/Q_{ij(A)}Q_{kl(B)}. \qquad (126)$$

Kimura further divides e_{ijkl} into its additive $\times$ additive, additive $\times$ dominance, and dominance $\times$ dominance components (Cockerham, 1954), and defines appropriate θ values for each. Thus using ε to represent any deviation from additivity, whether by dominance or epistasis, and θ the associated departure from random assortment, he writes finally

$$d\bar{M}/dt = V_A + \overline{\Sigma\varepsilon(d\log\theta/dt)} \qquad (127)$$

the equation being in this notation general for any number of loci.

Kimura here defines V_A (V_g in his notation) to be the total additive genic variance over all loci, which as will be shown is a non-orthogonal sum of the additive variances for all loci separately, excluding the additive $\times$ additive interaction variances. Thus the second term of Eq. (127) includes not only the phenomena for single loci already described, but changes in gametic excess due to selection and to recombination, and interaction of these changes with the single locus effects. The theorem is therefore very different in form from Eq. (120)–(124), where changes in gametic excess due to selection are included in the first term and those due to recombination are in the second term. The ingenious θ term of Eq. (127) sums up a wealth of phenomena, and Kimura's equation is much more general than Eq. (120)–(124), which are restricted to random mating, although the method of including inbreeding is clear enough.

Kojima and Kelleher (1961) show how the second term of Eq. (124) can be extended for any number of loci; the first term is already general. The equation becomes

$$d\bar{M}/dt = 2V_H - 2\Sigma M_{hh}[\Sigma R\{\Sigma(q_i q_l - q_j q_k)(M_i - M_j - M_k + M_l)\}] \qquad (128)$$

where M_{hh} is the fitness of a multiple heterozygote, R the appropriate recombination fraction, V_H is summed over all gametes, the term in q is a determinant and the term in M half of the partial derivative of $\bar{M}$ on that determinant. The equation is spelled out for three alleles by Kojima and Kelleher (1961) and by Turner (1967c).

Quasi Linkage Equilibrium

Fisher (1930) extended his theorem to multiple loci simply by adding the effects due to the single loci; in other words he wrote

$$d\bar{M}/dt = 2\Sigma q_0 q_1 a\alpha = \Sigma V_A \tag{129}$$

where the summation is over all loci and indicates total additive variance. This is obviously approximately true with weak dominance, weak epistasis, and loose linkage (Li, 1967b; Turner, 1967b), so that changes in mean fitness due to changes in D are negligible compared with those resulting from changes in gene frequency, especially in the case when D is always zero.

Kimura (1965) shows that Eq. (129) is approximately true under much more general conditions. We can see that $\varDelta \bar{W}$ would equal ΣV_A if the loss in fitness due to recombination exactly balanced the gain due to the epistatic variance. Ignoring dominance, and using Eq. (112, 115), rewrite Eq. (122)

$$\varDelta \bar{W} = \{\Sigma V_A + \tfrac{1}{2}(d\bar{W}/dD)^2/J - RDW_h(d\bar{W}/dD)\}/\bar{W}. \tag{130}$$

(See below for a note on ΣV_A.) To get $\varDelta \bar{W} = \Sigma V_A/\bar{W}$ we need the condition giving

$$d\bar{W}/dD = 2RDW_h J . \tag{131}$$

Kimura shows that under selection and random mating, using only linear approximations,

$$\varDelta \log X = \tfrac{1}{2}(d\bar{W}/dD - 2RDW_h J)/\bar{W} \tag{132}$$

(see Eq. (109) for the cross-product-ratio X) so that $\varDelta \bar{W} = \Sigma V_A/\bar{W}$ if X is constant. He further argues that with weak epistasis, high recombination and slow changes in gene frequency, populations will settle down into a state in which X is roughly constant for many generations, although gene frequencies and D are changing. In these circumstances, Eq. (129) is true, and because this is also a property of a population travelling in linkage equilibrium, Kimura calls the state of constant X "quasi linkage equilibrium". If X remains exactly constant, $\bar{W}$ cannot decrease, for Kimura (1966) shows that the stable equilibrium point is at the maximum value of $\bar{W}$ in the plane of constant X.

As the argument depends on a number of approximations, Kimura has checked it by simulating slow selection with epistasis on a computer. Initially the change in population fitness may be roughly equal to $2\ V_{H(2)}$ if the population starts with $D = 0$, or if it starts with a high value of D there may be rapid decreases from recombination (all in accord with Eq. (124)), but after about 20 generations X becomes roughly constant, and $\Delta \bar{W} = \Sigma V_A$ to a high degree of approximation.

An important point, overlooked before (Turner, 1967 b), is that ΣV_A is not simply an orthogonal sum. Kimura's partition of $V_{H(2)}$ separates an epistatic component (Eq. (115)) and the term V_{AC} or V_g which is the total additive genic variance. Kimura shows that if we fit, by least squares, the additive values $\mu + \alpha, \mu, \mu - \alpha$ for the A locus, and $\mu + \beta, \mu, \mu - \beta$ for the B locus (as in Eq. (26)), then if we consider the variance of the additive values of the two-locus genotypes (for example, $AaBB$ is $\mu + \beta$, $aaBB$ is $\mu - \alpha + \beta$) we find it is

$$V_g = 2q_A q_a \alpha^2 + 2q_B q_b \beta^2 + 4\alpha\beta D \qquad (133)$$

which is the value called ΣV_A in this account, whereas the individual values of V_A for the single loci are obviously

$$2V_{H(1)} = 2q_A q_a \alpha^2 \quad \text{and} \quad 2V_{H(1)} = 2q_B q_b \beta^2. \qquad (134)$$

There is thus a further important difference between the assumption that the loci can be treated independently, and Kimura's treatment: the cross product term $4\alpha\beta D$ is included in the additive genic variance. Extensions to non-random mating and more than two loci have not yet been made; but Kimura (1965) includes differences, here ignored, in viability between coupling and repulsion heterozygotes.

Changes in Fitness of Genotypes

Having found the change in the rate of population growth resulting from natural selection, Fisher (1930, 1958) added two terms to his equation, one allowing for decrease in population growth resulting from a change in the environment or by recurrent mutation – which would both normally decrease the mean fitness – and the other allowing for a check to population growth as the population becomes dense; using some simplifying assumptions he solved the final equation for the steady state. Kimura (1958) adds to his own equation a very general term representing the mean of all changes in the genotypic fitness values resulting from all causes and writes

$$d\bar{M}/dt = V_A + \overline{dM/dt} + \overline{\Sigma\varepsilon(d\log\theta/dt)}. \qquad (135)$$

Obviously inclusion of this middle term allows fitness to decrease, even when all other circumstances make it a nondecreasing parameter.

Like the third term of this equation, the second embraces many phenomena. We may note that changes in the M_{ij} may be secular ones, due to a change in the environment $(dM_{ij}/dE \times dE/dt)$, may be frequency dependent ones, due to changes in the genetic make-up of the population $(dM_{ij}/dQ_{kl} \times dQ_{kl}/dt)$, may be density-dependent ones, due to changes in population density (say $dM_{ij}/dN \times dN/dt$), or may depend on some joint function of genetic make-up and density, a change which Turner and Williamson (1968) call population-dependent (say $dM_{ij}/dQ_{kl}N \times dQ_{kl}N/dt$). We may also distinguish those changes which alter all absolute fitnesses (M_{ij}) in equal proportion, thus affecting neither the relative fitnesses (m_{ij}) nor the genetic composition of the population, but only the rate of growth, and those which affect the M_{ij} differentially, thus affecting the m_{ij} and also the evolution of the population. Turner and Williamson suggest a similar partition of the value of the M_{ij}, and obviously everything said here applies also to W_{ij}.

As, in general, changes in the genetic make-up of the population affect its rate of growth and its density may affect its genetic composition, altering the first and third terms, as well as the second, Eq. (135) describes a very complicated system.

Predictive Models

Wright (1932) proposed that adaptive value, or $\overline{W}$, could be used as a sort of topographic map plotted against gene frequencies; as $\overline{W}$ tends to increase, populations will move to the highest local point in this "adaptive topography". By considering the interaction between selection and the random genetic drift caused by restricted population size in determining the movement of populations on this surface, he proposed his "three-phase multiple peak" theory of evolution, in which selection, random drift, and interpopulation selection determine the course of evolution (Wright (1964) for a recent account). The adaptive topography may also be used in a more restricted way to discuss the movement and equilibria of particular populations under selection (Lewontin and White, 1960). (See also Chapter 1.)

Mean fitness is not strictly a potential function, as the change in fitness per generation is not the maximum possible change (Lewontin and Kojima, 1960); the deviation of the actual population path from the steepest path is, however, slight. I take this to be the meaning of Kimura's *maximum principle in the genetical theory of natural selection* (Kimura, 1958): *For a given short time interval δt, natural selection causes gene fre-*

quency changes $\delta q_0, \delta q_1, \ldots, \delta q_n$, *in such a way that the increase in population fitness,* $\bar{M}$, *shall be maximum under the restriction*

$$\Sigma_i(\delta q_i)^2/q_i = V_H(\delta t)^2 \,. \tag{136}$$

Kimura extends this theorem, proved initially for a single locus under random mating with constant genotypic fitnesses, to multiple loci and to nonrandom mating and variable fitness.

That mean fitness maximizes only for single loci, under random mating and with constant fitness, might seem to destroy the validity of the adaptive topography for the general case of multiple loci under non-random mating, with variable fitness, in which the equilibrium genotype frequencies are not given by the maximum value of $\bar{W}$. But Wright's three-phase theory only requires us to imagine that there are many different local points in the field of genotype frequencies to which populations may move under selection and other systematic pressures; if it helps us to imagine this as movement across a topography we may do so; for the most general purpose it matters little whether we know the formula for the parameter which maximizes. For more specific purposes, especially when we want accurately to predict population changes and equilibria, we must know the maximizing parameter. Changes in genotypic fitness resulting from changes in density can be removed, provided they affect all genotypes equally, by using $\bar{w}$ and $\bar{m}$ instead of the absolute parameters. This deprives the mean fitness of its physical meaning as the rate of population growth, but makes it easier to use for prediction. With inbreeding, the parameter $\bar{w} + F\bar{w}_I$ maximizes, provided F is constant (Wright, 1949; Li, 1955 a); if selection differs in the sexes, the product of mean fitness in males and females maximizes to a high degree of approximation (Campos Rosado and Robertson, 1966; Turner, 1968) and a slightly more elaborate parameter maximizes if the locus is X-linked (Li, 1967 c). Wright (1955) has defined an integral "fitness" function which, if it exists, always maximizes; this is conceptually helpful, but rather too abstruse for practical use.

With two loci it is very instructive to consider changes under selection (first term of Eq. (120, 122, 123)) separately from those due to recombination (second term), as this meets the difficulties felt by Moran (1964). $\bar{w}$ can be plotted against the four gamete frequencies by using a tetrahedron as the graph; it is most easily drawn by distorting the tetrahedron and then drawing $\bar{w}$ as a series of contour maps in slices of this model (Fig. 3). Selection is then thought of as pulling the population up the gradient in $\bar{w}$, and recombination as dropping the population back; equilibrium is a balance between these two forces, the stable equilibria falling short in a predictable way of the maxima of $\overline{w}$ and the unstable equilibria of points of minimum fitness. This is most easily seen by studying graphs of $\overline{w}$ against D (curve of Eq. (110)); in this way many of the properties of two locus polymorphisms are explained. One possibly important phenomenon is added to the multi-peak theory as the position of equilibria is very sensitive to changes in recombination and selection, and it is possible for a stable and unstable equilibrium to meet, thus annihilating each other and causing the population at the stable equilibrium to undergo rapid evolution.

Most of these rather elaborate methods of predicting equilibria can be generalized. It is obvious from Eq. (8) and (36) that equilibria, stable or unstable, will be points where V_H is zero, where the W elements in V_H may be functions of F or of genotype frequency; the system can be generalized to some more elaborate cases. Thus we can plot V_H instead of $\overline{W}$ in the topography. Whereas the maximizing parameters have stable equilibria at maxima, unstable at minima, both types are at zero points of V_H, and when using a V_H topography one must also use an approximate equation for ΔV_H (Turner, 1969) to determine the direction of movement at a few points on the surface. The maximizing parameters are in fact

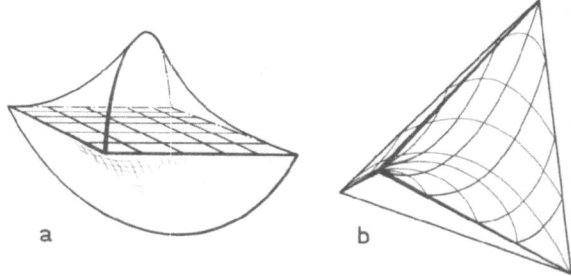

Fig. 3 a–d. Adaptive topographies for two loci. (a, b) Three-dimensional bodies in which the topography can be plotted. In the tetrahedron (b), the saddle surface represents gametic equilibrium $(D = 0)$ and the four homogeneous co-ordinates of the tetrahedron are the gamete frequencies. In (a) gametic equilibrium is the square, the two horizontal dimensions are the gene frequencies, and the vertical dimension is D (gametic excess). These models can be converted one to the other simply by distortion. (c) Contours of $\overline{W}$ drawn in horizontal and vertical slices of model (a); to guide the eye the same contour intervals have been shaded in the same way in all sections; vertical sections are a different scale from horizontal. $+, -$ indicate maximum and minimum $\overline{W}$ in the plane of the section. The matrix of figures is the fitness of the nine genotypes, as in Table 2, and the graph of $\overline{W}$ is drawn along a vertical line through the center of the model. (d) Main shapes of graph of $\overline{W}$ against D, as in (c), showing stable (closed circles) and unstable equilibria (open circles). From top to bottom – no recombination, moderate recombination, recombination of 50%. Redrawn after Turner (1967a, b) from the American Naturalist **101**, 195–221 (Univ. of Chicago Press) and Proceedings of the Royal Society London B **169**, 31–58

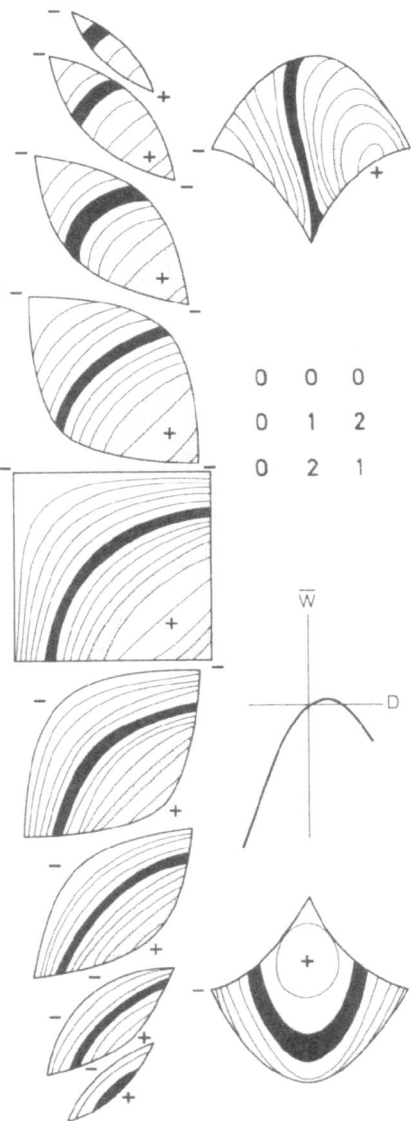

Fig. 3c.

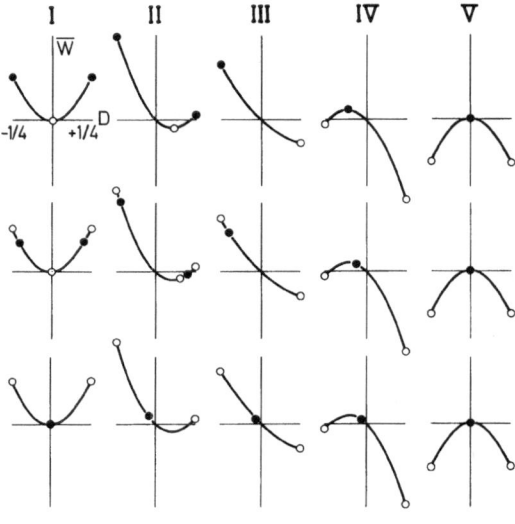

Fig. 3 d.

special cases of this theorem about V_H; for any parameter Y such that

$$V_H = \Sigma_i q_i (\partial Y / \partial q_i)^2 / n, \tag{137}$$

where n is a number, will be maximum or minimum at equilibrium (see Sacks (1967) for a case where $\bar{W}$ is minimum at stable equilibrium). The theorem of minimum V_H does no more than summarize the Δq equations; it has no physical meaning in population ecology. But then $\bar{w}$ and $\bar{m}$ have no physical meaning either.

Is Mean Fitness Maximized at Equilibrium?

We have seen that with separate generations, constant selection, and random mating, $\bar{W}$ always increases for single loci and is therefore maximum at a stable equilibrium. The mean fitness is unlikely, except in special cases, to be maximized in this way under nonrandom mating however the generations overlap. What of overlapping generations under random mating? At equilibrium there will be, by definition of equilibrium, no haploid variance in the Malthusian parameters. Similarly there will be no haploid variance in the probability that a zygote will survive and reproduce, for if there were such a variance the population composition would change. In other words

$$\Sigma_i q_i (M_i - \bar{M})^2 = \Sigma_i q_i (W_i - \bar{W})^2 = 0. \tag{138}$$

As under random mating $\overline{W}$ conforms to condition Eq. (137), being measured in zygotes, $\overline{W}$ is maximum at equilibrium. On the other hand M is measured in the whole population and does not conform to Eq. (137); thus $\overline{M}$ is not maximum at equilibrium. Therefore $\overline{M}$ can decrease under random mating and selection.

These surprising results raise the question — what is the effect of generation structure on the outcome of selection? The equations for overlapping generations often purport to describe a continuously breeding population, but they are usually used in such a way that they are merely approximations to equations for separate generations. One can be led into error by imagining that certain of the differential equations describe a real population. For instance, by assuming that the survival probability of a randomly picked individual is the same as the survival probability of a zygote, one obtains Eq. (21), as the Malthusian parameter is the log of the survival probability of a randomly picked individual. If one then assumes, incorrectly, that the population is in Hardy-Weinberg equilibrium, one obtains Eq. (20) and (23) and concludes that $\overline{\log W}$ is maximized with overlapping generations, that $\log \overline{W}$ is maximized if generations are separate (Eq. (22)); thus the generation system would alter the equilibrium point. In fact in both instances it is $\overline{W}$ or $\log \overline{W}$ which maximizes. Similarly, by assuming that $\overline{M}$ maximized with continuous generations, I found a difference in the behaviour of systems of two loci, according to the degree of overlap of the generations (Turner, 1967c); I now believe this conclusion to be wrong for obvious reasons.

The Fundamental Theorem in Ecology

Fisher's Fundamental Theorem was an attempt to weld population genetics and population ecology; there have been very few such attempts since (Sokal, 1962).

The Malthusian parameter has certain great disadvantages as a general parameter of population fitness. Relative fitness ($\overline{m}$, $\overline{w}$) has no physical meaning, and the variance theorem (above) has only a genetical meaning; it makes the banal statement that evolution continues until all the bits of genetical material are being handed on in equal proportions. Thus it is a true but not very interesting basic theorem. However, it has an ecological extension — community structure goes on changing until all competing species are equally successful or until there is no variance in competitive ability; this is another way of stating the theory of balanced competition (of species or genes) proposed by Williamson (1957). As an alternative to plotting the variance, we can plot the fitnesses of individual genes or species against population or community structure; the course of change and position of equilibria are easily determined (Fig. 4).

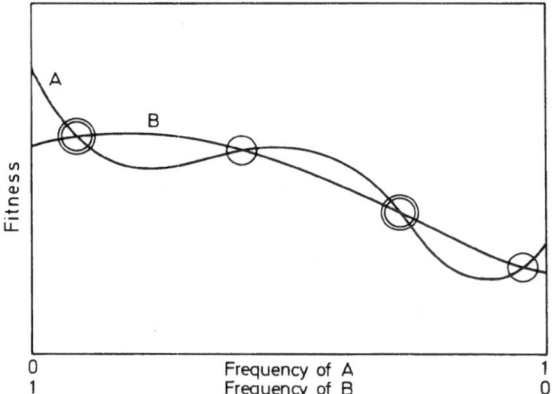

Fig. 4. Fitness of two components of a population (e.g. genes, genotypes or species) plotted against the composition of the population. Where the lines cross they produce stable (double) or unstable (single circle) equilibria. Adapted from Williamson (1957)

The absolute rate of population growth ($\bar{M}$, $\bar{W}$) is probably not often an important component of the competitive ability of a population; only when a species colonizes a new habitat or reproduces rapidly in early spring is its mean rate of growth likely to influence its success. Otherwise the short term success of the population is probably much more influenced by the way it copes with its limiting factors than by its innate rate of growth. It is most unfortunate that "fitness" means so many different things.

The theorem of increasing mean genetical fitness is thus not very successful as a fundamental theorem; the relative fitness, as mean or variance, is useful for genetical prediction but says nothing about adaptation (also Cain and Sheppard, 1954, 1956; Li, 1955a); the absolute mean fitness does not say much about adaptation. In addition, the rate of change of absolute fitness is by no means always positive, or equal to the variance: dominance, inbreeding, changes in the breeding system, frequency-dependent selection, and recombination all may invalidate this theorem. They are all aspects of the breaking up of genotypes discussed at the beginning of the chapter. Changes in mean fitness are also caused by changes in population density (C) and in environment (D), and recognizing these factors Fisher wrote his theorem

$$d\bar{M}/dt = V_A - \bar{M}/C - D .   \tag{139}$$

It is possible to save the form of the theorem by regarding the above nonadditive genetical factors affecting fitness as part of the "genetic environment" and therefore included in D (Kimura, 1958) or to say that *mean fitness increases subject to the restraints of the genetic system* (Turner, 1967b); neither of these approaches strikes me as very helpful.

Perhaps the fundamental theorem of evolution is one of increasing energy-flow or biomass, this following from the idea (MacArthur, 1962) that efficiency in using limiting resources maximizes, or of increasing homeostasis (Slobodkin, 1964), or of increasing ability not to make a mistake (Blakley, 1967); perhaps all these are the same theorem. Certainly it must involve the ability to cope with a changing environment (Levins, 1962 and after). On all these counts it must involve interpopulation selection (Lewontin, 1965) and some kind of multipeak process.

We may ask the humbler but complex question: what is the relation between genetic change and population density? Fisher's (1958) solution of Eq. (139):

$$d\bar{M}/dt = C(V_A - D) \qquad (140)$$

where $\bar{M}$ is conceived as constant at roughly zero and changes in the term within brackets are absorbed as changes in population size, is really much too simple a description. If $\bar{M}$ is zero, the equation says nothing about the constant population size attained. If $\bar{M}$ is not zero, and the population size fluctuates, as most real populations do (Ehrlich and Birch, 1967), then the equation says nothing about the fluctuations. Putting it another way, suppose V_A causes a genetical change in the population, then does the population size go up or down? The equation does not say. Changes in V_A are complex even with constant selection and the simplest genetic conditions, and contrary to popular opinion they are by no means a monotonic decrease (Warburton, 1967) (equations are given by Kimura, 1958; Nei, 1963; Turner, 1969). In general, both environmental changes and density-dependent death will alter V_A and hence the course of genetic change by changing the relative fitness of genotypes, and the genetic changes in turn affect population size (see Lerner (1965) for a discussion). The same restrictions apply to Kimura's equation, in which the genetical changes are more accurately stated than they are by Fisher. A general model, probably best produced by computer, including the interactions of genetic and ecological factors, and also changes in age-structure (i.e. partial overlap of generations) would certainly give us significant insight into both genetics and ecology; it might well generate fluctuations both in population size and gene frequency. In genetic terms an interesting question is: what is the relation between the production, consumption, and conservation of genetic variance?

Pimintel (1965) has proposed an integrated eco-genetic theorem of population regulation by genetic feedback. Consider two competing species. The individuals of the rarer species encounter members of the other species more frequently than of their own, and thus selection in the rarer species favours genes conferring success in interspecific competition; in the commoner species, whose individuals more frequently encounter each other, selection will favour genes aiding intraspecific competition. In this way genetic changes cause the two competing species to come to an equilibrium where neither is exterminated — they are in balanced competition as proposed by Williamson (1957) — as the commoner species is always at a disadvantage. Pimintel supports his theory with experiments on cage populations of house-flies and blow-flies, and supports a similar theory about predator-prey systems, which are examples of competition in Williamson's sense, with experiments on house-flies and a parasite species. The two populations evolve a balanced equilibrium after many generations of selection.

Here are some examples of other observations relevant to an integrated theory of ecological genetics. *Drosophila pseudoobscura* carries more than one allele at thirty per cent or more of its genetic loci (Lewontin and Hubby, 1966); the high rate of selective death needed to maintain these polymorphisms probably occurs when the population is numerous and is being thinned (Turner and Williamson, 1968); both selection and overall death-rate must have decreased when a population of the butterfly *Melitaea aurinia* was expanding rapidly (Ford, 1964); in *Drosophila melanogaster*, *Bar* genotypes which have the same viability at low density differ in viability at high density (Bentvelzen, 1963); all this supports the idea of density or population dependent selection, as do the observations of Kojima and Yarbrough (1967) on selection in *D. melanogaster*. In the Cothill colony of the moth *Panaxia dominula* (Ford, 1964) population size and gene frequency seem to have fluctuated out of phase up to 1946, while gene frequency was on the whole decreasing, but to have fluctuated in phase since then while the gene frequency has remained roughly constant (Fig. 5). Thoday (1963) found similar parallel fluctuations in the frequency of *white eye* and the density of a culture of *D. melanogaster* (Fig. 6).

Sokal and others, working on relationships between population genetics and population dynamics have found that in laboratory culture of *Tribolium*, survivorship, dry-weight, and speed of development, all of them important components of fitness (in the sense of W_{ij} or M_{ij}), are often significant functions of population density, of the genotypic composition of the population, and of the interaction of these two factors (Sokal and Huber, 1963; Sokal and Karten, 1964). Some of the functions appear to contain sharp discontinuities. Similar results were obtained for *Drosophila* (Lewontin, 1955). The great complexity of the results discussed by these authors gives the lie to any simple applications of equations about mean fitness to population ecology. For example, at the *sooty* locus in *T. castaneum*, the wild-type homozygote is the fittest genotype at most densities and gene frequencies. We might therefore expect populations with a low

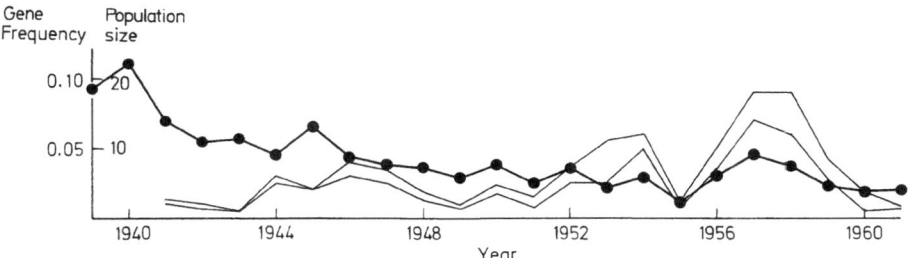

Fig. 5. Frequency of the *medionigra* gene (heavy line) and population size in thousards (thin lines, indicating probable upper and lower limits of the estimated size) in an English colony of the Scarlet Tiger Moth from 1939 to 1961. Data from Ford (1964)

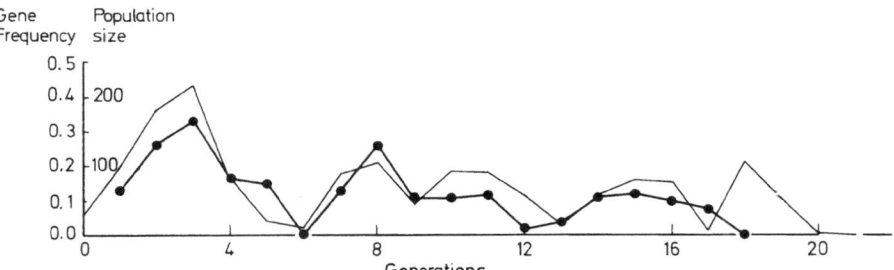

Fig. 6. Frequency of *white-eye* gene (heavy line) and population size (thin line) in a culture of *Drosophila melanogaster*. After Thoday (1963)

frequency of the wild-type allele to have a low fitness, but Schlager (quoted by Sokal and Huber, 1963) found that populations with this allele at a frequency of only 25% had the fastest growth rate and the greatest asymptotic density. Similarly Pimintel (1965) reports a decline in reproductive capacity in a cage population of parasites of house-flies, apparently as a result of selection. On the other hand there is a steady increase in the size of cultured populations of *Drosophila serrata* (Ayala, 1968). Also Beardmore, Dobzhansky, and Pavlovsky (1960) found a greater biomass and greater homeostasis in polymorphic than in mono-morphic populations of *Drosophila pseudoobscura;* in this experiment, natural selection maintained the polymorphism. In theory, as we have seen, there is no obvious reason why a selectively maintained poly-morphism should give a "fitter" population in this sense, and the example from *Tribolium* shows that the "fittest" population composition is not necessarily the one favoured by selection. Polymorphic populations prob-ably exploit the environment more efficiently than monomorphic ones (Dobzhansky, 1965), in the same way that, as Darwin knew, mixed stands

of crop species or varieties sometimes give a higher yield than single
stands (Roy, 1963; Harper, 1965); the connection of this with natural
selection is tenuous.

So while the correct equations for changes in mean fitness under selection will probably play an important part in sections of theoretical population genetics (and they can be extended to include other characters —
Kempthorne, 1957; Falconer, 1966; Turner, 1969), in the wider field of
ecological genetics the simple component of change in mean under constant selection must be welded as one factor into a complex model. Such
a part hardly deserves the name fundamental theorem. In the human
population, where the genetic load of human misery is the converse of
fitness, and where density may be regulated by contraception (we hope),
mean fitness and its changes are an important part of medical genetics.

Summing Up

The very interesting questions about the generation, conservation and
consumption of genetic variance, particularly the additive and haploid
variances in fitness, although introduced by Fisher (1930) in his demonstration that Mendelian inheritance maintained variance for longer than
did blending inheritance, have received little attention; but equations
have been developed for changes in the mean population fitness. Fisher,
equating two parameters which are in general different, claimed that his
"fundamental theorem of natural selection" (increase in mean fitness
equals additive variance in fitness) is more general than it is. Only for
a single locus, under random mating, with constant fitness values the
same in both sexes and not showing dominance, is this theorem strictly
true. But under these conditions with separate generations the change in
mean is roughly equal to the variance, and always positive, with any
degree of dominance. Although the equations describing populations
with overlapping generations tend to be unrealistic, as they take no
account of age-structure (see Hasofer (1966) for a simplified realistic version), Kimura (1958) has derived a general theorem which overcomes the
errors in Fisher's version. In this system the mean fitness (Malthusian
parameter) may decrease even under random mating.

The fact that $\bar{W}$ maximizes under random mating when generations
are separate and when they overlap can be used in the construction of
geometrical analogues (adaptive topographies) for the investigation of
evolutionary changes. These can be used for two loci by including recombination in the equations and adding a dimension to the topography;
alternatively one can use an approximation based on the condition of
quasi linkage equilibrium (Kimura, 1965). Under more general condi-

tions when $\overline{W}$ does not maximize, one may use other parameters which do so, although they are in no sense "fitness", or one may use a topography based on changes in variance.

Except in so far as "fitness" in the above sense is related to the genetic load and human tragedy, the above theorems have little application in population biology, beyond predicting genetical evolution. The changes in mean population growth, size, and biomass are not covered by any adequate or realistic theory; the experimental evidence shows that the theory would need to be elaborate.

The simple view of evolution that mean fitness will increase because the less fit genotypes die is not adequate because sexual reproduction shuffles the genes and the fitnesses of genotypes may change with changes in the population.

Acknowledgements

My grateful thanks go to C. Cannings, A. W. F. Edwards, J. Felsenstein, E. B. Ford, J. L. Harper, P. O'Donald, and M. H. Williamson for discussions on topics related to this review and to A. J. Cain, B. C. Clarke, J. D. Currey, P. J. Hogarth, and J. A. Metcalfe for criticizing various sections of the draft, and to C. C. Li for his comments on multiple alleles.

References

* indicates a relevant paper not discussed in the text.

Atkinson, F. V., G. A. Watterson, and P. A. P. Moran: A matrix inequality. Quart. J. Math. (Oxford) (2) 11, 137 – 140 (1960).

Ayala, F. S.: Evolution of fitness. II. Correlated effects of natural selection on the productivity and size of experimental population of *Drosophila serrata*. Evolution 22, 55 – 65 (1968).

Beardmore, J. A., Th. Dobzhansky, and O. Pavlovsky: An attempt to compare the fitness of polymorphic and monomorphic experimental populations of *Drosophila pseudoobscura*. Heredity 14, 19 – 33 (1960).

Bentvelzen, P.: Some interrelations between density and genetic structure of a *Drosophila* population. Genetica (Netherlands) 34, 229 – 241 (1963).

Blakley, G. R.: Darwinian natural selection acting within populations. J. Theor. Biol. 17, 252 – 281 (1967).

Bodmer, W. F., and J. Felsenstein: Linkage and selection: theoretical analysis of the deterministic two locus random mating model. Genetics 57, 237 – 265 (1967).

Boyle, R.: A Disquisition about the final Causes of natural Things. London: H. C. for John Taylor 1688.

Cain, A. J., and P. M. Sheppard: The theory of adaptive polymorphism. Amer. Natur. 88, 321 – 326 (1954).

— — Adaptive and selective value. Amer. Natur. 90, 202 – 203 (1956).

Campos Rosado, J. M., and A. Robertson: The genetic control of sex ratio. J. Theor. Biol. **13**, 324 – 329 (1966).

Cannings, C., and A. W. F. Edwards: Natural selection and the de Finetti diagram. Ann. Hum. Genet. (Lond.) **31**, 421 – 428 (1968).

Cockerham, C. C.: An extension of the concept of partitioning hereditary variance for analysis of covariances among relatives when epistasis is present. Genetics **39**, 859 – 882 (1954).

*Crow, J. F., and M. Kimura: Some genetic problems in natural populations. Proc. 3rd. Berkeley Symp. Math. Stat. Prob. **4**, 1 – 22 (1956).

Dobzhansky, Th.: Genetic diversity and fitness. Proc. XI Int. Cong. Genet. Hague **3**, 541 – 552 (1965).

*Edwards, A. W. F.: Fundamental Theorem of natural selection. Nature (Lond.) **215**, 537 – 538 (1967).

– On the fundamental theorem of natural selection. Duplicated paper available from the author at Gonville & Caius College, Cambridge (1968).

Ehrlich, P. R., and L. C. Birch: The "balance of nature" and "population control". Amer. Natur. **101**, 97 – 107 (1967).

Ellegård, A.: Darwin and the general reader. The reception of Darwin's theory of evolution in the British periodical press, 1859 – 1872. Gothenberg Studies in English VIII Ed. F. Behre. Göteberg 1958.

Falconer, D. S.: Introduction to quantitative genetics. Edinburgh-London: Oliver and Boyd 1961.

– Genetic consequences of selection pressure. In Genetic and environmental factors in human ability. pp. 219 – 232. Ed. by J. E. Meade and A. S. Parkes. Edinburgh – London: Oliver and Boyd 1966.

Feller, W.: On fitness and the cost of natural selection. Genet. Res. (Camb.) **9**, 1 – 15 (1967).

Fisher, R. A.: The genetical theory of natural selection. Oxford: Clarendon Press 1930.

– Average excess and average effect of a gene substitution. Ann. Eugen. (Lond.) **11**, 53 – 63 (1941).

– The genetical theory of natural selection. 2nd ed. New York: Dover Books 1958.

Ford, E. B.: Ecological genetics. London: Methuen; New York: John Wiley & Sons 1964.

Harper, J. L.: The nature and consequence of interference amongst plants. Proc. XI Int. Cong. Genet. Hague **2**, 465 – 482 (1965).

Hasofer, A. M.: A continuous-time model in population genetics. J. Theor. Biol. **11**, 150 – 163 (1966).

Jain, S. K., and P. L. Workman: Generalized F-statistics and the theory of inbreeding and selection. Nature (Lond.) **214**, 674 – 678 (1967).

Kempthorne, O.: An introduction to genetic statistics. New York: John Wiley & Sons 1957.

Kimura, M.: On the change of population fitness by natural selection. Heredity **12**, 145 – 167 (1958).

– Shudan iden-gaku gairon (Introduction to population genetics). Tokyo: Baihukan 1960.

– Attainment of quasi linkage equilibrium when gene frequencies are changing by natural selection. Genetics **52**, 875 – 890 (1965).

– Two loci polymorphism as a stationary point. Ann. Rep. Natl. Inst. Genet. Japan **17**, 65 – 67 (1966).

Kingman, J. F. C.: On an inequality in partial averages. Quart. J. Math. (Oxford) (2) **12**, 78 – 80 (1961).

Kojima, K.: Role of epistasis and overdominance in stability of equilibrium with selection. Proc. Natl. Acad. Sci. (Wash.) **45**, 984 – 989 (1959).

–, and T. M. Kelleher: Changes of mean fitness in random mating populations when epistasis and linkage are present. Genetics **46**, 527 – 540 (1961).

– and K. M. Yarbrough: Frequency-dependent selection at the *esterase 6* locus in. *Drosophila melanogaster*. Proc. Natl. Acad. Sci. (Wash.) **57**, 645 – 649 (1967).

Lerner, I. M.: Ecological genetics-synthesis. Proc. XI Int. Cong. Genet. Hague **2**, 489 – 494 (1965).

Levins, R.: Theory of fitness in a heterogeneous environment. I. The fitness set and adaptive function. Amer. Natur. **96**, 361 – 373 (1962).

Lewontin, R. C.: The effects of population density and composition on viability in *Drosophila melanogaster*. Evolution **9**, 27 – 41 (1955).

– Selection in and of populations. Ideas in modern biology. Proc. XVI Int. Cong. Zool. **6**, 299 – 311 (1965).

–, and J. L. Hubby: A molecular approach to the study of genetic heterozygosity in natural populations. II. Amount of variation and degree of heterozygosity in natural populations of *Drosophila pseudoobscura*. Genetics **54**, 595 – 609 (1966).

–, and K. Kojima: The evolutionary dynamics of complex polymorphisms. Evolution **14**, 458 – 472 (1960).

–, and M. J. D. White: Interaction between inversion polymorphisms of two chromosome pairs in the grasshopper, *Moraba scurra*. Evolution **14**, 116 – 129 (1960).

Li, C. C.: The stability of an equilibrium and the average fitness of a population. Amer. Natur. **89**, 281 – 295 (1955 a).

– Population genetics. Chicago-London: Chicago Univ. Press 1955 b.

– Fundamental theorem of natural selection. Nature (Lond.) **214**, 505 – 506 (1967 a).

– Genetic equilibrium under selection. Biometrics **23**, 397 – 484 (1967 b).

– The maximization of average fitness by natural selection for a sex-linked locus. Proc. Natl. Acad. Sci. (Wash.) **57**, 1260 – 1261 (1967 c).

– Increment of average fitness for multiple alleles. Proc. Natl. Acad. Sci. (Wash.) **62**, 395 – 398 (1969).

MacArthur, R. H.: Some generalized theorems of natural selection. Proc. Natl. Acad. Sci. (Wash.) **48**, 1893 – 1897 (1962).

Mandel, S. P. H.: Fundamental theorem of natural selection. Nature (Lond.) **220**, 1251 – 1252 (1968).

*Mode, C. J.: A stochastic calculus and its application to some fundamental theorems of natural selection. J. Appl. Prob. **3**, 327 – 352 (1966).

Moran, P. A. P.: The statistical processes of evolutionary theory. Oxford: Clarendon Press 1962.

– On the nonexistence of adaptive topographies. Ann. Hum. Genet. (Lond.) **27**, 383 – 393 (1964).

Mulholland, H. P., and C. A. B. Smith: An inequality arising in genetical theory. Am. Math. Monthly **66**, 673 – 683 (1959).

Nei, M.: Effects of selection on the components of genetic variance. In: Statistical genetics and plant breeding, pp. 501 – 515. Ed. by W. A. Hansen and H. F. Robinson. Washington, D. C.: Natl. Acad. Sci. – Natl. Res. Council, Publ. 982, 1963.

Peckham, M.: The origin of species, by Charles Darwin. A variorum text. Philadelphia: Univ. Penna. Press 1959.

Pimentel, D.: Population ecology and the genetic feedback mechanism. Proc. XI Int. Cong. Genet. Hague **2**, 483 – 488 (1965).

Roy, Subodh Kumar: Intra-specific interaction in rice. Proc. XI Int. Cong. Genet., Hague **7**, 145 (1963).

Sacks, J. M.: A stable equilibrium with minimum average fitness. Genetics **56**, 705 – 708 (1967).

Scheuer, P. A. G., and S. P. H. Mandel: An inequality in population genetics. Heredity **13**, 519 – 524 (1959).

Slobodkin, L. B.: The strategy of evolution. Amer. Sci. **52**, 342 – 357 (1964).

Sokal, R. R.: Some stages in the development of the concept of natural selection. Univ. Kansas Sci. Bull. Supp. **42**, 129 – 151 (1962).

—, and I. Huber: Competition among genotypes in *Tribolium castaneum* at varying densities and gene frequencies (the *sooty* locus). Amer. Natur. **97**, 169-184 (1963).

—, and I. Karten: Competition among genotypes in *Tribolium castaneum* at varying densities and gene frequencies (the *black* locus). Genetics **49**, 195 – 211 (1964).

Thoday, J. M.: Correlation between gene frequency and population size. Amer. Natur. **97**, 409 – 412 (1963).

Turner, J. R. G.: On supergenes. I. The evolution of supergenes. Amer. Natur. **101**, 195 – 221 (1967a).

— Mean fitness and the equilibria in multilocus polymorphisms. Proc. Roy. Soc. London B **169**, 31 – 58 (1967b).

— Why does the genotype not congeal? Evolution **21**, 645 – 656 (1967c).

— The effect of mutation on fitness in a system of two coadapted loci. Ann. Hum. Genet. (Lond.) **30**, 329 – 334 (1967d).

*— Fundamental theorem of natural selection. Nature (Lond.) **215**, 1080 (1967e).

— Natural selection for and against a polymorphism which interacts with sex. Evolution **22**, 481 – 495 (1968).

— The basic theorems of natural selection: a naive approach. Heredity **24**, 75 – 84 (1969).

—, and M. H. Williamson: Population size, natural selection, and the genetic load. Nature (Lond.) **218**, 700 (1968).

Warburton, F. E.: Increase in the variance of fitness due to selection. Evolution **21**, 197 – 198 (1967).

Williamson, M. H.: An elementary theory of interspecific competition. Nature (Lond.) **180**, 422 – 425 (1957).

Wright, S.: The roles of mutation, inbreeding, crossbreeding, and selection in evolution. Proc. 6th Int. Cong. Genet. **1**, 356 – 366 (1932).

— Adaptation and selection. In: Genetics, palaeontology and evolution,, pp. 365 – 389. Ed. by G. L. Jepsen, G. G. Simpson, and E. Mayr. Princeton, N. J.: Princeton Univ. Press 1949.

— Classification of the factors of evolution. Cold Spr. Harb. Symp. Quant. Biol. **20**, 16 – 24 D (1955).

— Stochastic process in evolution. In: Symposium on stochastic models in medicine and biology, pp. 199 – 244. Ed. by J. Gurland. Madison: The University of Wisconsin Press 1964.

Models and Analyses of Dispersal Patterns[*]

R. H. RICHARDSON

Introduction

For a given set of gametes, population structure is the controlling factor determining the array of genotypes exposed to selective forces. This array reflects the combined effects of genotype frequencies, mating preferences, and mate availability, as well as any changes in age distributions of fertile adults or availability of breeding sites. Because the extent of the array of genotypes may either enhance or render ineffective a particular mode of selection, population genetic theory always incorporates a statement specifying the population structure. Models developed from "conditions of random pairing of gametes" are useful as a first step for understanding selection forces, but ultimately such oversimplifications of population structure must be revised to better agree with the complexities of actual situations in nature. Evidence available from a variety of sources indicates that existing theory does not adequately account for non-random mating of several forms.

Among the multitude of factors affecting the genetic structure of a population, some are conspicuous and others subtle. It is obvious that summer and winter annual plants do not randomly interbreed, but, is it equally obvious that a few hours difference between male and female Drosophila in reaching sexual maturity can contribute materially to a reduction of inbreeding? It might be obvious, if one also knew for a given case that dispersal capabilities were developed sooner than sexual maturity and that a culture site was primarily a result of heavy egg laying of a few females over a short period of time. Many factors of behavior and ecology enter into a determination of population structure, most of which are, at best, only poorly understood, but dispersal is often thought to be a behavioral factor contributing an important component to the population structure of animals.

Dispersal models have historical beginnings before the Hardy-Weinberg formulations, dating at least to Karl Pearson's work around 1900. Many empirical functions have been used to describe data on dispersal phenomena. Public health studies describe the spread of epidemics in

* This paper is dedicated to Professor Th. Dobzhansky in honor of his long leadership in evolutionary genetics and of his pioneering efforts in ecological genetics of Drosophila.

time and space; biogeographers consider rates of extensions of species ranges; agronomists formulate models of pollen contamination in cross fertilized crops; entomologists study effectiveness of reinfestation of insect pests after localized pesticide application; and population geneticists formulate systems of joint effects of gene flow and selection. [An extensive review of data, derived primarily from the agricultural literature, is given by Wolfenbarger (1946).] Most of the models incorporate the assumption of random movement. Skellam (1951) presents several theoretically interesting models for random dispersal systems, and Wright (1969) discusses a number of other distributions, but critical data necessary for evaluation of these models are usually lacking.

Diffusion Models

Dispersal phenomena may be approached theoretically either from a discrete process of movement by "jumps" or by a continuous process by the solution of differential equations. Ordinarily for the limiting case of a discrete process there can be found a corresponding solution of a differential equation. Furthermore, most biological data may be equally well approximated by either the discrete or continuous models. Consequently, the ease of manipulation of the continuous cases is ordinarily considered sufficient justification for their use, although the summary of a development from discrete models may be useful in gaining an insight into the nature of the biological parameters incorporated in the model.

The general form of the continuous model is

$$\xi(r) = C \exp - (\alpha r^\kappa) \tag{1}$$

where C is a scaling constant determined by units of the data, α is a constant which controls the "spread" of the distribution, r is the radius from the origin, and κ is a constant which will change, depending upon the assumptions concerning the dispersal process. Kurtosis, β_2, for this case was given by Wright (1968 b) as

$$\beta_2 = \frac{\Gamma(1/\kappa)\,\Gamma(5/\kappa)}{[\Gamma(3/\kappa)]^2} . \tag{2}$$

This is a useful parameter for describing the "peakedness" around the mean and the "flatness" of the tails of the theoretical distributions, although it results in an inefficient comparison between shapes of theoretical and observed distributions.

Dispersal is usually on a plane, while Eq. (1) is on the real line. So long as movement from the origin is symmetrical, the model may be either for $r > 0$, or may be generalized to a transect through the origin, or may be rotated about the origin to describe planar cases. The value of C varies with these alternatives such that the areas under observed and expected distributions are equal. The model form of Eq. (1) with $r > 0$ also describes an average expectation of dispersal when movement is random, which is convenient for comparisons with data. However, the mean is assumed to be at the origin and is not calculated from the model. If movement is asymmetrical, more complicated general models are necessary. However, for preliminary comparisons with asymmetrical data, we can consider individual radii, rather than their average, and use the form of Eq. (1) as an approximate expectation.

Model of Brownian Motion

Homogeneous Population. The classical Brownian diffusion equation of physics may be used in developing a model of dispersal of a fixed number of particles (individuals) from a single point beginning at an initial time, t_0. This type of model might be appropriate for an experiment in which individuals are released at a point and disperse (diffuse) outward in a specified fashion. Also, this model might apply if individuals in a population simultaneously reach a dispersal phase of their life cycle.

The "particle density" function may be developed theoretically from several points of view, but, for present purposes, it was paraphrased from three references: Feller (1966, 1968) and Papoulis (1965). Let us assume that individuals are participating in a random-walk process (also referred to as Wiener-Levy or Wiener-Bachelier process). If an individual is placed at $X(0)$ and subsequently moves one step of a fixed size, γ, every t seconds, the direction either forward ($x = 1$) or backward ($x = -1$) chosen randomly, one may calculate the probability of the individual being at location $X(n)$ after n steps (in nt seconds).

Let $X(0) = 0$. After n random steps, if k forward steps were made and $n - k$ specified the number of backward steps, the position is

$$X(n) = \gamma \sum_{i=1}^{n} x_i$$
$$= [k - (n - k)] \gamma \qquad (3)$$
$$= [2k - n] \gamma .$$

The step-length, γ, may be a function of time. For example, such an additional complication might be the result of the life cycle involving both sedentary stages and mobile stages or could arise simply because older individuals tend to be less capable of effective movement. Furthermore, the actual step-size at any given time may randomly vary, in which case γ is the expected (or average) step-size for that time. So long as γ is time-independent, it is a scaling factor which may be set to $\gamma = 1$. From Eq. (3), it is then apparent that $X(n)$ may take values n, $n-2$, $u, \ldots, -n$, and that the function (a probability density function in this special case) is

$$P[X(n) = u] = \binom{n}{k} p^k (1-p)^{n-k} \tag{4}$$

where p is the probability of a forward step (or $P(x=1)$) and $(1-p)$ is the probability of a backward step (or $P(x=-1)$). If $p \neq 1/2$, then individuals are more likely to move in one direction than in the other on the real line, and their density function will be skewed and the mean of the function will be shifted from the initial starting point. This situation would result in the mean of the dispersal function being time-dependent, and also, for planar cases, would mitigate against applying the model without reference to θ, the angle of rotation about the mode. Also, even if $p = 1/2$, but γ, the step-size, were a function of θ, the model for the dispersal pattern would be modified such that the two effects, $p \neq 1/2$ and γ a function of θ, would be confounded.

For purposes of simplicity, assume $p = 1/2$ and γ is constant. Then, for the i-th step

$$E(x_i) = 0 \tag{5}$$

and

$$\begin{aligned} E(x_i^2) &= \gamma^2 \\ &= 1 \\ &= \sigma^2, \text{ the variance.} \end{aligned} \tag{6}$$

However, $X(n) = \sum_{i=1}^{n} x_i$ and we assume that the x_i are independent Bernouli trials. Thus,

$$E[X(n)] = 0 \tag{7}$$

and

$$\begin{aligned} E[X(n)^2] &= \sum_{i=1}^{n} E(x_i^2) \\ &= n. \end{aligned} \tag{8}$$

Note that n is a measure of time, for example, each unit being t seconds. As we allow both γ and t to decrease to very small values, by the de Moivre Laplace theorem (or by the Central Limit theorem) the limiting form is the normal (Gaussian) distribution. From Eq. (3),

$$k = (n + \sum x_i)/2$$

so that, from Eq. (4)

$$P[X(n) = u] \cong \left(\frac{1}{\sqrt{2\pi n}} \right) \exp - (\sum x_i)^2/2n, \quad -\infty < x < \infty \qquad (9)$$

Since n (the variance) is the measure of time, this normal density is a function of time. The distribution continually spreads with time after t_0, but the mean remains at $X(0)$ as a result of the assumptions that $p = 1/2$ and γ is independent of θ (Fig. 1). For the normal density, Eq. (9), κ of Eq. (1) equals 2, and from Eq. (2), $\beta_2 = 3$. The model for dispersal on a plane is the bivariate normal density with zero correlation.

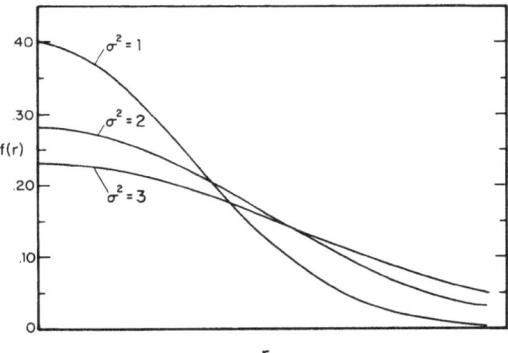

Fig. 1. Gaussian distributions with variances of one, two, and three

Heterogeneous Populations. Although no biological population is phenotypically homogeneous, sometimes homogeneity may be mimicked by effects of different factors cancelling. When this is not the case, the nature of the impact of heterogeneity upon distribution may be questioned.

For the Brownian movement model, Wright (1968a) summarized the effects of the pooling of individuals from populations with a common mean but different variances. When individuals from two such populations are pooled, the composite population has a density function with $\beta_2 > 3$, which Wright calls a "compound distribution" and which is leptokurtic relative to the normal. Those individuals taken from populations with relatively large variances contribute greatly to the tails of the composite density, while those from populations with small variances constitute most of those found near the mean. The tails are more sensitive to distortion than the modal region of the curve. The examples used by Wright. (Fig. 2) where the standard deviations differ by a factor of four, show considerably greater kurtosis for a mixture consisting mostly of individuals from a population with a density having the smaller variance. Only a small addition of individuals from a population with a larger variance has a marked impact on the value of β_2.

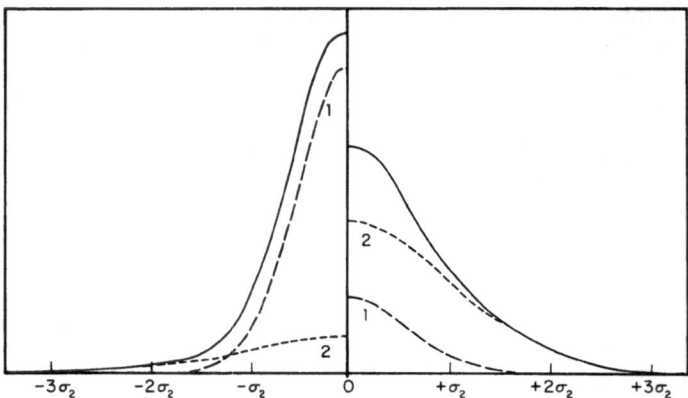

Fig. 2. Composite (compound) distributions: Distribution on the left obtained by adding Gaussian distributions, 20% from one with variance of 4 (distribution 2) and 80% from one with variance of 1 (distribution 1). Conversely, on the right the composite distribution is formed by adding 80% from one with variance of 4 and 20% from one with variance of 1 (modified from Wright, 1968a)

The compounding of distributions could occur in a variety of ways. For example, if there is an overlap of generations, each having a phase of dispersal, the variances of densities would differ among generations. Generations would be at different intervals from their respective t_0's (t_0 is defined as the initial point of a dispersal phase). Also, if there is individual variation in behaviour, such that some individuals have larger values of γ than others ("travelers" *versus* "home-bodies"), the variances for their respective densities would differ. For this derivation, Eq. (4) would be $P[X(n) = \gamma u]$, where the appropriate values of γ would be supplied. The differences in scaling, from Eq. (6), would differ by the differences in the respective γ's.

Another statistically less predictable, but biologically more meaningful, possibility involves ecological heterogeneity as a factor reflected in heterogeneous behavior. If a small region is ecologically attractive, individuals may tend to move with a small value for γ. However, if individuals happen to enter an adjacent, less attractive region, they may move with a larger γ, and, consequently, disperse to greater distances with higher than expected frequencies. Such a disturbance is detectable, since the second derivative of the density function would change in sign at ecological interfaces between attractive and unattractive regions. However, it is unlikely that such a set of critical data could be readily obtained. Wright (1968a) illustrates how a bimodal distribution could have $\beta_2 < 3$, or, depending upon the effects on behavior, it is easy to visualize cases where $\beta_2 < 3$. Thus, kurtosis would offer no assistance in detecting many types of ecological effects on the dispersal distribution.

Since a variety of biological effects may result in the same compound density function, they are confounded until additional information is obtained. Knowledge of the shape of the distribution, or the way it deviates from expected, tells very little about the underlying reasons for the deviation. This situation leads to controversies in interpretation of empirical data, and individual preferences of investigators, rather than analyses, frequently account for the particular interpretations presented.

Random Dispersion with Replacement at the Origin

A localized population may have a continual replacement by birth and immigration of those individuals lost by death or emigration. Physics again furnishes a useful analogy from which to begin the derivation of a descriptive model (Glasstone and Sesonske, 1963). If N is the number of individuals in a unit area of the plane and t is time, the change in N per change in t may be written

$$\frac{\partial N}{\partial t} = \text{birth} + \text{immigration} - \text{death} - \text{emigration}. \tag{10}$$

The biological factors are analogous to neutron production (birth), absorption (death), and leakage (emigration − immigration). The neutron density at any place in the vicinity of a thermal neutron point source with diffusion in a plane is such an example.

Under steady-state conditions, $\partial N/\partial t = 0$; that is, the number of individuals at any given point is constant.

The assumptions of the model are as follows.

(1) The rate of movement is assumed to be independent of age, spatial position, or other changing environmental variables. This is similar to the case presented for Brownian motion, where step-size, γ, and time between steps are assumed to be constant. The consequences of the fact that rate of movement is a function of the age of individuals is analogous to γ being a function of time, since older individuals tend to be found farther from the origin (point source). However, if rate of movement declines with age, the effect would be to reduce the expected dispersal to distant positions from the origin.

(2) It is assumed that the net number of individuals in a unit area always declines with increasing distance from the origin. Individuals tend to move randomly, and, since there is a higher density of individuals on the side of an area toward the origin, there is a greater probability that a specified number will move into the area from the side toward the origin than a specified number will move into the area from the side of of the area away from the origin. We might say, simply, that we assume the density gradient is away from the origin, and, from the first assumption, that the steepness of the gradient is a function of the rate of movement.

The situation may be formulated as Fick's law of diffusion as follows. Rate of movement, or velocity, is measured in units of distance traveled per unit time. For a specified unit time scale, the distance traveled may be used to measure velocity. Since the movement is random, but the gradient is negative, the *net* (or vector) velocity away from the origin, may be defined as D, the diffusion coefficient, which is assumed to be constant. Then the net flow of individuals (current) for one direction, r, may be written as

$$J_r = -D \frac{\partial N}{\partial r}. \tag{11}$$

This assumption is also implicit in the case for Brownian motion and is reflected by the symmetrical planar model being a bivariate normal density.

(3) It is assumed that there is a constant probability of death. If most deaths are "accidental" from predation on adults (for example, flies caught in spider webs), the assumption is likely to be acceptable. However, for many biological systems, the probability of death is a function of age, sex, density, or spatial position. Unless the details are known for a specific case, the best first approximation is that of a constant probability. This assumption differs from that of Brownian motion, where it was assumed there was no death.

(4) It is assumed that individuals entering the population arise at a common point, the origin, at a continuous rate. In defining a freely interbreeding population (a deme) one might consider only individuals arising (being born) at a point and lost from the breeding population by death or emigration. Maintenance of the population would result from continual reproduction only at the "point source". A pulsating source might be a better biological analog, but, due to a lack of relevant data, an examination of this possibility will be deferred to some future time. (The model would be intermediate between that for Brownian motion and that being developed.) If immigrants were numerous, they would greatly increase the census in the tails of the distribution. However if they could be distinguished from the "natives", only that single distribution for natives

would be considered, rather than the compound distribution of natives and immigrants.

Under these assumptions the solution of Eq. (10) is

$$\zeta(r) = C \exp - (\alpha r) \tag{12}$$

(Glasstone and Sesonske, 1963), which is the form of the exponential probability distribution. The variance of Eq. (12) is

$$\sigma^2 = \frac{2}{\alpha^2} \tag{13}$$

and, thus, Eq. (12) may be rewritten as

$$\zeta(r) = C \exp - (\sqrt{2}\ r/\sigma). \tag{14}$$

Note that we are considering the steady-state solution and ζ is not a function of time. Since, from Eq. (1), $\kappa = 1$, then

$$\beta_2 = 6$$

from Eq. (2).

Population Dynamics and Genetic Structure

Without a formal mathematical relationship between dispersal phenomena and interbreeding, some comment may be made concerning the uses of dispersal data to infer amounts of gene flow. From previous discussions of inferences concerning behavioral characteristics, it is apparent that multiple sets of factors may generate a particular dispersal distribution. Thus, ordinarily additional parameters (dispersal rate functions, for example) must be incorporated into models for a unique set of factors to correspond to an observed distribution. The estimation of the additional parameters has been technologically handicapped.

So long as one considers distributions of the form of Eq. 1, the distribution with the greater value of β_2 will intersect the one with lesser value of β_2 in two places – one near the mode and another in the tail. If "genetically significant dispersal" need be only an occasional immigrant reaching a specified area forming a breeding site, then those populations whose dispersal behavior is best described by a more kurtic distribution likely will be greater contributors to adjacent demes than populations whose behavior pattern is less kurtic. This relationship between relative kurtosis and gene flow clearly will depend on many factors, but assumes, for example, that the mating activity is localized, while dispersal activity is essentially uniform, a different model from Wright (1969, for example), where both breeding and dispersal are covariates in time and space. Whether situations described by the model given here are common remains largely for future studies to evaluate, but from the consideration of a few examples, one may get the impression that this is a model which is closer to the rule than one for the exceptions.

Suppose Drosophila living in a decaying cactus represents the breeding population of a deme. The individuals may tend to remain very near to a particular "rot pocket" as would be expected if it were highly attractive. However, the life span of the rot pocket is limited to a few generations of the Drosophila, and when it dried and no longer was attractive, the flies supposedly would disperse, more or less at random. Depending on their dispersal pattern, they might contribute more or less to the genetic composition of some other deme at another rot pocket still supporting an active Drosophila culture. Of course, the gene flow would not take place until the dispersing individuals reached the vicinity of the rot pocket, since individuals of that deme would not be dispersing at that time. Analogous situations would be expected any time a vital component of the habitat periodically was missing, with a frequency less than the generation interval of the organism, and thereby forced emigration. Historically, for example, wars and famine resulted in this pattern of dispersal and gene flow in humans.

But there is still more information needed than accurate dispersal theory before gene flow may be accurately inferred. The genetic arrays in various portions of the distribution may vary. Thus the dispersal behavior may be well known, but the genetic composition of emigrants remain obscure. Even if emigrants are random samples of populations, inferences about gene flow remains tenuous.

Dispersal processes may be sufficiently infrequent in the life history of a population to make the necessary measurements impossible. If important movements of individuals are sporadic, the fact may go undetected, although there may result large amounts of gene flow among demes. Heterogeneity of behavior varying with time presents the same difficulties of interpretation of empirical data as those deriving from heterogeneities discussed earlier, except for the infrequent occurrence measured by the time scale of the investigation. What may superficially appear to be great isolation in a desert species of Drosophila exploiting widely spaced rotting cacti (Heed et al., 1968) may, in actuality, be subject to extensive genetic mixing when new colonies are formed by individuals dispersing from several surrounding colonies.

The genetic impact of dispersal may be greatly affected by the mating patterns involving immigrants, as suggested by works such as those of Ehrman (1966). Gene flow may be greater than predicted from measures of the frequency of immigration. By contrast, a territorial breeding structure would have the opposite effect. For instance, the tribe structure in house mice with a single dominant male is likely to reduce or eliminate the mating effectiveness of an immigrant male, thus diminishing gene flow (Reimer and Petras, 1967).

Dispersal is an important part of the dynamics of any population, and, given relevant additional information, may be related to the genetic structure of a population, to the degree of isolation, and to variations in behavior caused by ecological heterogeneity. For studies of population dynamics, technology has developed slowly, and most data were acquired by the use of inadequate, laborious, and inefficient techniques. In addition, the techniques themselves tend to bias behavioral responses. As new techniques develop (e.g. Richardson et al., 1969; Stern and Mueller, 1968) more precise estimates of the usual parameters will become possible. These new techniques make it possible to uniquely and simultaneously mark individuals at several places. The combined study of several dispersal distributions in an area make it possible to map attractive and unattractive areas by comparing ecological effects on symmetry and variances in the distributions rather than extra "dips and humps" (sign changes in second derivatives, see the section *Heterogeneous Populations* on page 84, 85) in a single dispersal distribution. Nevertheless, studies using other techniques have revealed a number of factors affecting dispersal distributions, and some examples are presented in the following section.

Selected Comparisons of Data with Models

Either of the two models described in *Model of Brownian Motion* or *Random Dispersion with Replacement at the Origin* involves the assumption of random movement. Also, the diffusion coefficient, or average radial vector velocity, remains constant for either case. However, effects of the failure to meet those assumptions may be mutually nullified by behavioral patterns of organisms. Random movement includes random orientation from the origin for a given individual at any given point in time or at any position on the surface. Non-random orientation changes the average vector velocity of an individual without a change in the absolute velocity. Orientation toward the origin reduces, while orientation away from the origin increases the average vector velocity. Thus, the dispersal distribution could be developed in terms of the probability distribution of vector velocities of individuals for a specified time, t, after t_0. This basis of derivation has intuitive appeal for some of the biological cases to be discussed.

By the variety of derivations from different premises possible for the same solution of the diffusion equations, it should be intuitively clear that there is likely to be a wide variety of biological ways in which the situation may differ from these particular assumptions which would result in similar changes in the distributions. Some possibilities were mentioned in *Model of Brownian Motion*. It is the purpose of this section to present several biological examples taken from the literature in order to illustrate some of the ways in which the expectations under a simple model may not be realized.

Insect Spread of a Virus Disease

Frampton *et al.* (1942) described the pattern of spread of insect borne virus diseases into fields of potatoes, endive, and lettuce. In the case of aster-yellows virus disease in endive, it was known that the invading leaf hoppers were mostly viriferous. Only the small percentage not already carrying viruses could become carriers by feeding on diseased plants in the field. This further production of viriferous leaf hoppers could be ignored with negligible bias of the analysis.

The model for the "wave front" of the disease as it spreads from the source is time-dependent because plants which are reinfected cannot be counted. The form of the model is that of Eq. (1) where $\kappa = 1$, except that there are time-dependent terms in the exponent. These terms increase the value of the (negative) exponent above that of Eq. (14), although the form remains that of Eq. (12), so that the shape of the distributions is similar for a specified time value, and assumptions concerning movement of insects are the same as for Eq. (14). Disease incidence represents the feeding sites of the leaf hoppers as they move into the field, and thus their dispersion.

Fig. 3 shows the disease incidence at a particular time, together with the expected frequencies based on the model from Eq. (1) with $\kappa = 1$. The correspondence between observed and expected results is

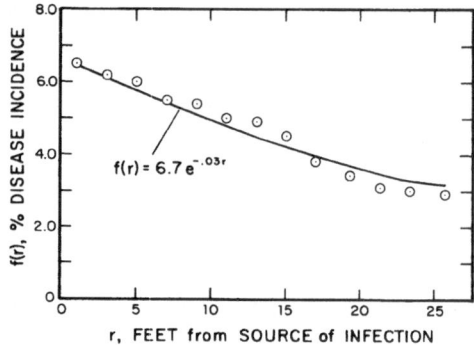

Fig. 3. Spread of infection (percentage infected plants) of aster-yellows virus into endive (from Frampton *et al.*, 1942, Fig. 4c)

fairly close (except for a possible discontinuity at about $r = 15$), and probably indicates an absence of major disturbing ecological complexities affecting leaf hopper dispersal behavior. The relatively simple experimental situation might be anticipated because the area was a homogeneous cultivated field, rather than a more heterogeneous uncultivated area.

Although for the spread of a virus disease, this model is analogous to that for colonization of new environments with unoccupied niches. If we assume that, once occupied, a niche is no longer available to new colonizers, then the filling of a niche is analogous to the infecting of a plant. For example, this is the continuous form of the model used by Hamilton and Rubinoff (1963) when considering the colonization opportunities of insular species of Darwin finches in the Galapagos Islands.

Pollen Dispersal by Bees

Bateman (1947a, b; 1948) performed several experiments with bees, in which a central plot of radishes of one variety was planted, with "stringers" of another variety consisting of one to several rows radiating to the four points of the compass. Two examples have been chosen from Bateman's work because they illustrate non-random pollination behavior of the bees. Bateman (1951) has discussed the widespread occurrence of dispersal distributions with $\beta_2 > 3$. A part of the unexpected kurtosis is the result (as is true in his analyses) of the common application of an inappropriate model, Eq. (9) (Brownian movement) when the biological expectations are more like those obtained from Eq. (14) (dispersion from a point source). The bees cannot spread contaminant pollen without first visiting the central plot. The supply of bees bearing contaminant pollen would be continuously replenished, as well as depleted, which would be analogous to birth and death.

Since the supply of bees is approximately constant, as is the pollen during the experiment, they would be expected to move pollen each visit from the central "contaminant" plot out along the stringers according to the model given by Eq. (14). Because the distribution of bees first visiting the central plot and subsequently the stringers is time-independent, so (except for possible weather or other environmental pertubations of behavior) the data on contamination *versus* distance from the contaminant pollen source legitimately may be pooled. The data are, in actuality, cumulative since seed harvested were from pollinations over a span of time.

The first example is for a single row stringer 45 feet long (Fig. 4, left). There is superficially good agreement with the exponential curve with variance equal to that estimated from the data. However, closer examination reveals a large excess of contaminant seed obtained from the extreme end of the stringer, with smaller but consistent excesses for the 20 feet adjacent to the end sample. Furthermore, the samples most adjacent to the contaminant pollen source are somewhat more contaminated than expected. In general, the observed distribution is leptokurtic to the expected distribution. (Estimated kurtosis is about 13. The theoretical kurtosis of 6 is reduced to 4.95 by truncation of the distribution at 45 feet.)

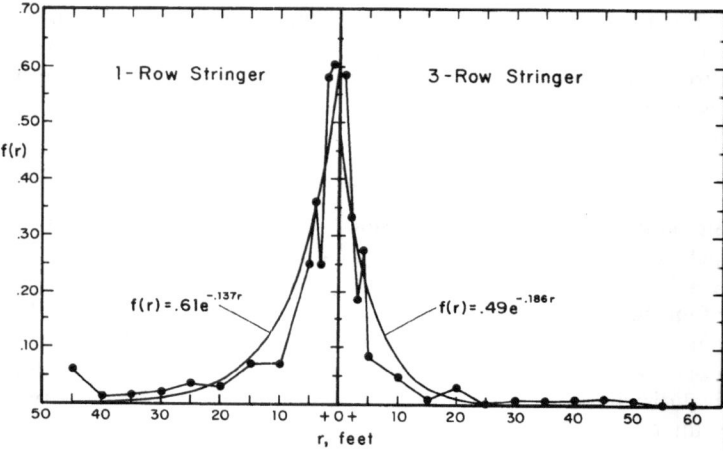

Fig. 4. Percentage of seed from contaminant pollen in radish *versus* distance from contaminant pollen source; on left, single-row stringer from central contaminant plot, and on right, three-row stringer from central contaminant plot (from Bateman, 1948, Table 4)

The explanation for the samples taken one and two feet from the contaminant pollen source might be due to some wind pollination, but this does not account for the extremes, especially the most distant sample at 45 feet. A likely explanation for this pattern, especially in light of the next example, might be that a significant proportion of bees reached the end of the stringer and then doubled back in the opposite direction. Because there was no pollen to feed upon beyond 45 feet, they in effect reached an "ecological interface" and were "reflected" back from it.

In the second example from Bateman's work (Fig. 4, right), the stringer was sixty feet long and consisted of three rows. Kurtosis was 23 in the data, with the expected value for the truncated exponential of 5.93. The disturbance, although having greater impact upon the value of kurtosis, is much less apparent to an observer examining the form of the distribution. There is no interface effect at the end of this stringer. Possibly the greater length of this stringer, combined with a greater pollen supply, is sufficient to explain the lack of an observed "interface effect" – fewer individuals reached the end. However, since the distribution is still highly leptokurtic, one might suspect that movement was not random, even apart from interface complications.

Dispersal Patterns in Drosophila

Most studies of Drosophila dispersal have been conducted by releasing marked laboratory-reared flies in nature and subsequently recapturing them. One of the earliest was that of the Timofeeff-Ressovskys (1940), but their data does not lend itself to detailed analysis. The extensive work by Dobzhansky and Wright (1943, 1947) is well suited to the comparison of theoretical and observed distributions, and furthermore, their results have recently been the topic of renewed discussion in the literature (Wallace, 1966, 1968a, b; Wright, 1968a, b).

Especially during the first few days after the release of their marked flies, the observed dispersal distributions were non-Gaussian, contrary to expectations from the model (Eq. (9)). Kurtosis was between 7.6 and 10.4 for the first day and had declined to a range of 4.4 to 5.9 on the second day (Wright, 1968b). Wallace (1966) reanalyzed these data and found a highly descriptive empirical model for the data in the form of Eq. (1) with $\kappa = 1/2$. Wright (1968b) pointed out that this model would be theoretically expected if the vector velocity were the average absolute velocity, i.e., that for a uniform value of κ to be $1/2$ throughout the experiment, one must assume that, "... daily flights of each fly were perfectly correlated in direction. It seems unlikely that each fly would remember the direction it had flown and keep on in the same direction throughout the experiment." It certainly is not characteristic of Drosophila to exhibit behavioral attributes such as prolonged unidirectional flight. Moreover, this model assumes a constant, rather than declining, density at the origin. However, at the time of release, the natural density at the spot was greatly increased, and it then decreased as flies dispersed. By the end of the first day, the ratio of released to native flies at the origin was about 9.1 (Fig. 5, from Wright, 1968a). For that particular, very transitory stage of the experiment, there may have been sufficient overcrowding to have effected orientation of dispersal by increasing the frequency of movements along radii leaving the release point and by agitated activity increasing general mobility. (Wallace (1966) discusses such a possibility of experimental alteration of normal behavior.) Together these effects might lead to the model Wallace arrived at empirically. However, as pointed out by Wright (1968b), the compounding of distributions over days causes a rapid decline in β_2, and the compound distribution approaches the Gaussian, although the general shape of the distribution may not change

in a spectacular fashion. Fig. 5 is data for the first day only and is not
such a compound distribution, but in general, interpretation of behavior
patterns measured by the shape of the distribution must be approached
cautiously.

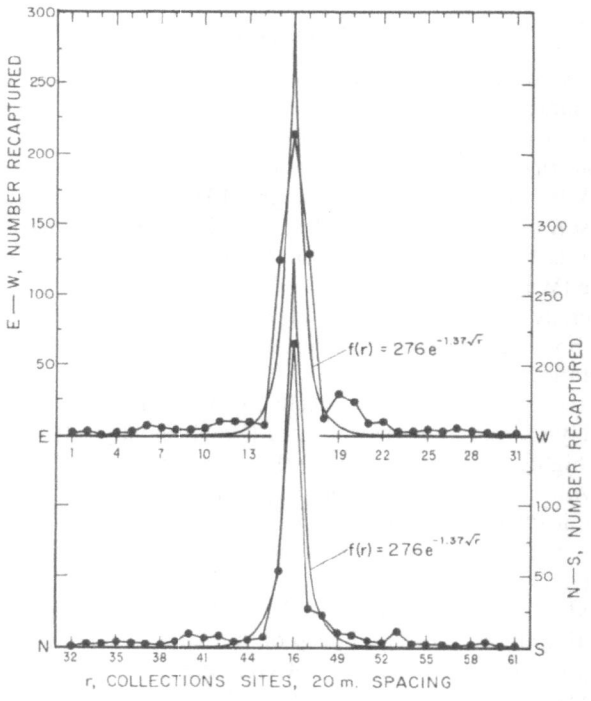

Fig. 5. Spread of marked *Drosophila pseudoobscura* after release previous day; top, East
West transect, and bottom, North-South transect (from Dobzhansky and Wright, 1943,
Fig. 1). Curve was fitted to average of four radii of transects

These data suggest a more general phenomenon of non-random
behavior. Wright (1968b) reiterates the importance of ecological heterogeneity, which was also mentioned in earlier publications (Dobzhansky
and Wright, 1943, 1947). In one experiment, the release site was in a
region of low density of native flies, which suggests that the area was
unattractive to *D. pseudoobscura*. The resulting behavior was rapid
movement to adjacent areas of higher native density, and presumably
a more attractive environment. If the usual pattern of movement results
in a clustering in an attractive area, one would expect this pattern. That
is, one would predict a higher (more peaked) mode around the attractive

site. Among those individuals which happened to reach the relatively unattractive areas between nearby attractive areas, some would continue to the next area. Thus, there would be larger than expected numbers in the tail portion of the expected distribution, which included the nearby attractive area. From these data, however, one may still argue that the movement was a result of the release of a large number of flies in an area normally only having a very few inhabitants.

In support of the supposition of an ecological attractiveness mechanism are the data from an experiment using *ad libitum* feeding of wild flies on a bait containing a trace of a rare earth, dysprosium, in which case the ones having fed there were distinguishable (by neutron activation analysis) from the ones which had not (Richardson, 1969a; Richardson *et al.*, 1969b). A dispersal distribution of marked individuals around this spot was estimated by captures in the vicinity and the identification of the marked individuals. The appropriate model would be of the form of Eq. (14), since there was a continuous production of marked flies. The pooled results shown for *D. aldrichi* in Fig. 6 have a kurtosis of about 8 compared with the expected value of less than 6.

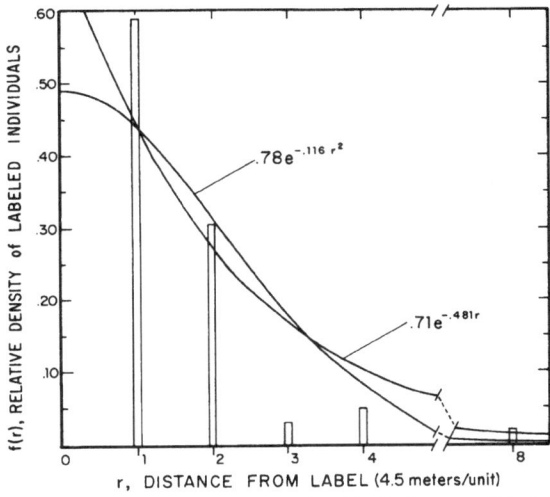

Fig. 6. Average radius of spread of *Drosophila aldrichi* from small bait labeled with dysprosium acetate (from Richardson, 1969)

In this experiment, the labeling site was intentionally selected to be one with a high natural density of flies in order to label as many individuals as possible. Thus, the origin was probably at a site of relative ecological attractiveness, and some that were collected from an arc with a 35 meter radius, came from nearby sites of higher natural density. Based

on the previous reasoning, one might expect this sample to contain somewhat greater than expected numbers of marked individuals.

From the experiments of two types utilizing two quite different
species, it is clear that the higher than expected values obtained for
kurtosis may result from two different situations. First, the disturbance
of density from a "release experiment" resulted in the measurement
of an effect probably not of general genetical importance, but which
obscured another disturbance of random behavior. There is distinct non
randomness of movement present in all cases; this is clearly seen only
in the results of the *ad libitum* feeding experiment, which is related to
ecological variability. This behavior pattern probably is of genetical
importance, affecting gene flow between adjacent populations and in-
fluencing adaptation to micro-environmental conditions.

Stable Distribution of Daphnia

Brownlee (1911) released *Daphnia pulex* in a shallow porcelain dish,
and described the situation as follows:

> (Daphnia) usually chose to distribute itself from one corner
> along one side of the dish. ... The corner being an impene-
> trable boundary may be taken as representing a centre of
> diffusion, so that the simple fundamental integral should
> apply and the grouping should conform to the exponential.

His data and fitted curves are given in Fig. 7. He points out that there
is good agreement between the data and the exponential model, except
for a single group of four individuals observed when only 1.7 individuals

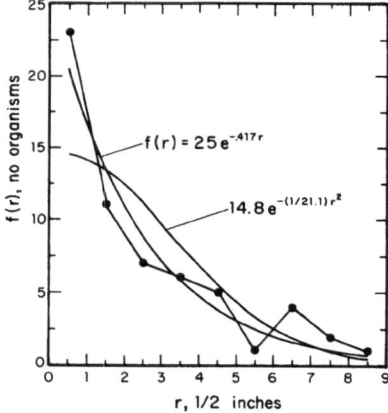

Fig. 7. Distribution of *Daphnia* from edge of dish (from Brownlee, 1911, p. 279)

were expected. This might have been a result of a transitory secondary disturbance of activity, such as local clustering. But, considering the small sample size, the explanation is best left as an open question.

The estimate of kurtosis from these data is 3.4, a value just slightly greater than that expected for a Gaussian distribution and not at all like that of 6 expected for an exponential distribution. However, the expected values for these distributions, calculated over the range of the data, are 2.8 and 3.5, respectively. Truncation of the range for the data is a primary reason for the small value of the estimate of kurtosis, and is an extreme case of inefficiency of fitting data to distributions by comparing moments. As can be seen in Fig. 7, the Gaussian distribution is much less descriptive than the exponential, in spite of the contradiction from kurtosis comparisons.

These data illustrate a special case of the exponential (where $\kappa = 1$) even though there is no birth-death process and Eq. (11) equals zero. The dispersion is away from an attractive interface, beyond which the individuals may not pass, but those leaving the interface are replaced by an equal number of individuals returning to the interface. If an ecological generalization can be made from this effect, it concerns the nature of distortion of a Brownian motion model by the presence of an attractive site at the mode. A stable distribution may be reached, whereby the variance is no longer increasing with time, but becomes independent of time. However, for a symmetrical distribution to be obtained, members of the population must be attracted to a common spot accessible from all directions. For many populations, even those with discrete generations, whereby a Brownian motion model might be appropriate, the exponential model may be a much better empirical approximation of dispersal behavior.

Dispersion of Juvenile Lizards

Blair (1960) presents an extensive account of the natural history of the rusty lizard (Sceloporus olivaceus), which includes data on dispersal obtained by marking members of a brood of newly hatched lizards and subsequently identifying the individuals when they reach sexual maturity. The lizards disperse as juveniles and eventually establish a restricted "home range". Being territorial in nature, they can occupy only certain locations within their home range. A limited number of desirable sites results in new occupations primarily reflecting the death of the resident lizard or the expulsion of the resident lizard by the new occupant of a site.

Males are more aggressive than females and are more strongly territorial, excluding other males from an area. The females may share portions of a territory. It was also found that the female occupies a smaller home range than does the male and thus had reduced space requirements for a given census number.

Blair's data (1960) is given in Fig. 8, together with both Gaussian and exponential distributions fitted to the sample variances for females and males. Estimates of kurtosis are 6.9 for females and 9.7 for males, and corresponding expectations for the truncated exponential distributions are 4.5 and 4.6, respectively.

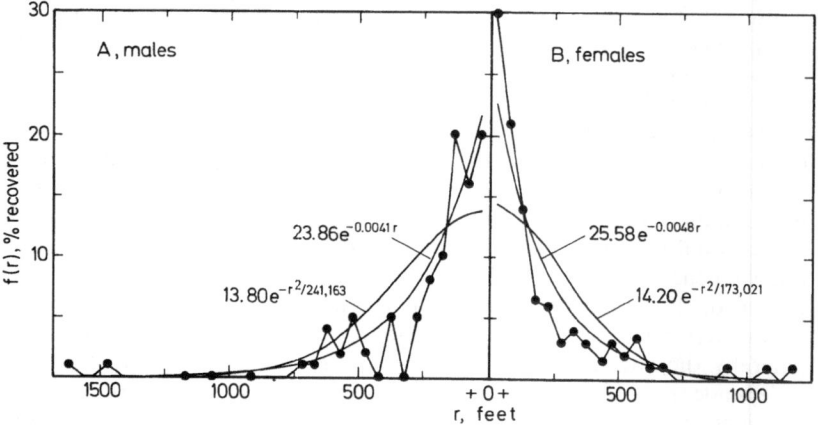

Fig. 8. Percentages of newly matured male and female lizards at various distances from hatching site (from Blair, 1960, Fig. 25)

Individuals were hatched and marked during a given season. The time of dispersal varied: some dispersed very soon after hatching, some established a juvenile home range and dispersed later, and others remained near their place of birth. Within wide limits this behavior fits the model form of Eq. (14).

However, territorality will modify the expected results. The greater range of males might be a result of at least two components. Greater aggressiveness between non-relatives would tend to increase the male dispersion as they moved away from a birth site, while such an effect might be reduced in less aggressive females.

Another effect is likely to be the result of both ecological interfaces and of gradations of ecological diversity around the birth sites. Descriptions indicating both of these ecological variables are given by Blair. He found a highway bordering one side of the study area to be a partial barrier to dispersal, which would affect the more dispersive males to a greater extent than the females. Some of the observations were noted as follows:

> "Among 9 lizards hatched along the north fence of the ten acres and later recovered as adults, none dispersed across the road. ... Among the lizards marked as hatchlings and later recovered as adults, at least 7 dispersed initially northward to or near the north fence and subsequently reversed the direction and dispersed back onto the ten acres before becoming sexually mature. Some lizards other than the ones mentioned above were known to cross the highway ..."

An asymmetrical pattern of dispersal was also noted and was thought to reflect the attractiveness of the ecology to the organism, although a complete explanation was not obvious. For example,

> "The hatchling lizards in all three of these broods tended to move initially toward open terrain with scattered posts, mesquites, and other basking sites, and with rather sparce ground cover. The area immediately south and southeast of the hatching site was a dense thicket of mesquite, some of it with a dense cover of grass. It is easy to see why the lizards did not disperse in this direction, for both juveniles and adults tend to avoid dense thickets and dense ground cover. It is less easy to explain the failure to disperse westward and southwestward."

Since the data are dispersal records pooled over a variety of environmental variables, likely there were variations in dispersal rates. The appropriate expected distribution would have a greater kurtosis than that of Eq. (14). Again, since males range further than females, the impact of dispersal rate heterogeneity would be increased.

In addition to discussing the complexities of these data, consider the consequencies of not separating a heterogeneous population into homogeneous portions. The general patterns for males and females are similar, except for greater mean distance of movement of males than for females. Kurtosis for the distribution formed from the two fitted exponential distributions is 4.63, while that for females is 4.47, and for males is 4.61. Kurtosis of the compound distribution is increased over that of the greater of the components. The same type of effect may be seen in the other situations where sexes differ in dispersal characteristics or where other groups with different rates of dispersal are inadvertently pooled. For the Gaussian distribution, Wright's discussion has been presented. However, as seen in this case, even for truncated exponential distributions the result of pooling distributions of unequal variance is an increased kurtosis.

Summary

Models of dispersal are of two general types. The first, one of Brownian motion, applies where a constant number of individuals begins moving simultaneously from a common point. Application of this model to biological problems leads to a density function of individuals which is time-dependent, the variance being proportional to the length of time dispersal has been taking place. The second model is developed for cases where there is a continuous "production" of individuals at a given point, replacing those which die or are otherwise lost. Application of this model leads to a density function of individuals which is independent of time. An example is given, however, where a stable distribution results for a constant number of individuals at an attractive interface, illustrating the possible distortion by non-random movement of a density from the first model to the density described by the second model. For both models, the underlying assumptions concerning the behavior of individuals are the same; both models assume random movement. It is also assumed that the expected velocity of movement of individuals is constant, being neither a function of intrinsic individual differences or the spatial position of individuals. Improper distinction and incorrect choice of model by a number of authors has resulted in confusing the literature. Many words were wasted explaining differences between observations and expectations derived from inappropriate models.

Application of models by population and ecological geneticists is directed primarily to three objectives: first, the identification of dispersal behavior patterns and the estimation of dispersal functions; second, the evaluation of effects of dispersal on colonization and on population and gene pool compositions; and, third the determination of ecological factors affecting dispersal behavior. The first objective is primarily the description of nature; the second leads to more realistic genetic and evolutionary models, or at least allows the shortcomings of simplified models to be determined; and the third objective is primarily related to generalizations of empirical interpretations and theoretical predictions – devising the guidelines for "reasonable" extrapolation beyond the facts.

Attainment of the first objective has been hampered by many difficulties. Some techniques of studying behavior modified the behavior. Consequently, study techniques for behavioral characteristics are notoriously prone to measure artifacts! In the absence of empirical data, attempts directed toward the other two objectives have often been based upon the intuition of the theoretician and his concept of nature.

A re-examination of selected data emphasizes some of the major biological variables at odds with the model assumptions and some of the problems to be encountered when attempting refinement.

(1) Individuals may tend to move haphazardly, and their dispersal distributions may be closely approximated by models assuming random movement. If by chance this agreement between data and expected distribution reflects the use of a realistic model, the situation would suggest that there exists some set of conditions whereby variation of individual responses are the same. In other words, under some conditions behavior patterns are essentially uniform. The limits of modification of this set of conditions to which average behavior closely describes individual behavior would be useful in applying a number of genetic, ecological, and sociological models.

(2) Regional variation in ecological "attractiveness" may be reflected in sharp modifications of dispersal activity, even to the extent of apparently changing the function from one assuming no birth or death to one which incorporates these processes. Obviously, neither model is based on appropriate premises. As a warning to empirical model building, it is clear that the appropriate models are likely to evolve only by an examination of details of behavior rather than by the study of such gross descriptions as dispersal distributions.

(3) The statistical estimation of parameters of models is complicated by the nonlinear models, and biological interpretation of comparisons among parameter values, either observed or expected, are difficult. Unless the models are based on biological premises, meaningful interpretations are essentially impossible. It appears that the ultimate choice among models is likely to be based upon comparisons of sets of premises among models, not upon elimination of all but one model by comparison with overall observed results. Critical information is confounded by the summarization of individuals' activity into the manageable distribution functions for the population.

Acknowledgements

The author is pleased to acknowledge the helpful discussions and comments of Dr. S. Wright during his visit to Austin (supported by GM 15769 to Kojima, et al.). Some further expansion of *Dispersal Patterns in Drosophila* to better illustrate the model of population structure presented here was a direct consequence of these discussions, as was the correction of an error in Heterogeneous Populations. Of course, any errors remaining are strictly those of the author. The editorial assistance of several colleagues, especially that of Drs. H. E. Schaffer and R. K. Selander, is gratefully acknowledged. Financial support for this work was derived from N.S.F. Grant GB-7252 and A.E.C. Contract AT-(40-1)-4023 to the author.

References

Bateman, A. J.: Contamination of seed crops. II. Wind pollination. Heredity 1, 235–246 (1947a).
— Contamination of seed crops. III. Relation with isolation distance. Heredity 1, 303–336 (1947b).
— Contamination of seed crops. I. Insect pollination. J. Genet. 48, 257–275 (1948).
— Is gene dispersion normal? Heredity 4, 253–263 (1951).

Blair, W. F.: The Rusty Lizard. Austin: University of Texas Press 1960.

Brownlee, J.: The mathematical theory of random migration and epidemic distribution. Proc. Roy. Soc. Edinburgh 31, 262–289 (1911).

Dobzhansky, Th., and S. Wright: Genetics of natural populations. X. Dispersion rates in *Drosophila pseudoobscura*. Genetics 28, 304–340 (1943).
— — Genetics of natural populations. XV. Rate of diffusion of a mutant gene through a population of *Drosophila pseudoobscura*. Genetics 32, 303–339 (1947).

Ehrman, L.: Mating success and genotype frequency in Drosophila. Anim. Behav. 14, 332–339 (1966).

Feller, W.: An introduction to probability theory and its applications. Vol. II. New York: John Wiley & Sons 1966.
— An introduction to probability theory and its applications. Vol. I. 3rd ed. New York: John Wiley & Sons 1968.

Frampton, V. L., M. B. Linn, and E. D. Hansing: The spread of virus diseases of the yellows type under field conditions. Phytopathology 32, 799–808 (1942).

Glasstone, S., and A. Sesonske: Nuclear reactor engineering. Princeton, N. J.: D. van Nostrand Co. 1963.

Hamilton, T. H., and I. Rubinoff: Isolation, endemism, and multiplication of species in the Darwin finches. Evolution 17, 388–403 (1963).

Heed, W. B., J. S. Russell, and B. L. Ward: Host specificity of cactiphilic Drosophila in the Sonoran Desert. Drosophila Information Service 43, 94 (1968).

Papoulis, A.: Probability, random variables, and stochastic processes. New York: McGraw Hill Co. 1965.

Reimer, J. D., and M. L. Petras: Breeding structure of the house mouse, *Mus musculus*, in a population cage. J. Mammalogy 48, 88–99 (1967).

Richardson, R. H.: Migration, and enzyme polymorphism in natural populations of Drosophila. Jap. J. Genet. Suppl. 1, 172–179 (1969a).
—, R. J. Wallace, Jr., S. J. Gage, G. D. Bouchey, and M. Denell: Neutron activation techniques for labeling Drosophila in natural populations. In: Studies in genetics. Ed. by M. R. Wheeler. Austin: University of Texas Press 1969b.

Skellam, J. G.: Random dispersal in theoretical populations. Biometrika 38, 196–218 (1951).

Stern, V. M., and A. Mueller: Techniques of marking insects with micronized fluorescent dust with especial emphasis on marking millions of *Lygus hesperus* for dispersal studies. J. Econ. Ento. 61, 1232–1237 (1968).

Timofeeff-Ressovsky, N. W., and E. A. Timofeeff-Ressovsky: Populationsgenetische Versuche an Drosophila Z. indukt. Abstamm.- u. Vererblehre 79, 28–49 (1940).

Wallace, B.: On the dispersal of Drosophila. Amer. Natur. **100**, 551–563 (1966).

— Topics in population genetics. New York: W. W. Norton Co. 1968a.

— On the dispersal of Drosophila. Amer. Natur. **102**, 85–87 (1968b).

Wolfenbarger, D. O.: Dispersion of small organisms. Amer. Midland Natur. **35**, 1–152 (1946).

Wright, S.: Evolution and the genetics of populations. Vol. I. Genetic and biometric foundations. Chicago: University of Chicago Press 1968a.

— Dispersion of *Drosophila pseudoobscura*. Amer. Natur. **102**, 81–84 (1968b).

— Evolution and the genetics of populations. Vol. II. The theory of gene frequencies. Chicago: University of Chicago Press 1969.

Avoidance and Rate of Inbreeding[*]

C. C. COCKERHAM

Introduction

Until Kimura and Crow (1963) discovered to the contrary, it was generally accepted that the cousin mating systems (Wright, 1921) which avoid the mating of relatives as much as possible had maximum avoidance of inbreeding during all generations. They gave examples of mating close relatives, circular half sib and circular (pair) first cousin mating systems, which in early generations led to more inbreeding but in late generations to less inbreeding than cousin systems with the same population size. Robertson (1964) contrasted the cousin and circular systems. Utilizing the coancestry of mates and the average coancestry for the line, he deduced that the more a mating plan makes the coancestry of mates greater than the average the more the final rate of approach to homozygosity is reduced. Wright (1965) also compared the two types of mating systems and others, coming to the same conclusions. He provided much insight into the situation by defining various correlations, F statistics, corresponding to the structuring of the relatives in the population, and showed in certain cases that some were negative relative to the mean.

Robinson and Bray (1965) compared four methods, random number and two offspring per family, each with and without the avoidance of mating full sibs. Reduction in variance of family size reduced the rate of inbreeding in a manner well known (Wright, 1938). The effect of avoidance of mating full sibs, however, was to reduce the rate of inbreeding when the number of offspring per family was random and to increase it when the number was exactly two.

These clues appear to suggest a picture. Involved are population size, variance of number of gametes per parent, structuring of the population, and the avoidance of the mating of relatives. Some mating systems specify exact combinations of these factors. In the cousin systems, for example, the gametic variance is always zero and the structuring of the

* Paper number **2763** of the Journal Series of the North Carolina State University Agricultural Experiment Station, Raleigh, North Carolina. This investigation was supported in part by Public Health Service Research Grant GM 11546 from the Division of General Medical Sciences.

population is related to the population size, as is also the avoidance of the mating of relatives. The factors are interconnected. One cannot structure a population or avoid the mating of relatives, for example, without affecting the gametic variance.

Some fairly general results relating these factors to avoidance of inbreeding and close inbreeding and rate of inbreeding are to be presented. Also, measures of relationships of structured groups, as well as of the population as a whole, other than the inbreeding coefficient will be utilized. These correspond in many respects to Wright's (1965) F statistics, but instead of correlations of various categories of genes, they are based on the probabilities of genes being "identical by descent" (Malécot, 1948). With these measures and a few simple rules, one can work with pedigrees of groups, structured populations, with almost the same ease as one does with pedigrees of individuals (Cockerham, 1967).

Definitions

The following list of symbols will be used throughout. For more complete definitions and rules of operation for structured populations see Cockerham (1967).

Symbol	Definitions
N (n, m, s, u)	is the total number of individuals in the sample population and remains constant from generation to generation. Subdivisions of the population will be into $2^s = m$ subdivisions such that there are $n = N/m$ individuals in each subdivision. $s = 0, 1, 2, ..., u = \log_2 N$. For $s = u, n = 1$ and the mating system is of individuals.
F	is the inbreeding coefficient corresponding to the probability of alleles being identical by descent in individuals. Its average value over individuals is to be studied.
θ	is the probability of a random allele in one individual being identical by descent to a random allele in another individual, called the coancestry herein. For two individuals, it is the inbreeding coefficient of the offspring, $\theta_{ij} = F_{i \cdot j}$. For an individual with itself, it is $\theta_{ii} = (1 + F_i)/2$.
$\bar{\theta}$	is the average of the coancestries among individuals of a group and applies to a random pair of individuals from the group.

F_l is the average of the probabilities of identity by descent of all the genes in a group of individuals. For the i-th group, $F_{li} = \dfrac{F_i}{2n_i - 1} + \dfrac{2(n_i - 1)}{2n_i - 1}\,\bar{\theta}_i\,.$

θ_l will be used for groups of individuals in the same context as θ is for individuals. For two groups, it is the average of the coancestries of individuals in one group with those in the other and related to the inbreeding coefficient of the offspring from random union of gametes as

$$\theta_{lij} = F_{i \cdot j}\,.$$

In the case of the coancestry of a group with itself,

$$\theta_{lii} = \frac{1 + F_i}{2n_i} + \frac{(n_i - 1)}{n_i}\,\bar{\theta}_i = \frac{1}{2n_i} + \frac{(2n_i - 1)}{2n_i}\,F_{li}\,.$$

ψ, Ψ
$(n_g, n_e, \sigma_{kn}^2)$
is the probability of two random alleles being identical by descent in a set of gametes from the same group.

The general form of the gametic variance effective number is

$$\frac{\bar{k}(\bar{k}n - 1)}{\bar{k}(\bar{k} - 1) + \sigma_{kn}^2}$$

where $\bar{k}n$ is the number of gametes. Since only integral values of $\bar{k}$ are to be considered, the following relationship among gametic variances will be assumed

$$\sigma_{kn}^2 = \bar{k}\sigma_{1n}^2\,,$$

which is zero for a constant number of gametes per parent, and

$$\sigma_{kn}^2 = \frac{\bar{k}(n - 1)}{n}$$

for equal chance of each parent contributing any gamete. Two effective numbers will be distinguished

$$n_g = \frac{n - 1}{\sigma_{1n}^2}$$

for $\bar{k} = 1$, and

$$n_e = \frac{2(2n - 1)}{2 + \sigma_{2n}^2}$$

for $\bar{k} = 2$. Also, two corresponding gametic probabilities will be distinguished.

For independent sampling of gametic sets, the gametic probability for the set from parental group j to offspring group i is

$$\psi_{ij} = \frac{1 + F_j}{2n_g} + \frac{n_g - 1}{n_g} \bar{\theta}_j.$$

The number of gametes in a set is always n. With independent sampling of gametic sets from the same group, the gametic probability is the same for all sets, e.g. $\psi_{ij} = \psi_{i'j}$, and the probability of a random allele in one set being identical by descent to a random allele in the other set is θ_{ljj}. For equal chance of each parent contributing any gamete in a set $\psi_{ij} = \theta_{l:j}$ also.

For combined sampling of $2n$ gametes from a parental group the gametic probability is

$$\Psi_{ij} = \frac{1 + F_j}{2n_e} + \frac{n_e - 1}{n_e} \bar{\theta}_j.$$

If the allocation of gametes to the two offspring groups, i and i', is at random, then $\Psi_{ij} = \Psi_{i'j} = \Psi_{\bar{i}j}$. Also, the probability of a random allele in one set being identical by descent to a random allele in the other set is the same gametic probability $\Psi_{jj} = \Psi_{\bar{i}j}$, in contrast to θ_{ljj} for independently sampled gametic sets. With equal chance of any parent contributing any gamete there is no difference between the sampling plans,

$$\psi_{ij} = \Psi_{ij} = \Psi_{jj} = \theta_{ljj}.$$

$\theta_{lP}, F_{lP}, \bar{\theta}_P, \Psi_P$ — When applied to the entire sample population, the notations θ_{lP}, F_{lP}, $\bar{\theta}_P$ and Ψ_P will be used.

t — measures time in generations. $t = 0, 1, 2, \ldots$. Initial members ($t = 0$) of the sample population are always assumed to be noninbred and unrelated.

Random Mating and Avoidance of Mating Relatives

A single population without subdivisions is envisioned. Generations are nonoverlapping. Individuals are bisexual. Mating is random subject to the conditions imposed by avoiding the mating of all relatives to degree v. The degree, v, of relatives is defined such that if they were not inbred $\theta = (\frac{1}{2})^{v+1}$. Thus, $v = 0$ for an individual with itself, $v = 1$ for parent offspring and full sibs, $v = 2$ for half sibs, and double first cousins, and so on.

For the avoidance of mating of v and lower degree relatives, an individual in generation t must trace to 2^v different ancestors in generation $t - v$, 2^{v-1} of these ancestors being for each of the uniting gametes received by the individual. Thus, genes in uniting gametes cannot stem from a common ancestor before the $t - v - 1$ previous generation. For a gene in one of the uniting gametes of each individual in generation t there are $2(N - 2^{v-1})$ possible pairings of ancestral genes in generation $(t - v - 1)$, giving a total of $2N(N - 2^{v-1})$ possible pairings for the N individuals. Expressing the number of pairings of genes from the same individual in terms of the mean $= 2$ and variance of the number of gametes per parent, the probability of two uniting genes in generation t being from a common ancestor in generation $t - v - 1$ is

$$\frac{1}{N_{ev}} = \frac{\sigma_{2n'}^2 + 2}{2(2N - 2^v)}. \tag{1}$$

When each parent contributes exactly two genes, $\sigma_{2n'}^2 = 0$,

$$N_{ev} = 2N - 2^v. \tag{2}$$

With equal chance of each parent contributing to the next generation, $\sigma_{2n'}^2$ is affected by the number of steps of avoidance. To carry out the randomization plan one must restrict the number of matings (gametes) per individual to be no greater than $2N/2^v = 2n'$ for v steps of avoidance. Since each individual in generation t traces to 2^v distinct ancestors in generation $t - v$, with equal chance of parenthood (hereafter termed just "equal chance") the probability of an ancestral individual being one of these (it can be no more than one) is $2^v/N = \frac{1}{n'}$. Further, the ancestor can be involved in only $N/2^v$ such ancestral groups. The distribution of k, the number of gametes per parent, is then

$$f(k) = \binom{2n'}{k} \times \left(\frac{1}{n'}\right)^k \left(1 - \frac{1}{n'}\right)^{2n'-k},$$

$$k = 0, 1, 2, ..., 2n', \tag{3}$$

$$\bar{k} = 2, \sigma_{2n'}^2 = 2\left(\frac{n'-1}{n'}\right) = 2\left(\frac{N - 2^v}{N}\right).$$

Substitution of $\sigma_{2n'}^2$ (3) into N_{ev} (1) gives $N_{ev} = N$ for any degree of avoidance. Thus, N_{ev} varies from N with equal chance to $2N - 2^v$ for $\sigma_{2n'}^2 = 0$.

For formulating the recurrence relation for F, refer to Diagram 1. Given a random gene (gamete), a_1, we have developed the probability, $1/N_{ev}$, that its mate, a_2, descended from the same common ancestor, A_{t-v-1}, in which case the probability of a_2 being indentical by descent to a_1 is

$$\theta_{A\,A\,t-v-1} = \frac{1 + F_{t-v-1}}{2}.$$ (4)

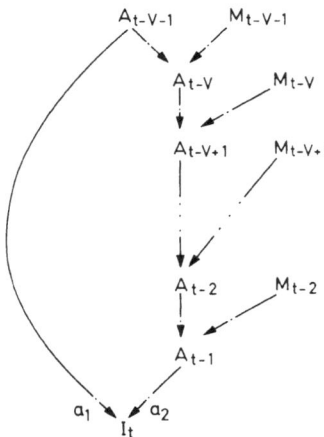

Diagram 1. Pedigree of uniting gametes from a common ancestor

Note that a_2 must have descended through the series of ancestors indicated in Diagram 1. Whatever the probability that a_2 descended from A_{t-v-1}, there is the same probability, $1/N_{ev}$, that it descended from the mate, M_{t-v-1}, of the ancestor, in which case, the probability of a_2 being identical by descent to a_1 is the same as between the two uniting genes in $t-v$, i.e. $\theta_{A\,M\,t-v-1} = F_{t-v}$. By random segregation arguments, the probability that a_2 came from a mate, M_{t-v}, of the ancestor in the next generation is just twice that from A_{t-v-1}, and the probability of a_2 being identical by descent to a_1 is F_{t-v+1}. Carrying the argument forward, we find the probability that a_2 came from a mate of an ancestor in the $t-v-1+x$ generation to be $2^x/N_{ev}$, $x=0, 1, 2, \ldots v-1$, with a corresponding probability of the genes being identical by descent of F_{t-v+x}. Summing up the probabilities of the gene having come from the common ancestor or mates of an ancestor, we obtain

$$\frac{1}{N_{ev}} + \frac{1}{N_{ev}} + \frac{2}{N_{ev}} + \frac{4}{N_{ev}} + \cdots + \frac{2^{v-1}}{N_{ev}} = \frac{2^v}{N_{ev}}.$$ (5)

For the remaining probability, $1 - 2^v/N_{ev}$, the gene must come from an individual sufficiently unrelated to A_{t-2} to be eligible as a mate, and for which the probability of the genes being identical by descent is F_{t-1}. In summary, the inbreeding coefficient is

$$F_t = \frac{1 + F_{t-v-1} + 2F_{t-v} + 4F_{t-v+1} + \cdots + 2^v F_{t-1}}{2N_{ev}} + \left(1 - \frac{2^v}{N_{ev}}\right) F_{t-1}$$

$$= F_{t-1} + \frac{1 + F_{t-v-1} + 2F_{t-v} + \cdots + 2^{v-1} F_{t-2} - 2^v F_{t-1}}{2N_{ev}} \tag{6}$$

which reduces to the appropriate ones for self-fertilization and avoidance of self-fertilization. From an alternative form of Eq. (6),

$$F_t = F_{t-1} + \frac{1 - F_{t-v-2}}{4N_{ev}} + \frac{F_{t-1} - F_{t-2}}{2}\left(1 - \frac{2^v}{N_{ev}}\right), \tag{7}$$

it is easy to arrive at a more simple recurrence form for maximum avoidance, $N = 2^v$, $N_{ev} = N$,

$$F_t = F_{t-1} + \frac{1 - F_{t-v-2}}{4N}. \tag{8}$$

We may arrive at this last result in a simple alternative way. For maximum avoidance, each of a pair of mates must trace to each of the $2^v = N$ individuals in the v-th previous generation. The coancestry of mates is always equal to the inbreeding coefficient of the offspring, but in this case it is equal to the average of all the coancestries, including individuals with themselves, in the v-th previous generation. In other words,

$$F_t = \theta_{lPt-v-1} = \frac{1 + F_{t-v-1}}{2N} + \frac{N-1}{N} \bar{\theta}_{Pt-v-1}. \tag{9}$$

Taking note that maximum avoidance can be accommodated only if $\sigma_{2n'}^2 = 0$, and utilizing $F_{lPt} = \Psi_{Pt-1}$, we have

$$\frac{F_t}{2N-1} + \frac{2(N-1)}{2N-1} \bar{\theta}_{Pt} = \frac{1 + F_{t-1}}{2(2N-1)} + \frac{2(N-1)}{2N-1} \bar{\theta}_{Pt-1}. \tag{10}$$

Substituting for the $\bar{\theta}$'s their equivalences in F's from Eq. (9) leads to

$$F_{t+v+1} = F_{t+v} + \frac{1 - F_{t-1}}{4N} \tag{11}$$

or

$$F_t = F_{t-1} + \frac{1 - F_{t-v-2}}{4N} \tag{12}$$

as before.

With v steps of avoidance the inbreeding coefficient is $1/2N_{ev}$ in the $v + 1$ generation and accrues according to Eq. (6) or an alternative form.

Subdivision of Gametes

Sexuality of gametes has been ignored, meaning they were at random with respect to the parents subject only to the restriction that there be an equal number of each. This procedure is termed combined gametic sampling. When subdivision of the gametes into male and female sets is taken into account, and each is independently sampled, the results are different from Eq. (6) for $v = 0$ and σ_{fN}^2 less than for equal chance (Cockerham, 1967). In this case

$$F_t = F_{t-1} + \frac{1 - F_{t-2}}{4N} + \frac{1 + F_{t-2} - 2F_{t-1}}{4N_g}.$$ (13)

With equal chance $N_g = N$ and F_t has the same recurrence form as in Eq. (6). With no gametic variance $N_g = \infty$, and

$$F_t = F_{t-1} + \frac{1 - F_{t-2}}{4N},$$ (14)

starting with $F_1 = 1/2N$.

With avoidance of self-fertilization or further early avoidance of inbreeding, subdivision of the gametes into sexual sets has no effect on the recurrence form for F, Eq. (6), as long as individuals are mated at random, subject only to the gametic sampling plans. To use only certain individuals as progenitors of a type of gamete or a subset of gametes brings about subdivisions of the population.

Subdividing the Population

Avoidance to the Limit of the Subdivision

The population is subdivided into $2^s = m$ aliquots. Offspring aliquots are produced by crossing parental aliquots in a manner to avoid the mating of related groups as much as possible, $v = s$. With 2^v subdivisions, inbreeding can be avoided initially for v generations as in the case of individuals. The mating pattern does not really matter with maximum avoidance permitted by the number of subdivisions, but is probably easiest visualized with the cousin type avoidance system. An example of this system corresponding to double first cousins is given in Diagram 2. In each group are $n = N/4$ individuals. For independent sampling of the two gametic sets from each group,

$$F_t = \theta_{IABt-1} = \frac{1}{4}(\theta_{IAB} + \theta_{IAD} + \theta_{IBC} + \theta_{ICD})_{t-2}$$

$$= \frac{1}{16}(\theta_{IAA} + \theta_{IBB} + 2\theta_{IAB} + 2\theta_{IAC} + 2\theta_{IAD}$$

$$+ 2\theta_{IBC} + 2\theta_{IBD} + 2\theta_{ICD} + \theta_{ICC} + \theta_{IDD})_{t-3}$$

$$= \theta_{IPt-3} = \frac{1 + F_{t-3}}{2N} + \left(1 - \frac{1}{N}\right)\bar{\theta}_{Pt-3}.$$ (15)

C. C. Cockerham

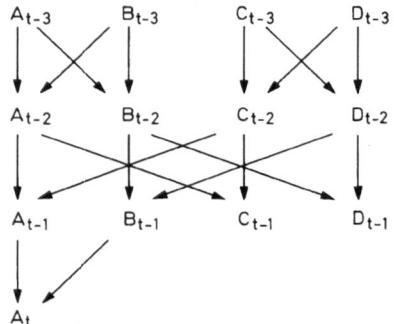

Diagram 2. Double first cousin mating system of groups of N/4 individuals

This averaging process demonstrates the equality of the inbreeding coefficient and the coancestry of the population with itself in the third previous generation for this gametic sampling plan, regardless of variations in the gametic variance. To find the recurrence form for F, note that

$$\theta_{IADt-2} = \theta_{IBCt-2} = \theta_{IACt-2} = F_{t-1},$$

$$\theta_{IABt-2} = \theta_{IBCt-2} = (\theta_{IAA} + \theta_{IBB} + 2\theta_{IAB})_{t-3}/4$$

$$= (\theta_{It-3} + F_{t-2})/2,$$

$$\theta_{It-3} = \frac{1+F_{t-3}}{2n} + \frac{(n-1)}{n}\bar{\theta}_{t-3}, \tag{16}$$

$$\bar{\theta}_t = \frac{F_t}{2} + \frac{\psi_{t-1}}{2},$$

$$\psi_{t-1} = \frac{1+F_{t-1}}{2n_g} + \frac{(n_g-1)}{n_g}\bar{\theta}_{t-1}.$$

Other helpful equivalences are,

$$2\bar{\theta}_t - F_t - \bar{\theta}_{t-1} = \frac{1+F_{t-1}-2\bar{\theta}_{t-1}}{2n_g} \tag{17}$$

and

$$\frac{(N-4)}{N}(1+F_{t-3}-2\bar{\theta}_{t-3}) = 1+F_{t-3}+2F_{t-2}+4F_{t-1}-8F_t, \tag{18}$$

and the recurrence form for F is

$$F_t = F_{t-1} + \frac{1-F_{t-4}}{4N} + \frac{1+F_{t-4}+2F_{t-3}+4F_{t-2}-8F_{t-1}}{16n_g}. \tag{19}$$

The argument for v steps of avoidance with 2^v equal subdivisions of the population will be sketched. Since each of a pair of mate groups traces to all of the ancestral groups in the v-th previous generation, by successive averaging of the θ_i's as in Eq. (14) we find

$$F_t = \theta_{lPt-v-1}. \tag{20}$$

The inbreeding coefficient is the coancestry of the least related groups in the $(t-1)$-th generation. By successive averaging, the coancestry of the next least related groups in $t-2$ is $F_{t-1}/2$, of the third least related groups in $t-3$ is $F_{t-2}/4$, and so on to the common ancestral groups in $t-v-1$ with a coancestry with themselves of θ_{lt-v-1}. Summarizing,

$$F_t = \frac{F_{t-1}}{2} + \frac{F_{t-2}}{4} + \cdots + \frac{F_{t-v}}{2^v} + \frac{\theta_{lt-v-1}}{2^v}. \tag{21}$$

The other probability measures have the same form as before, θ_l, $\bar{\theta}$ and ψ in Eq. (16). Utilizing $n = N/2^v$,

$$\frac{\theta_{lt-v-1}}{2^v} = \frac{1 + F_{t-v-1}}{2N} + \frac{N-2^v}{2^v N} \bar{\theta}_{t-v-1}. \tag{22}$$

Of the equivalences, the one corresponding to Eq. (17) remains unchanged and the general form of Eq. (18) is

$$\frac{N-2^v}{N}(1 + F_{t-v-1} - 2\bar{\theta}_{t-v-1})$$
$$= 1 + F_{t-v-1} + 2F_{t-v} + \cdots + 2^v F_{t-1} - 2^{v+1} F_t, \tag{23}$$

and

$$F_t = F_{t-1} + \frac{1 - F_{t-v-2}}{4N} + \frac{1 + F_{t-v-2} + 2F_{t-v-1} + \cdots + 2^v F_{t-2} - 2^{v+1} F_{t-1}}{2^{v+2} n_g}, \tag{24}$$

which reduces to

$$F_t = F_{t-1} + \frac{1 - F_{t-v-2}}{4N} \tag{25}$$

for $\sigma_{1n}^2 = 0$, and to

$$F_t = F_{t-1} + \frac{1 + F_{t-v-1} + 2F_{t-v} + \cdots + 2^{v-1} F_{t-2} - 2^v F_{t-1}}{2N} \tag{26}$$

for equal chance. It may be noted that Eq. (24) includes the case of $v = 0$ when the gametes are subdivided Eq. (13).

The inbreeding coefficient is $1/2N$ in the $v+1$ generation and from then on accrues according to Eq. (24).

Turning now to combined gametic sampling, one need only to substitute Ψ for θ_t and for ψ in the previous formulae, where

$$\Psi_t = \frac{1 + F_t}{2n_e} + \frac{n_e - 1}{n_e}\,\bar{\theta}_t\,. \tag{27}$$

From Eq. (21)

$$F_t = \frac{F_{t-1}}{2} + \frac{F_{t-2}}{4} + \cdots + \frac{F_{t-v}}{2^v} + \frac{\Psi_{t-v-1}}{2^v}\,. \tag{28}$$

By substituting the F expression of Ψ from Eq. (28) into

$$\bar{\theta}_{t-v-1} = \frac{F_{t-v-1}}{2} + \frac{\Psi_{t-v-2}}{2}\,, \tag{29}$$

and then of $\bar{\theta}_{t-v-1}$ into Ψ_{t-v-1}, the recurrence form is found to be

$$F_t = F_{t-1} + \frac{1 + F_{t-v-1} + 2F_{t-v} + \cdots + 2^{v-1}F_{t-2} - 2^v F_{t-1}}{2^{v+1}n_e}\,. \tag{30}$$

Since $n = N/2^v$ and $n_e = 2(2n - 1)/(2 + \sigma_{2n}^2)$,

$$2^{v+1}n_e = \frac{4(2N - 2^v)}{2 + \sigma_{2n}^2} = 2N_{ev}\,, \tag{31}$$

and the recurrence form for F is the same as that Eq. (6) for a nonsubdivided population and v steps of avoidance.

The two recurrence forms Eq. (24, 30) are the same with equal chance and also with maximum avoidance, $2^v = N$. With less than maximum avoidance and gametic variance other than for equal chance, a subdivision of the gametic sets leads to slightly different results.

With maximum avoidance, by varying $N = 1, 2, 4, 8, \ldots$, the recurrence forms are generated for many of the classical mating systems of individuals, i.e., self-fertilization, full sib, double first cousin, and higher order cousins.

It is interesting to note that the situation with exactly one male and one female offspring per family and random mating (Robinson and Bray, 1965) corresponds to $v = 1$, independent sampling of gametic sets and $\sigma_{1N}^2 = 0$, while the avoidance of mating full sibs corresponds to $v = 2$, combined sampling of gametic sets and $\sigma_{2N}^2 = 0$, or just plain avoidance, $v = 2$, no gametic variance, and random mating. There are only two subdivisions in the population, the two sexes. The subdivision of gametic sets has an effect only when gametic variance is less than for equal chance and only when compounded on the subdivisions of individuals. If avoidance is of a greater degree than the subdivisions of individuals, then the subdivisions of gametes lose their identity.

Avoidance and Subdivisions General

To distinguish between subdivisions and avoidance, let $m = 2^s = N/n$. For avoidance of mating related subdivisions of degree v, we shall use an effective number, m_{ev}, corresponding to N_{ev} (1),

$$m_e = \frac{2(2m-1)}{2 + \sigma_{2m}^2},$$

$$m_{ev} = \frac{2(2m - 2^v)}{2 + \sigma_{2m'}^2} = \frac{2(2^{s+1} - 2^v)}{2 + \sigma_{2m'}^2}, \tag{32}$$

$$m' = \frac{m}{2^v}.$$

When $v = s$, avoidance is maximum, and v can be no larger than s. Following analogous arguments that led to Eq. (5) and Eq. (6), we find for v steps of avoidance,

$$F_t = F_{t-1} + \frac{\theta_{lt-v-1} + F_{t-v} + 2F_{t-v+1} + \cdots + 2^{v-1}F_{t-2} - 2^v F_{t-1}}{m_{ev}} \tag{33}$$

for independent gametic sampling, and it is unrealistic to consider combined gametic sampling. Making use of the relationships for structured populations,

$$F_t = F_{t-1} + \frac{1 - F_{t-v-2}}{4nm_{ev}} + \frac{1 + F_{t-v-2} + 2F_{t-v-1} + \cdots + 2^v F_{t-2} - 2^{v+1}F_{t-1}}{4n_g m_{ev}}$$
$$+ \frac{F_{t-1} - F_{t-2}}{2}\left(1 - \frac{1}{n_g}\right)\left(1 - \frac{2^v}{m_{ev}}\right). \tag{34}$$

The form of Eq. (34) is the same as Eq. (24) except for the addition of the last term which disappears when $v = s$, in which case $\sigma_m^2 = 0$ and $m_{ev} = 2^s = 2^v = m$. For equal chance of subdivisions, $m_{ev} = m$ and of individuals, $n_g = n$, and

$$F_t = F_{t-1} + \frac{1 + F_{t-v-1} + 2F_{t-v} + \cdots + 2^{v-1}F_{t-2} - 2^v F_{t-1}}{2N}$$
$$+ \frac{F_{t-1} - F_{t-2}}{2} \frac{(N - 2^s)}{N} \frac{(2^s - 2^v)}{2^s}. \tag{35}$$

The factor $(1 - 2^s/N)(1 - 2^v/2^s)$, which goes to zero when $2^s = N$ or $2^v = 2^s$ as it should, has a maximum at $2^s = \sqrt{N2^v}$ and is larger for smaller v. The choosing among subdivisions at random increases the total gametic variance and the rate of inbreeding. Even with $n_g = \infty$, the last term of

$$F_t = F_{t-1} + \frac{1 - F_{t-v-2}}{4nm_{ev}} + \frac{F_{t-1} - F_{t-2}}{2}\left(1 - \frac{2^v}{m_{ev}}\right) \tag{36}$$

does not disappear unless $v = s$.

Mating of Close Relatives

The counterpart of the avoidance of mating of relatives is for mates to be more related than random individuals. A few examples will illustrate the main features. Consider first self-fertilization, i.e., the mating of relatives of degree $v = 0$. Let the probability of a random pair of offspring being from the same parent be

$$\frac{1}{N_g'} = \frac{\sigma_{1N}^2}{N-1}, \tag{37}$$

which is similar to $1/N_g$ except σ_{1N}^2 is for the number of offspring per parent and thus is for sets of two gametes. Inbreeding increases at the rate for self-fertilization

$$F_t = \frac{1 + F_{t-1}}{2} = 1 - \left(\frac{1}{2}\right)^t. \tag{38}$$

The average coancestry is

$$\bar{\theta}_t = \frac{F_t}{N_g'} + \frac{N_g' - 1}{N_g'} \bar{\theta}_{t-1} \tag{39}$$

which is always zero if $\sigma_{1N}^2 = 0$, and for equal chance, $N_g' = N$, and

$$\theta_{1Pt} = \bar{\theta}_{t+1}. \tag{40}$$

Since F approaches one very rapidly, the rate of increase of $\bar{\theta}$,

$$\frac{\bar{\theta}_t - \bar{\theta}_{t-1}}{1 - \bar{\theta}_{t-1}} \cong \frac{1}{N}, \tag{41}$$

is a result of random sampling and loss of progenitor groups.

To mate relatives of at least degree v that all stemmed from the same group of ancestors in the v-th previous generation, one must divide the population into permanent sublines of $n = 2^v$ individuals. The inbreeding and coancestry of individuals within groups will depend entirely on the effective number, n_e within groups, i.e., one may specify random mating subject to the avoidance of selfing, full sibbing, etc. If avoidance is up to $v - 1$, and $n = 2^v$, then the mating is between relatives with 2^v common ancestors in the v-th previous generation – full sibs for $v = 1$, double first cousins for $v = 2$, quadruple second cousins for $v = 4$, etc. Variation in the production of groups can be accommodated by

$$m_g' = (m-1)/\sigma_{1m}^2,$$
$$m = N/n. \tag{42}$$

The average population coancestry will be an average of $\bar{\theta}_w$ within groups and of $\bar{\theta}_b$ between groups,

$$\bar{\theta}_{Pt} = \frac{n-1}{N-1}\bar{\theta}_{wt} + \frac{N-n}{N-1}\bar{\theta}_{bt} \tag{43}$$

where

$$\bar{\theta}_{bt} = \frac{\bar{\theta}_{wt-1}}{m_g} + \frac{m_g-1}{m_g}\bar{\theta}_{bt-1}. \tag{44}$$

The within group coancestry is that for the mating system of the 2^v individuals in the group. Included in Eq. (43) and Eq. (44) is the previous example of self-fertilization for which $n = 1$.

If there is some variation allowed in the production of progeny groups from parental ones, the whole population eventually goes to fixation. On the other hand, if the identity of the groups is maintained, i.e., permanent sublines, $m_g = \infty$,

$$F_\infty = 1, \quad \bar{\theta}_{b\infty} = 0, \quad \bar{\theta}_{w\infty} = 1,$$

$$\bar{\theta}_{P\infty} = \frac{n-1}{N-1}, \quad \theta_{lP\infty} = \frac{n}{N}, \tag{45}$$

the sublines go to fixation but remain unrelated.

If there is an occasional interchange among subdivisions, the population measures, $\bar{\theta}_P$ and θ_{lP}, will approach one, but very slowly. As an example, consider N lines of individuals which are selfed t' generations and then mated at random for one generation, after which the process is repeated. Let the random mating generations be measured in t so that the actual number of generations is $t(t'+1)$. Let the gametic variance be zero for the random mating generations since it is for the selfed generations. We may note that $\bar{\theta}_{Pt-1}$ does not change during the subsequent selfed generations, so that with random mating,

$$F_t = \bar{\theta}_{Pt-1}. \tag{46}$$

The inbreeding coefficient does change,

$$F_{t-1+t'} = 1 - \left(\frac{1}{2}\right)^{t'}(1 - F_{t-1}), \tag{47}$$

which leads to an average coancestry of,

$$\bar{\theta}_{Pt} = \frac{1 + F_{t-1+t'}}{4(N-1)} + \left(1 - \frac{1}{2(N-1)}\right)\theta_{Pt-1}, \tag{48}$$

and

$$F_t = F_{t-1} + \frac{2 - \left(\frac{1}{2}\right)^{t'}(1 - F_{t-2}) - 2F_{t-1}}{4(N-1)}. \tag{49}$$

A plot of F is now in units of $t' + 1$ generations with F in the intervening t' generations increasing according to $1 - (\frac{1}{2})^{t'}(1 - F_{t-1})$ and dropping to F_t with random mating. These intervening upward bulges become more and more damped in time. A corresponding type of result, but with different details, is obtained for larger subdivisions.

There are systems of mating relatives with various proportions of their ancestors common which do not bring about permanent sublines. Examples are the circular and circular pair mating schemes of Kimura and Crow (1963). A class of such systems may be generated in the following way. Let the type of rotation in the mating of parents or parental groups be defined by $y = 0, 1, 2, \ldots s \leqq \log_2 m$. For $y = 0$ there is no rotation and each parental group is mated to itself. For all other values of y each parental group is mated to its two 2^{y-1} neighbors, i.e. $2^{y-1} - 1$ groups separate mate groups. We are now in a position to consider rotations of degree $r = 1, 2, \ldots, s$. To perform a rotation of degree r the initial parental groups are mated in type $y = 1$ fashion, and in each successive generation y is increased by 1 for the type of rotation until the desired degree has been reached, i.e. $y = 1, 2, \ldots, r$ in successive generations. The groups in the r-th generation are mated in $y = 1$ fashion, and the cycle is repeated leading to recurring cycles in multiples of r successive generations.

A third degree rotation is illustrated in Diagram 3 for eight groups. Had the rotation been of the first degree, the first rotation would have been repeated every generation, and of second degree the first two rotations would have been repeated every two generations.

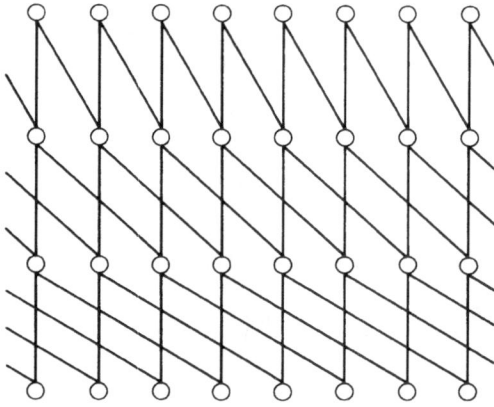

Diagram 3. A third degree circular system of mating of eight groups or individuals. (Easiest to visualize as drawn on the surface of a cylinder)

With these systems, mate groups in the r-th generation have $2^r - 1$ common ancestral groups in the initial generation if $m > 2^r$ and 2^r common ancestral groups if $m = 2^r$, when the circle is completed. Thus inbreeding starts in the $r + 1$ generation, and F_{r+1} has the following values:

	Combined gametic sampling	Independent gametic sampling
$m > 2^r$	$\dfrac{2^r - 1}{2^{2r+1} n_e}$	$\dfrac{2^r - 1}{2^{2r+1} n}$
$m = 2^r$	$\dfrac{1}{2 m n_e}$	$\dfrac{1}{2N}$

Thus we have a class of systems which includes all of the maximum avoidance systems, avoidance up to the limit of the subdivisions, plus that of mating relatives of varying degrees, the greatest degree being for $r = 1$, $N = m$ in which case the system is the circular (half-sib) one of Kimura and Crow (1963), and $v = 2$ if $N > 2$. No permanent sublines are formed, although for the special case of $\log_2 m =$ integer, rotations for $y > 1$ do divide the groups into 2^{y-1} divisions of $m/2^{y-1}$ groups which are intermated.

The accounting problems for F and $\bar{\theta}$ are very tedious for all of these systems except for maximum avoidance of mating individuals or groups. Robertson (1964) contrasted circular mating with maximum avoidance in terms of F and $\bar{\theta}$. Wright (1965) also studied circular and circular pair mating in some detail. Critical in the comparison of mating systems is how F relates to $\bar{\theta}$. In the mating of individuals, $N = m$, it appears that $F > \bar{\theta}$ when $N > 2^{r+1}$ while $F < \bar{\theta}$ for $2^r \leqq N \leqq 2^{r+1}$, and $F - \bar{\theta}$ is increased as r is decreased or N increased.

Rate of Inbreeding

We shall use as a measure of rate of inbreeding

$$\delta_t = \frac{F_t - F_{t-1}}{1 - F_{t-1}}. \tag{50}$$

This rate is constant only for the idealized monoecious population,

$$\delta_t = \frac{1}{2 N_e}. \tag{51}$$

Asymptotic rates for the avoidance of mating relatives may be arrived at in the following manner. For one of the recurrence forms Eq. (6) for the avoidance of mating relatives,

$$\delta_t = \frac{F_t - F_{t-1}}{1 - F_{t-1}} = \frac{1}{2 N_{ev}} - \frac{\displaystyle\sum_{i=0}^{v-1} (2^{v-i} - 1)(F_{t-i-1} - F_{t-i-2})}{2 N_{ev}(1 - F_{t-1})}. \tag{52}$$

In time δ_t approaches a constant, δ. At the same time

$$\frac{F_{t-i-1} - F_{t-i-2}}{1 - F_{t-1}} \to \delta, \tag{53}$$

so that an approximation δ_u to δ is

$$\delta_u = \frac{1}{2N_{ev}} - \frac{(2^{v+1} - v - 2)}{2N_{ev}} \delta_u,$$

$$\delta_u = \frac{1}{2N_{ev} + 2^{v+1} - (v+2)}. \tag{54}$$

Note that the approximations are such that this rate is on the high side of the true rate, i.e.,

$$\delta_u \geqq \delta. \tag{55}$$

Proceeding in the same manner for the alternative recurrence form Eq. (7), we find

$$\delta_l = \frac{1}{4N_{ev}} + \frac{\sum\limits_{i=1}^{v+1}(F_{t-i} - F_{t-i-1})}{(1 - F_{t-1})4N_{ev}} + \frac{(F_{t-1} - F_{t-2})}{(1 - F_{t-1})2}\left(1 - \frac{2^v}{N_{ev}}\right),$$

$$\delta_l = \frac{1}{4N_{ev}} + \frac{(2N_{ev} - 2^{v+1} + v + 1)}{4N_{ev}} \delta_l, \tag{56}$$

$$\delta_l = \frac{1}{2N_{ev} + 2^{v+1} - (v+1)}.$$

The approximations in Eq. (56) are such that $\delta_l \leqq \delta$, and we have δ bounded

$$\frac{1}{2N_{ev} + 2^{v+1} - (v+1)} \leqq \delta \leqq \frac{1}{2N_{ev} + 2^{v+1} - (v+2)}. \tag{57}$$

With equal chance subject only to the avoidance of mating relatives, $N_{ev} = N$,

$$\frac{1}{2N + 2^{v+1} - (v+1)} \leqq \delta \leqq \frac{1}{2N + 2^{v+1} - (v+2)} \tag{58}$$

and the rate decreases as v increases. For zero gametic variance, $N_{ev} = 2N - 2^v$,

$$\frac{1}{4N - (v+1)} \leqq \delta \leqq \frac{1}{4N - (v+2)}, \tag{59}$$

and there is a slight decrease in asymptotic rate with an increase in v.

Let δ^* be the true asymptotic rate for Eq. (24) corresponding to independent sampling of gametes and subdivisions. For equal chance, or for maximum avoidance, $\delta^* = \delta$. It is for gametic variance less than for equal chance and less than complete avoidance that the two differ, and the greatest difference is for no avoidance and $\sigma^2_{1N} = 0$. For $v = 0$ and $\sigma^2_{1N} = 0$,

$$\delta \equiv \delta_u = \frac{1}{4N - 2}, \tag{60}$$

and

$$\delta^*(1 - \delta^*) \equiv \frac{1}{4N}. \tag{61}$$

Since

$$\delta^* \leqq \frac{1}{4N - 1 - 1/(4N - 3)}, \tag{62}$$

to an extremely close approximation, particularly for large N,

$$\delta^* \cong \delta_l = \frac{1}{4N - 1}. \tag{63}$$

Thus, for no gametic variance, δ^* and δ are at the lower and upper bounds, respectively, for $v = 0$ and converge together for maximum avoidance. Increasing the gametic variance decreases the difference between them, and they are the same for equal chance.

When subdivisions and individuals are chosen and mated at random, the asymptotic rate of inbreeding may be increased considerably. For the recurrence form Eq. (35),

$$\delta_u = \frac{1}{2N + 2^{v+1} - (v+2) - \dfrac{(2^s - 2^v)(N - 2^s)}{2^s}} \tag{64}$$

which reduces to δ_u in Eq. (58) when $s = v$ or $N = 2^s$. The rate is considerably larger for v much less than s. For example, when $v = 0$ and $2^s = \sqrt{N}$, the rate is maximum,

$$\delta_u = \frac{1}{N - 1 + 2\sqrt{N}}. \tag{65}$$

When the system of mating of close relatives causes a sublining of the population, the rate of inbreeding is that for the system with the dimension, n, of the subgroup. The coancestry, $\bar{\theta}_w$, within subgroups will have the same asymptotic rate as F, but that, $\bar{\theta}_b$, between subgroups will depend on the variance of the number of progeny groups from parental groups. In the case of complete sublining, $\bar{\theta}_b = 0$.

If sublines are occasionally intermated, the asymptotic rates of F, $\bar{\theta}_w$ and $\bar{\theta}_b$ will be the same. For the example of selfed lines and occasional random mating (49), in terms of t or units of $t' + 1$ generations,

$$\delta \cong \frac{1 - (\frac{1}{2})^{t'+1}}{2(N-1) + (\frac{1}{2})^{t'+1}}, \tag{66}$$

and the per generation rate is

$$\delta \cong \frac{1 - (\frac{1}{2})^{t'+1}}{(t'+1)\,[2(N-1) + (\frac{1}{2})^{t'+1}]} \tag{67}$$

which can be made arbitrarily small by adjusting t', a point made by Robertson (1964).

The asymptotic rates for large N were given as $\pi^2/4(N+2)^2$ for circular and $\pi^2/(N+12)^2$ for circular pair mating systems by Kimura and Crow (1963), which are much less than for any of the other systems except for variations of Eq. (67). To see that the degree of rotation, r, relative to N plays a role in the asymptotic rate note that for $N = 4$, the asymptotic rate of 0.0727 given by Kimura and Crow (1963) for circular mating, $r = 1$, is greater than δ Eq. (60) and δ^* Eq. (62) for no avoidance and zero gametic variance. Also for circular pair which is slightly different from $r = 2$, when $N = 8$ the asymptotic rate is greater than for no avoidance and zero gametic variance.

Discussion

Most of the developments in this paper are concerned with solving a complex puzzle. Many facets, the most important of which were already known, do relate to the maintenance of control and experimental populations. Extension of the treatment of population subdivisions might offer an alternative approach to the study of partially isolated populations in nature.

Involved in the transition from one generation to the next for any system is the sampling of gametes and the manner in which they are put together. The measure taking into account the sampling plan is the population gametic probability,

$$\Psi_{Pt} = \frac{1 + F_t}{2N_e} + \frac{N_e - 1}{N_e}\,\bar{\theta}_{Pt}, \tag{68}$$

and of the storage of the genes is the population inbreeding coefficient

$$F_{iPt} = \frac{F_t}{2N-1} + \frac{2(N-1)}{2N-1}\,\bar{\theta}_{Pt}. \tag{69}$$

For any system $F_{lPt} = \Psi_{Pt-1}$ since they are just two different expressions of the same average of the probabilities of identity by descent. A description of the storage system is given by how F and $\bar{\theta}$ deviate from the mean. Using

$$F_{lPt} - \Psi_{Pt-1} = \frac{F_t - \Psi_{Pt-1}}{2N-1} + \frac{2(N-1)}{2N-1}(\bar{\theta}_{Pt} - \Psi_{Pt-1}) = 0 \qquad (70)$$

we find the exact relationship between these two deviations,

$$F_t - \Psi_{Pt-1} = -2(N-1)(\bar{\theta}_{Pt} - \Psi_{Pt-1}). \qquad (71)$$

The increase in average probability,

$$\Psi_{Pt} - \Psi_{Pt-1} = \frac{1 - \Psi_{Pt-1}}{2N_e} + \frac{F_t - \Psi_{Pt-1}}{2N_e} + \frac{N_e - 1}{N_e}(\bar{\theta}_{Pt} - \Psi_{Pt-1}), \qquad (72)$$

with substitution for $F_t - \Psi_{Pt-1}$ given in Eq. (71)

$$\Psi_{Pt} - \Psi_{Pt-1} = \frac{1 - \Psi_{Pt-1}}{2N_e} + \frac{N_e - N}{N_e}(\bar{\theta}_{Pt} - \Psi_{Pt-1}) \qquad (73)$$

reflects both the effects of storage and sampling.

Of primary importance in Eq. (73) is N_e. First let us consider the effects of subdividing the population and gametes and of the sampling plan on N_e. With m groups and exactly two sets of n gametes per group, the probability of a random pair of gametes being from the same set is $(n-1)/(2N-1)$ and from the same group but different sets is $n/(2N-1)$. To be determined is the probability, $1/N_e$, of a random pair of gametes being from the same parent. For combined gametic sampling when from the same group, whether from the same gametic set or not, the probability of a random pair of gametes being from the same parent is $1/n_e$. Thus,

$$\frac{1}{N_e} = \frac{n-1}{2N-1}\frac{1}{n_e} + \frac{n}{2N-1}\frac{1}{n_e} = \frac{2n-1}{2N-1}\frac{1}{n_e}$$

$$= \frac{2n-1}{2N-1}\frac{2+\sigma^2_{2n}}{2(2n-1)} = \frac{1+\sigma^2_{1n}}{2N-1}, \qquad (74)$$

since $\sigma^2_{2n/2} = \sigma^2_{1n}$.

With independent gametic sampling,

$$\frac{1}{N_e} = \frac{n-1}{2N-1}\frac{1}{n_g} + \frac{1}{2N-1}\frac{1}{n}$$

$$= \frac{n-1}{2N-1}\frac{\sigma^2_{1n}}{n-1} + \frac{1}{2N-1} = \frac{1+\sigma^2_{1n}}{2N-1}. \qquad (75)$$

The two sampling plans give the same overall effective number which varies from $(2N - 1)/(2 - m/N)$ for equal chance to $2N - 1$ for no gametic variance. Subdividing the population with exactly two sets of gametes per subdivision automatically causes a reduction in the gametic variance and an increase in the effective number. With maximum subdivision, the effective number is maximum.

To account for an arbitrary distribution of the number of gametic sets per subdivision we make use of the probability $1/m_e$ of a random pair of sets being from the same group. The probability of a random pair of gametes being from different sets is $n(2m - 1)/(2N - 1)$. Only independent gametic sampling is considered in this case. The effective number,

$$
\begin{aligned}
\frac{1}{N_e} &= \frac{n-1}{2N-1} \frac{1}{n_g} + \frac{n(2m-1)}{2N-1} \frac{1}{m_e} \frac{1}{n} \\
&= \frac{\sigma_{1n}^2}{2N-1} + \frac{2m-1}{2N-1} \frac{2+\sigma_2^2 m}{2(2m-1)} \\
&= \frac{1+\sigma_{1n}^2+\sigma_{1m}^2}{2N-1},
\end{aligned}
\tag{76}
$$

now shows the additional effect of sampling of gametic sets. With exactly two sets of gametes per group, $\sigma_{1m}^2 = 0$, and the result is the same as in Eq. (75). With equal chance within sets and of sets from the same group, $\sigma_{1n}^2 = (n - 1)/n$ and $\sigma_{1m}^2 = (m - 1)/m$, and

$$
\frac{1}{N_e} = \frac{1}{N} + \frac{(m-1)(n-1)}{N(2N-1)}.
\tag{77}
$$

Because of redundance of the situations $N_e = N$ when m or $n = N$, and N_e is a minimum when $n = m = \sqrt{N}$.

With sublines and variation in the number of offspring sublines per parental subline the result is slightly different from Eq. (76) because the two gametic sets in each offspring subline of necessity came from the parental subline. A random pair of genes will be from the same subline with probability $(2n - 1)/(2N - 1)$ and from different sublines with probability $2n(m - 1)/(2N - 1)$. In the former case the probability of being from the same parent is $1/n_e$ and in the latter case is $(1/m_g')(1/n)$. Thus,

$$
\begin{aligned}
\frac{1}{N_e} &= \frac{2n-1}{2N-1} \frac{1}{n_e} + \frac{2n(m-1)}{2N-1} \frac{1}{m_g'} \frac{1}{n} \\
&= \frac{1+\sigma_{1n}^2+2\sigma_{1m'}^2}{2N-1}.
\end{aligned}
\tag{78}
$$

The coefficient of two for $\sigma_{1m'}^2$ in contrast to one in Eq. (76) corresponds to the sets of gametes in each sample.

Most of the effects of avoidance of mating relatives and subdividing the population with and without avoidance of mating related groups are reflected in N_e. When individuals are sampled at random, $N_e = N$, but this can be accomplished only in the idealized monoecious population. The avoidance of mating relatives, the mating of close relatives or subdividing the population with equal sampling of gametic sets dictates that $N_e > N$ while sampling among subdivisions or sublines brings about $N_e < N$.

Of major importance also in the increase in average probability in Eq. (73) is how the genes are stored reflected by how $\bar{\theta}_P$ (or F) deviates from the mean. Avoidance of mating relatives up to degree v avoids any inbreeding in individuals until generation $v+1$, the increase in probability being stored in $\bar{\theta}_P$. Avoidance, then, makes F less than the mean and $\bar{\theta}_P$ greater than the mean, more so with more avoidance, and for a given N_e leads to a greater increase in average probability. It is in this respect that subdivisions of gametic sets with independent sampling of the sets has a differential effect. For example, the greatest effect is for two gametes per parent and no avoidance. The inbreeding in the first generation is $1/2N$ as compared to $1/2(2N-1)$ for no subdivision, and the asymptotic rates are reversed, $1/(4N-1)$ and $1/(4N-2)$, but with only a trivial difference.

The rate of inbreeding for the mating of close relatives which require sublining is of course that for the subline. With permanent sublines and no gametic variance among subdivisions the population measures approach their asymptotes with the same rate, but their asymptotes are different and less than one, with a final average probability of $F_{lP\infty} = (2n-1)/(2N-1)$. With gametic variance among the sublines, $N_e < N$, and as can be noted in (73), the increase in average probability is augmented by $\bar{\theta}_P > \Psi_P$.

The closest regular system of mating relatives which has no sublines is the first degree circular system for which N_e is maximum, and $\bar{\theta}_P$ and asymptotic rates are minimum. The asymptotic rates can of course be made arbitrarily small by sublining and occasionally intermating, which is really an averaging for systems some of which have asymptotes less than one.

Without sublining, we may use as a reference point the monoecious population for which $F = \bar{\theta}_P$ and for which all rates are constant. The striking differences in F and $\bar{\theta}_P$ for avoidance or the mating of close relatives occur in the early generations. Asymptotically, F and $\bar{\theta}_P$ for a particular mating system have the same rates. The outcome among systems is dominated by the $\bar{\theta}_P$'s, which have $2(N-1)$ times the storage space as does F, and which maintain the same ranks throughout the process.

Pair matings and full sib families have not been considered except in an incidental way at times. For the avoidance of mating relatives and otherwise pairing mates at random, an effective number can be defined in terms of the variance of the number of individuals per family (much the same as for the gametic variance among progenitor groups) which gives the same result as for a dioecious population and independent gametic sampling. The types of relatives involved differ because there can be no half sibs, for example, but the average probabilities are the same. Certain mating systems of close relatives, such as circular pair or circular subpopulation mating (Kimura and Crow, 1963) do not correspond exactly to the circular systems as outlined. For example, a second degree rotation of individuals is slightly different from circular pair. Of course in many natural populations the variance of the number of individuals per family reduces the effective number drastically below the census number, but only one natural population knows anything about recipe mating systems.

Summary

General recursion formulae for the inbreeding coefficient were derived for the avoidance of the mating of v degree relatives and otherwise random mating, which also accommodated variations in the gametic variance not dictated by the avoidance. It was developed also that subdividing the population into 2^v groups and mating among groups in a manner to avoid the mating of v degree relatives gave the same results. Further division of the gametes from each group gave slightly different results when the gametic variance was less than for equal chance of parents in the group; the main difference being that subdivision of the gametes led to higher initial rates of inbreeding and slightly less asymptotic rates. Subdividing the population more than necessary for the degree of avoidance and random sampling of the subdivisions increased the rate of inbreeding and all other probability measures considerably.

The systems of mating of close relatives were found to fall into two classes: those for which all ancestors are in a parental group or all ancestors are common in a previous generation and which dictate a sublining of the population, and those for which a portion of the ancestors are common in a previous generation such as the circular systems, and which do not subline the population. When sublining, the rate of inbreeding is the appropriate one for any subline. If progenitor sublines are sampled, the population measures of probability approach one at a rate depending on the rate of loss of sublines. If all sublines are maintained, the population measures approach asymptotes less than one.

The circular systems fit with the avoidance systems into a framework using population probability measures for which it was shown that early rates of inbreeding are reversed asymptotically. Bounds on asymptotic rates were found for all of the avoidance systems.

References

Cockerham, C. Clark: Group inbreeding and coancestry. Genetics, **56**, 89–104 (1967).

Kimura, M., and J. F. Crow: On the maximum avoidance of inbreeding. Genet. Res. **4**, 399–415 (1963).

Malécot, G.: Les Mathématiques de l'Hérédité. Paris: Masson et Cie. 1948.

Robertson, A.: The effect of non-random mating within inbred lines on the rate of inbreeding. Genet. Res. **5**, 164–167 (1964).

Robinson, P., and D. F. Bray: Expected effects on the inbreeding coefficient and rate of gene loss of four methods of reproducing finite diploid populations. Biometrics **21**, 447–458 (1965).

Wright, S.: Systems of mating. Genetics **6**, 111–178 (1921).

— Size of population and breeding structure in relation to evolution. Science **87**, 430–431 (1938).

— The interpretation of population structure by F-statistics with special regard to systems of mating. Evolution **19**, 395–420 (1965).

Genetic Loads and the Cost of Natural Selection* , **

J. F. CROW

Introduction

The basic ideas of genetic load and the cost of natural selection are both from J. B. S. Haldane. In his early papers on natural selection (1924–1932), Haldane was concerned with both the dynamics and the statics of evolution. He emphasized that, although evolution depends on changes of gene frequency, nevertheless at any one time the population is in approximate equilibrium for most factors.

In "The effect of variation on fitness" (1937) Haldane discussed such an equilibrium situation and inquired into the effect on fitness of recurrent deleterious mutations and inferior homozygotes that arise by Mendelian segregation from better adapted heterozygotes. He defined the fitness of a genotype as the average number of progeny left by an individual of that genotype (or half that number, if both mother and father are credited with each progeny). But he noted that:

"It is clear that the mean fitness of all members of a species must always be very close to unity, if we average over any length of time. If the fitness were 1.01 the population would increase 20,959 times in 1,000 generations. In almost all species the mean fitness over 1,000 generations must vary from unity by far less than one percent. But in any species some genotypes have a fitness less than unity, ranging to zero in the case of lethal genes and genes causing complete sterility. So it is clear that the fitness of the standard type containing no deleterious genes must exceed unity. A population composed of such a type would of course increase until, owing to its pressure on the means of subsistence, the fitness was again reduced to unity."

He also pointed out that the standard type may not be the most fit genotype. For example, if two or more individually deleterious mutants collectively increase the fitness beyond the prevailing type, they may not increase in frequency in a sexual population, and their net effect will be to decrease the average fitness. Later these may by

* Paper number **1303** from the Laboratory of Genetics, University of Wisconsin. Some of the work reported here was supported by the National Institutes of Health (GM–15422).

** This paper is dedicated to Professor Th. Dobzhansky for his seventieth birthday celebration and his long lasting leadership in experimental population genetics.

some means become the prevailing type, perhaps through the kinds of mechanisms that Wright (1931, 1949) suggests in a structured population, in which case this genotype becomes the standard.

In discussing deleterious mutants, Haldane goes on to say:

"It is at once clear that in equilibrium such abnormal genes are wiped out by natural selction at exactly the same rate as they are produced by mutation. It does not matter whether the gene is lethal or almost harmless. In the first case, every individual carrying it, or if it is recessive, every individual homozygous for it, is wiped out. In the second the viability or fertility of such individuals may only be reduced by one-thousandth. In either case, however, the loss of fitness to the species depends entirely on the mutation rate and not at all on the effect of the gene upon the fitness of the individual carrying it, provided this is large enough to keep the gene rare."

In "The cost of natural selection" (1957a) Haldane considered the effect on fitness of the natural selection required to carry out gene substitutions in evolution. He showed that the total cost over the whole period in which the substitution takes place is determined almost entirely by the initial frequency of the mutant and hardly at all by the selective advantage of the mutant.

What makes both of these ideas interesting, and totally new at the time they were first presented, is that the selection coefficient drops out of the equations. Whether the individual gene effect is large or small, the consequences are the same.

The change of average fitness associated with maintaining the variability in the population has come to be called the genetic load. The word "load" was first used by Muller (1950) in assessing the impact of mutation on the human population. He used the Haldane principle in an attempt to get around the fact that the effects of individual mutant genes were mainly of unknown magnitude. This principle provided a means of assessing the total affect of mutation on the average fitness of the human population from knowledge of the mutation rate alone. In Muller's thinking the load was clearly a burden, measured in terms of reduced fitness, but felt in terms of death, sterility, illness, pain, and frustration (see Wallace, 1968, pp. 267–280). In most evolutionary and population genetics considerations, however, it is used as a measure of the amount of natural selection associated with a certain amount of genetic variability. It is not necessarily a burden; on the other hand, it may be the *sine qua non* for further evolution without which the population could become extinct. Another frequently noted example where the load is not bad is the situation where a population, previously of genotype AA, undergoes a mutation to A' such that $A'A$ is superior to AA, but $A'A'$ is inferior. This creates variability and a genetic load, but the polymorphic popu-

lation has a higher average fitness than if it were monomorphic for the
original AA genotype. For a critical review of these points, see Brues
(1969).

Despite these semantic reservations, I shall use the words "load"
and "cost" for Haldane's two situations. The first refers to the consequences of genes that are in approximate equilibrium. The second refers
to genes in the process of changing frequency. Thus the load refers to
static properties of a population while the cost refers to the dynamic
aspects. Kimura (1960) has referred to the cost of evolution as the "substitution load". There is considerable justification for this, for in many
ways the load and cost concepts are very similar. However, I will use
the two words in order to emphasize that one is involved with equilibria
and the other with change.

The load and cost are consequences of genetic variability. If this
is the case, one might well ask why a relevant measure is not some function
of the genetic variance of the population rather than a function of the
mean fitness. It is often easier to measure the genetic component of the
population variance than it is to measure the difference between the
mean fitness of the population and that of a reference genotype; this is
especially true when fertility differences are an important consideration,
as they usually are, in determining total fitness. I believe that both kinds
of measures can be useful. They are more complementary than alternative and together can give a deeper understanding of the population
than either by itself.

The load and cost concepts have generated a great deal of discussion
and criticism. Some of this has to do with the appropriateness and meaning of some of the definitions and mathematical conventions employed
(Van Valen, 1963; Sanghvi, 1963; Li, 1963a, b, c; Brues, 1964; Feller,
1966, 1967; Wallace, 1968). These have been or can be resolved by
being more careful with definitions and working out mathematical
points more rigorously. Another serious question is the extent to which
theory based mainly on total fitness is applicable to individual components thereof (e.g. Levene, 1963; Haldane and Jayakar, 1965).

A major question has been whether cost and load arguments can
be used to set upper limits on such things as the number of selectively
maintained polymorphisms, the mutation rate that can be tolerated,
or the rate of gene substitution by natural selection. Haldane and Muller
have argued that such arguments can be useful, if only roughly. But
King (1966, 1967), Milkman (1967), and Sved, Reed, and Bodmer (1967)
independently pointed out the relevance of threshold models to this
situation. They think of the gene loci as acting more or less additively
on some underlying scale, leading to a monomodal distribution that
is roughly normal in shape. Then selection is assumed to act by trun-

cating those genotypes that are at one end of the scale. Genotypes that are above this point on the scale are retained by selection and therefore have fitness one; those at the point or below are rejected and have fitness zero. There is a quantitative difference between such models and those based on selection acting on the loci independently that is so great as to be almost qualitative.

The truncation point can be thought of as determined by the carrying capacity of the environment, which restricts the average survival and fertility to a level that keeps the population size roughly constant. Truncation selection is applied in artificial selection of livestock for quantitative traits, and models based on this theory have considerable predictive value. On the other hand, one of the first lessons the animal breeder learns is that he cannot select for several different characters at the same time. I have been reasonably sure that the kinds of epistasis that are observed are not such as to make any very great change in the conclusions based on independence. But, if truncation selection of the type envisaged by King, Milkman, Sved, Reed, and Bodmer is widely applicable, then these conclusions have to be revised. The question then becomes the empirical one of whether, in fact, gene action and the way selection acts are consistent with a threshold model.

In the sections to follow I will discuss only selected aspects of the subject, those with which I am most familiar. But I shall try to give at least an idea of the much greater breadth of the subject as it has grown by short discussions and references to other work.

Genetic Loads

Kinds of Genetic Loads

There are many kinds of genetic loads. Almost any factor that influences gene frequencies can lead to a change in the average fitness. Only a few have been investigated in any detail, but the general way in which some of them operate is already clear. I give here a list of some of the factors that may produce a genetic load, together with a short discussion of each. The first two were, of course, introduced by Haldane in his original paper (1937).

1. Mutation. Every population is exposed continuously to newly occurring mutations, the average effect of which is to lower the fitness in the environment where the population lives. This decrease in fitness, Haldane suggests, is the price paid by a species for its capacity to evolve.

2. Segregation. The segregation load, or balanced load (Dobzhansky, 1955, 1957), occurs when the fittest genotype is heterozygous. In a sexual population homozygous progeny are regularly produced by

Mendelian segregation, and the segregation load is the decrease in fitness compared to the most fit heterozygote – the type that would prevail if it were not for segregation.

3. Recombination. As pointed out long ago by Fisher (1930), if two chromosomes, say *AB* and *ab*, are on the average favored by natural selection in the various zygote combinations into which they enter, while the two types, *Ab* and *aB*, are in disfavor, the average effect of recombination will be to increase the inferior types. Thus if the four types are maintained in the population in some kind of stable equilibrium, recombination produces a decrease in fitness. Under such circumstances natural selection would favor a gene or chromosome rearrangement that reduces recombination in this region, and Kimura (1956 b) has discussed a specific model for this. The extent of the genetic load introduced by recombination has been discussed for some specific cases of 2-locus interaction by Nei (1965).

4. Heterogeneous Environment. If there are different environments and the various genotypes respond to these in different ways, it is advantageous if those genotypes that do best in an unusual environment are rare while those which prefer a prevalent environment are common. This means that there will be an optimum allotment of genotype frequencies corresponding to the numbers of available spaces in the various environments. Any actual population will depart from this ideal and, to the extent that this is true, there is a reduced fitness and therefore a load. Haldane (personal communication) suggested that such a load might be called the "dysmetric" load. As far as I know it has not been systematically studied. Conditions under which multiple niches maintain a polymorphism have been given by Levene (1953) and Prout (1968).

5. Meiotic Drive or Gamete Selection. A gene that is favored by meiotic drive or gamete selection, or genes that are closely linked to such a locus, may persist in the population despite being harmful in various zygotic combinations. The population fitness will therefore be lowered, and this again creates a genetic load.

6. Maternal-Fetal Incompatibility. A genotype may have a lowered fitness in association with certain maternal genotypes. A well known example is an Rh + child with an Rh − mother. The genetic load from this cause has been called the incompatibility load. For a discussion, see Crow and Morton (1960).

7. Finite Population. In any finite population the gene frequencies drift away from the values that they would have if they had reached equilibrium in an infinite population. The equilibrium proportions are usually the values that maximize the population fitness, although this is not always true if there are frequency dependent genotypic fitnesses or if there are changes in the mating system. The drift away from the equilibrium frequencies will usually reduce the fitness, thereby creating a genetic load.

8. Migration. Immigrants into a population, having evolved elsewhere under other environmental conditions, are less likely to be adapted to the local environment than the native population. Therefore migrants and the descendants of migrants usually lower the fitness and thereby create a migration load. In Wright's (1931, 1949) theory of evolution, migration between partially isolated subpopulations is a major factor in evolutionary progress, so again a genetic load may be an evolutionary benefit.

Definitions of Genetic Load

There are several different ways of defining the genetic load. I shall consider three. They are intended to be equivalent in that they ordinarily lead to the same numerical conclusions. Having more than one definition may be useful in that more procedures are thereby suggested, with increasing probability that one may be successful. Furthermore, the different definitions may deepen our insights by providing slightly different views that may reveal the situation more clearly.

The three definitions are:

I. The genetic load is the fraction by which the population mean differs from a reference genotype. The trait measured is usually fitness, or some component thereof, and the reference genotype is usually the maximum among the actually or theoretically available types.

II. The genetic load is the fraction by which the population mean is changed as a consequence of the factor under consideration in comparison to a population, supposed to be identical otherwise, in which the factor is missing.

III. The genetic load is the amount of reproductive excess, expressed as a fraction of the reproductive rate, required to maintain the size of the population when it is at equilibrium for the factors under consideration.

As illustrations of the three definitions, consider a very simple situation – a haploid population in which the less favored gene, a, is maintained by recurrent mutation at a rate u per generation.

Genotype	A	a
Frequency	p	$q = 1 - p$
Fitness	w	$w(1 - s)$

The equilibrium value of q, $\hat{q}$, is obtained immediately by writing the frequency of A in terms of the frequency in the previous generation and equating these:

$$\frac{pw(1-u)}{pw+qw(1-s)} = p$$

from which $\hat{q} = u/s$. The average fitness at equilibrium is

$$\overline{w} = w(1-s\hat{q}) = w(1-u).$$

To use definition I, we choose A as the reference genotype and, from the definition, the load in terms of fitness is

$$L = \frac{w-\overline{w}}{w}. \tag{1}$$

To use definition II, we note that a population without mutation would, after reaching equilibrium, be composed only of A individuals with fitness w. Thus the mutation load is the same as given in Eq. (1). In both cases the load L is equal to the mutation rate u.

The third definition can be applied in the following fashion. Let w be interpreted as the number of progeny which the genotype of maximum fitness must produce in order that the average number of progeny of all genotypes equals the number of parents. Parents and progeny must be counted at the same age. It is convenient if the count is at the zygote stage, for then the effects of differential mortality and of differential fertility are in the same generation. In order to have a stable population of size N in gene frequency equilibrium, the following relationships must hold:

$$Nwp(1-u) = Np,$$

$$Nwq(1-s) + Nwup = Nq.$$

From the first equation, $w = 1/(1-u)$. The necessary reproductive excess to maintain the population size is $w-1$, so the load is

$$L = \frac{w-1}{w} = u. \tag{1'}$$

Substitution of $1/(1-u)$ for w in the second equation gives the equilibrium value of q, u/s, as obtained previously.

The first definition (Crow, 1958) was intended to correspond to Haldane's original conception. The second was introduced (Crow, 1962) as a guide to the choice of the reference genotype (or, perhaps, group of genotypes). For example, the reference population when assessing the mutation load is one without mutation, which in the example already discussed would be entirely of genotype A at equilibrium. If we are measuring the load caused by Mendelian segregation, the natural reference population is one without segregation, that is to say, an asexual population. With a single heterotic locus, where AA' is fitter than either AA or $A'A'$, an asexual population at equilibrium would contain 100% AA'; hence the segregation load is measured from this reference genotype. Likewise, the load due to recombination is measured in comparison with a population that is segregating but has no recombination; a meaningful comparison could be made between a group of freely combining genes and the same group locked in an inversion.

The third definition lets us see the relation between the genetic load and the necessary reproductive potential of the species if this load is to persist. To support an amount of variability associated with a load, L, the reference genotype (or group of genotypes) must have survival and fertility sufficient to leave $1/(1 - L)$ descendants, or in other words, an excess survival and reproductive capacity equal to $L/(1 - L)$.

If this capacity does not exist, there are several alternatives, for example: (1) The population size may diminish, and perhaps finally drop to extinction. (2) The equilibrium gene frequencies may not be maintained. (3) The selection may occur in ways that do not contribute proportionately to the measured load – for example, gametic selection, selection acting on very young embryos, selection acting between litter mates when only a certain number are permitted to survive anyhow, or selective deaths allocated primarily among those individuals that were going to be eliminated for other reasons, such as overcrowding.

The genetic load of an equilibrium population, usually mating approximately at random, is the expressed load. In addition, the total load includes the effect of those recessive or partially recessive factors whose effect is ordinarily concealed but is revealed by inbreeding.

The total load is of secondary interest as far as the fitness or other measures of the population are concerned, but it is often useful in assessment of important parameters such as the accumulated mutant frequency, the average dominance of mutants, or the effect of a change in the mating system.

In principle the definition of genetic load applies to any measurable trait. If viability is being considered, it is sometimes convenient to measure the load in terms of lethal equivalents (Morton, Crow, and Muller, 1956). A lethal equivalent is a group of genes that, if dispersed in different individuals chosen at random from the population (or chosen according to the prevailing mating system), would cause an average of one death. Likewise, the reduction in fertility may be measured in sterile equivalents (Chung, 1962; Temin, 1966) and specific detrimental effects as detrimental equivalents (Morton, 1960).

The Mutation Load

Independent Loci

Consider an idealized model in which generations do not overlap and each generation is counted at the same age, most conveniently at the zygote stage. Let A be the normal gene with frequency p and a be the mutant gene with frequency q. F is a measure of departure from random genotypic proportions. The frequencies and fitnesses of the three genotypes are:

Genotype	AA	Aa	aa	
Fitness	$w_{AA} = w$	$w_{Aa} = w(1 - hs)$	$w_{aa} = w(1 - s)$	(2)
Frequency	$P_{AA} = p^2(1 - F) + pF$	$P_{Aa} = 2pq(1 - F)$	$P_{aa} = q^2(1 - F) + qF$	

The relations to follow apply strictly under the conditions of the model. However, the relations should be very good approximations to populations with overlapping generations since the formulae are to be applied to populations near equilibrium. The situation is not materially altered if the "normal" gene is really a population of isoalleles; in this case w_{AA} is the average fitness of all combinations of these weighted by their frequency in the population. If there is more than one mutant type, we make the simplifying assumption that heterozygotes between different mutants are the average of the two mutant homozygotes, as is often roughly true for Drosophila mutants; in this case s and h are average values.

If s and h are positive, the mutant factor, a, will tend to decrease until the rate of loss of a factors by selection balances the rate of occurrence by mutation. Let the rate of mutation from A to a be u per generation. If a is rare, as it will be if s is large relative to u, then reverse mutation from a to A will be infrequent enough that it can be ignored.

Starting with the frequencies in Expression (2), the frequency of the A factor one generation later will be

$$\frac{(w_{AA}P_{AA} + \frac{1}{2}w_{Aa}P_{Aa})(1 - u)}{\overline{w}} \tag{3}$$

where

$$\overline{w} = w_{AA}P_{AA} + w_{Aa}P_{Aa} + w_{aa}P_{aa}$$

and

$$P_{AA} + P_{Aa} + P_{aa} = 1.$$

Expression (3) is the proportion of A factors in the population after selection (which is the proportion of A among the genes transmitted to the next generation) corrected by $1 - u$, which is the proportion of A factors which have remained unmutated during this generation.

If the population has reached equilibrium, the gene frequencies are no longer changing, and we can equate Expression (3), the A frequency next generation, to p, its frequency in the present generation. Thus,

$$p\bar{w} = (w_{AA}P_{AA} + \tfrac{1}{2}w_{Aa}P_{Aa})(1-u). \tag{4}$$

Using the relations in Expression (2), Eq. (4) becomes

$$s(1-F)(1-2h)q^2 + [hs(1-F)(1+u) + sF]q - u = 0 \tag{5}$$

from which the equilibrium frequency, $\hat{q}$, of the deleterious gene can be calculated when h, s, u, and F are known.

It is clear that in this case the meaningful reference population is a population of AA homozygotes, for this is the genotype that would ultimately prevail in the absence of mutation. Thus, from the definition of the genetic load Eq. (1), substituting from Eq. (4) and using Expression (2) gives,

$$L = u + \hat{q}hs(1-F)(1-u). \tag{6}$$

From Eq. (6) we see immediately that for a recessive mutant ($h = 0$), $L = u$; that is to say, the mutation load is exactly equal to the mutation rate for any value of F. Also we see that $L = u$ when $F = 1$. Thus any highly inbred population where F approaches 1 will have a mutation load equal to the mutation rate.

For other values, we can solve the quadratic Eq. (5) for q and substitute this value into Eq. (6). Some numerical examples are given in Table 1.

If $F = 0$ and $h = 0.5$, Eq. (5) becomes $qhs = u/(1+u)$, and substitution of this into Eq. (6) gives

$$L = \frac{2u}{1+u},$$

or very nearly $2u$. Values of h larger than 0.5 give a larger value; values less than 0.5 give a smaller. Over a wide range of values the load is very nearly twice the mutation rate, as Table 1 shows.

If either h or F is appreciable, then q is approximately $u/s(h+F)$. For example, if $h = F = 0.05$, $s = 0.1$, $u = 0.00001$, then $q = 0.001$ by this formula and $L = 1.48 \times 10^{-5}$, in agreement with the exact value in Table 1.

It is possible for L to be considerably larger than $2u$. The most striking cases occur when s is very small and h is greater than 1. For example, if $h = 1.5$ and $s = 4u$, the load is $2.5u$ (Kimura, 1961). Such instances where the heterozygous mutant is worse than the homozygote are presumably unusual. In the more typical case of a fully dominant mutant, the load is very nearly $2u$; for a dominant lethal it is $2.00001u$ when $u = 0.00001$.

Table 1. *Values of C where Cu is the mutation load per locus. Diploidy is assumed, u is the mutation rate, s the homozygous selective disadvantage of the mutant, and hs is the disadvantage in heterozygotes. The mutation rate is taken to be 10^{-5}*

					h				
F	s	0	0.01	0.02	0.05	0.1	0.5	1.0	1.5
0	0.001	1.00	1.10	1.18	1.41	1.66	2.00	2.01	2.01
	0.01	1.00	1.27	1.47	1.78	1.93	2.00	2.00	2.00
	0.1	1.00	1.62	1.83	1.97	1.99	2.00	2.00	2.00
	0.5	1.00	1.86	1.96	1.99	2.00	2.00	2.00	2.00
	1.0	1.00	1.92	1.98	2.00	2.00	2.00	2.00	2.00
0.05	0.001	1.00	1.07	1.14	1.31	1.51	1.90	1.96	1.98
	0.01	1.00	1.13	1.24	1.45	1.63	1.90	1.95	1.97
	0.1	1.00	1.16	1.27	1.48	1.65	1.90	1.95	1.97
	0.5	1.00	1.16	1.27	1.49	1.65	1.90	1.95	1.97
	1.0	1.00	1.16	1.27	1.49	1.65	1.90	1.95	1.97

If the genotypic fitnesses of AA, Aa, and aa are w, $w(1-s)$, and $w(1-s)^2$, then the equilibrium frequency of a is u/s and the average fitness is,

$$\bar{w} = w(1-u)^2,$$

and the load is given by

$$L = 1 - (1-u)^2.$$

Thus, absence of dominance for fitness is perhaps most naturally defined as having the heterozygote with the geometric mean of the two homozygotes.

Thus we verify the essential correctness of Haldane's (1937) original statement that the load is between one and two times the mutation rate. For an X-linked recessive with equal mutation rates in the two sexes, Haldane showed that the mutation load is $1.5u$. If the mutation rates in the two sexes are not equal,

$$L = \frac{u_m + 2u_f}{2} \tag{7}$$

where u_m and u_f are the mutation rates in the two sexes. This assumes equal numbers of males and females in the population. A more general discussion of the load for an X-linked locus, including the above formula, is given by Kimura (1961).

That dominant mutants produce approximately twice the load that recessives do is not surprising. It can be seen intuitively by using the genetic extinction or "genetic death" argument of Muller (1950). In a diploid population of constant size N (therefore having $2N$ alleles), a genetic extinction caused by death or failure to reproduce eliminates one mutant among the $2N$ genes at this locus. If the mutant is recessive, two mutants are eliminated at once so there are 2 eliminations among the $2N$ alleles per genetic extinction. The $2:1$ ratio will be correct as long as the dominant mutant is rare enough that it is never eliminated as a homozygote. If the mutant is an X-linked recessive, practically all the eliminations are through genetic extinction of males. There are $3N/2$ genes in the population so the extinction eliminates one among the $3N/2$. Thus the number of genetic extinctions per gene eliminated for dominant, recessive, and X-linked recessives is in the ratio $2:1:1.5$. Likewise, if the inbreeding coefficient is high, the eliminations will be through homozygotes and therefore the situation is like that of a recessive mutant.

If the effects of different loci are independent, one minus the load for all loci combined is given by

$$1 - L = \Pi(1 - L_i) \approx e^{-\Sigma L_i} \tag{8}$$

where L_i is the load at the *i-th* locus. Since the individual L_i's are very small, this is very close to

$$L \approx \sum L_i .$$

The mutation load is then, to a very good approximation,

$$L = C\sum u_i = C U \tag{9}$$

where C is a constant between 1 and 2 (usually much nearer 2), u_i is the total mutation to all deleterious mutants at the *i-th* locus and the summation is over all relevant loci. In other words, $U = \sum u_i$ is the total mutation rate to harmful genes per gamete per generation.

The Mutation Load with Epistasis

Epistasis may increase or decrease the mutation load, depending on the direction of the interaction. If the double mutant is less deleterious than would be expected if the genes were acting independently, then the mutation load is somewhat increased. The situation is analogous to a single locus with h greater than $1/2$. On the other hand, the mutation load is decreased if the multiple mutant is more deleterious than with independent action.

That this is true is apparent from a simple example. Assume a classical epistasis situation where *aa bb* is harmful, but all other genotypes are normal. Then each genetic extinction eliminates four mutant genes. Thus the load per mutant is only one-fourth as large as for a dominant, or one half that of a recessive.

This can be demonstrated and extended as follows: I follow the procedure of King (1966). Let $\bar{x}$ be the average number of mutants per individual, u be the mutation rate per locus and $U = \sum u$ the total rate per gamete, n be the average number of mutants eliminated with each genetic death, $\bar{q}$ be the average frequency of mutant alleles per locus, and $\bar{w}$ be the average fitness. Then, the number of mutant alleles per individual in the following generation will be given by

$$\bar{x}' = \frac{\bar{x} - nL + 2U(1 - \bar{q})}{\bar{w}}$$

where L is the fraction of genetic deaths or the mutation load. Noting that $\bar{w} = 1 - L$ and equating $\bar{x}'$ to $\bar{x}$ to give the equilibrium condition, we obtain

$$L = \frac{2U(1 - \bar{q})}{n - \bar{x}}. \tag{10}$$

For a single locus with completely recessive mutants, $n = 2$, $\bar{x} = 2\bar{q}$, which leads to $L = U = u$, in agreement with the previous section. For two loci, both recessive and such that only the double homozygote is deleterious, $n = 4$, $\bar{x} = 4\bar{q}$, $L = U/2 = u$; and for N loci producing completely recessive mutants, the load is U/N.

It is likely, I think, that complete epistasis of this kind is as rare as complete recessivity, and that therefore the same considerations apply as with partial dominance; so the mutation load is probably nearer the total mutation rate than the above paragraph might imply.

However, one model has received particular attention in the context of genetic loads. This is a threshold, or truncation selection model. It is assumed that individuals with a certain number of mutants or more are eliminated while those with less than this number are retained. With strict truncation at a threshold of m mutants, then each eliminated individual has at least m mutants, and the considerations mentioned above become applicable. This kind of model has been discussed by Kimura and Maruyama (1966) and especially by King (1966) who emphasized that the mutation load can be considerably reduced in such a situation.

The total number of loci contributing to a threshold trait may be very large, in which case $\bar{q}$ is very much smaller than $\bar{x}$ in Eq. (10). As $\bar{q}$ becomes unimportant, we see that the mutation load is approximately equal to the total mutation rate divided by the difference between the average number of mutants in those individuals that are selectively eliminated (n) and the average number for the whole population ($\bar{x}$).

Table 2 gives some calculations from Kimura and Maruyama. I assume that an individual is eliminated if he has m or more mutants,

Table 2. *The mutation load with a threshold model. Each individual with m or more mutants is eliminated, while all with less than m are retained. The mean number of mutants per individual at equilibrium ($\bar{x}$), the mutation load (L), and the mutation load expressed as a fraction of what it would be if the genes were eliminated independently (L/L_I) are given. The total mutation rate per gamete for all relevant loci is assumed to be 0.1*

m	$\bar{x}$	L	L/L_I
1	0.20	0.181	0.91
2	0.56	0.108	0.54
3	1.00	0.080	0.40
4	1.49	0.065	0.33
5	2.03	0.056	0.28
10	5.07	0.034	0.17

and that individuals with any number less than this are normal. The mutation rate is taken to be 0.1 per gamete. The number of mutants per individual is assumed to have a Poisson distribution, as would be approximated if they occur independently and are not closely linked. The equilibrium distribution can then be calculated by the procedure given by Kimura and Maruyama (1966). The mutation load, L, is given in the table together with the fraction that L is of the load if the mutants were eliminated independently rather than by truncation selection.

As can be seen, the mutation load with a high threshold number (m) becomes considerably less than the mutation rate. The relevance of Eq. (10) is apparent in the last row of the table. If the mean number of mutant genes is 5.07, the mean number among those with 10 or more is almost exactly 11. If $n = 11$ and $\bar{x} = 5.07$, $1/(n - \bar{x})$ is 0.17, and since $\bar{q}$ is very small $L = 2U/(n - \bar{x}) = 0.034$ as the last line in the table shows.

Thus, we see that a model of truncation selection can lead to a substantial reduction of the mutation load, as King emphasized. On the other hand, Kimura (1961) showed that if the effect on fitness is proportional to the square of the number of mutants the mutation load is approximately half what it is with independent effects, which is much less of a decrease than with the threshold model. It becomes important, then, to inquire how mutations affecting fitness do, in fact, interact.

Figure 1 shows several curves in which fitness is plotted as ordinate against the number of deleterious mutants as abscissa. If the loci are independent, then the fitness of an individual with k mutants is $(1 - s_1)$ $(1 - s_2) \ldots (1 - s_k)$, where s_i is the reduction in fitness caused by the *i-th* mutant. This is roughly $e^{-k\bar{s}}$, where $\bar{s}$ is the average of the values. This is shown in 1A. 1B shows the simple relation if the genes are strictly additive in their effect on fitness. 1D illustrates a threshold or truncation selection model. All individuals to the left of the threshold point have fitness one; those to the right, fitness zero. Finally, 1C shows an inter-

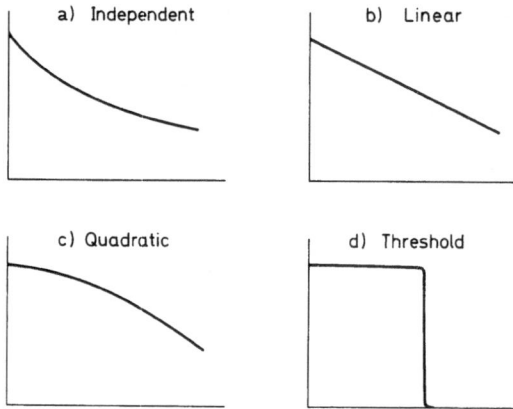

Fig. 1. Four models for the relationship of viability or fitness to the number of mutant genes. Abscissa: Number of mutant genes. Ordinate: Fitness

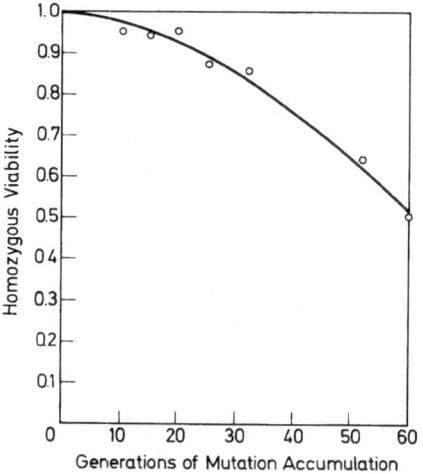

Fig. 2. The change in viability with accumulation of deleterious mutations. The mutations were allowed to accumulate in a chromosome which was maintained in the heterozygous condition for several generations; then viability was measured for the same chromosome when made homozygous. Data from Mukai (1969)

mediate situation where there is some synergistic epistasis, but not as extreme as with the threshold. Are there data which can discriminate among these possibilities?

The largest body of data of which I am aware that relates the number of mutants to their cumulative effect has been published by Mukai

(1969). His data are for homozygous effects of spontaneously arising mutants on the second chromosome of *Drosophila melanogaster*. These mutants have an estimated viability effect, when homozygous, of about 1–5%, those mutants with more drastic effects having been eliminated from the data.

When the viability is plotted against the number of mutants, there is clearly a non-linear relationship, as shown in Fig. 2. A quadratic equation gives a much better fit, and the improvement is statistically significant. However, fitting a cubic gives no significant improvement. So I shall assume that the data are adequately represented by a quadratic equation.

I should add that the distribution is for homozygous mutants, whereas we are really concerned mainly with the heterozygous effects. However, there is strong evidence for a very high level of dominance (Temin, 1966; Wills, 1966; Crow, 1968a; Morton, et al., 1968), so I shall assume that the curve for heterozygous effects parallels at least roughly the curve for homozygous effects, with lesser individual values of course.

To develop a procedure for estimating the mutation load when the relationship is quadratic, I follow rather closely the methods of Kimura and Maruyama (1966).

Let w_x be the fitness of an individual carrying x mutant genes. On a quadratic model, we let

$$w_x = 1 - ax - bx^2,$$

where for convenience the fitness of an individual with no mutants is taken to be one. (I am assuming that x never gets to be large enough for w_x to become zero or negative. This assumption is consistent with the data in Drosophila. If the assumption were not reasonable, a natural model would be to assume that the fitness, m, where $m = \log_e w$, is a quadratic function of x; that is $m_x = -ax - bx^2$.) The mutation load is given by

$$L = 1 - \overline{w} = \sum f_x (ax + bx^2),$$

where f_x is the frequency of individuals with x mutants ($\Sigma f_x = 1$). Then

$$L = a\overline{x} + b\overline{x^2}$$
$$= a\overline{x} + b(\overline{x}^2 + V_x),$$

(11)

where V_x is the variance of x.

If the mutants arise independently and there is free recombination, the number of mutants per individual, x, will have a distribution that is approximately Poisson, and thus $V_x = \overline{x}$. So we arrive at

$$L = (a + b)\overline{x} + b\overline{x}^2.$$

(12)

We now ask what is the average amount of selection against a mutant, for this and the mutation rate will determine the average number of accumulated mutants per individual. The average selection against a mutant will be given by the weighted average difference in fitness between an individual with x mutants and an individual with one more, $x + 1$, expressed as a fraction of the average fitness. Letting $\overline{hs}$ be the average selection against a mutant heterozygote, we obtain

$$
\begin{aligned}
\overline{hs} &= \frac{\sum f_x(w_x - w_{x+1})}{\overline{w}} \\
&= \frac{\sum f_x[1 - ax - bx^2 - 1 + a(x+1) + b(x+1)^2]}{\overline{w}} \\
&= \frac{\sum f_x(a + b + 2bx)}{1 - L} \\
&= \frac{a + b + 2b\overline{x}}{1 - (a + b)\overline{x} - b\overline{x}^2}
\end{aligned}
\tag{13}
$$

where $\overline{x} = \sum f_x x$.

From Eq. (5) when $F = 0$ and u is very small, we have approximately

$$
u = hsq. \tag{14}
$$

Summing over all relevant loci, and substituting $\overline{x}$ for $2\sum q$, U for $\sum u$, and $\overline{hs}$ for $\sum hsq / \sum q$, we obtain

$$
2U = \overline{hs}\,\overline{x} \tag{15}
$$

or

$$
\overline{hs} = \frac{2U}{\overline{x}}. \tag{15'}
$$

Equating the right halves of Eq. (13) and (15'), we eliminate $\overline{hs}$ and get the quadratic

$$
2b(U + 1)\overline{x}^2 + (2U + 1)(a + b)\overline{x} - 2U = 0 \tag{16}
$$

of which $\overline{x}$ is the positive root. We then compute L from Eq. (12).

Mukai's (1969) data in Fig. 2 give the relative viability of Drosophilae that are homozygous for a second chromosome which has been allowed to accumulate minor deleterious mutations while being held heterozygous for n generations. The regression equation for relative viability, w_n, as a function of n is

$$
w_n = 1 - 0.0014n - 0.00011n^2.
$$

The estimated mutation rate for mildly deleterious mutants is about 0.1 per chromosome per generation. Therefore, x, the number of accumulated mutants in n generations is $0.1n$ and the corresponding equation in x is

$$w_x = 1 - 0.014x - 0.011x^2. \tag{17}$$

The decreased fitness in heterozygotes is not known, but there is strong evidence for a large amount of partial dominance — perhaps 25% or more — as mentioned before. If we assume that heterozygous effects follow the same curve as the homozygous effects, the fitness of a fly with kx heterozygous mutant genes would be the same as that of one with x homozygous mutants, where $1/k = h$, the amount of dominance. The coefficients, a and b, in the equation $w_x = 1 - ax - bx^2$ for heterozygous effects, corresponding to 0.014 and 0.011 in Eq. (17) would become $0.014h$ and $0.011h^2$.

The average number of deleterious mutant genes per zygote ($\bar{x}$), the average heterozygous effect of a mutant in the equilibrium population ($\overline{hs}$), the mutation load (L), and the ratio of the mutation load to what it would be with independent gene action ($L/2U$) are given in Table 3. U is the total mutation rate per gamete.

The values of a and b in the first four rows are those of Eq. (17) for homozygous effects of newly occurring mutants, and for heterozygous effects 1/2, 1/4, and 1/10 as large. Rows 5–8 are corresponding numbers when the quadratic component is 1/10 times as large; rows 9–12 are for a quadratic component 10 times as large as for rows 1–4. Rows 13–15 give three values when there is no linear component at all. Rows 16–19 are the same as 1–4 except that the mutation rate is only half as large, and in rows 20–23 the mutation rate is twice as large. Finally rows 24–26 give the loads when there is no quadratic component for three mutation rates. The load in this case is always $2U/(2U+1)$ for moderate values of $\overline{hs}$.

Notice that for quite a range of values of a and b the load is roughly half that which would be expected with independent gene action. Even when $b = 0$, which means that the relationship is linear, the load is less than that for independence. This is because additivity on the fitness scale (Fig. 1 B) is not the same as independence (Fig. 1 A), which is additive on a log scale, and when there is a high mutation rate the difference is appreciable.

The load-reducing effect of synergistic epistasis becomes more important as the mutation rate gets larger (see rows 20–23). Here the mutation rate is 0.8 per zygote, which with partial dominance and the general level of epistasis that appears in Mukai's data, leads to an average loss of fitness of about 1/3 attributable to deleterious mutations.

Table 3. *The mean number of mutants per zygote ($\overline{x}$), the average heterozygous effect at equilibrium ($\overline{hs}$), the mutation load (L), and the ratio of the load to that expected with independent gene effects (L/2U). U is the mutation rate per gamete and a and b are the coefficients in the quadratic viability equation $w_x = 1 - ax - bx^2$*

Row	U	a	b	$\overline{x}$	$\overline{hs}$	L	L/2U
1	0.20	0.0140	0.01100	3.3	0.122	0.201	0.50
2		0.0070	0.00275	6.8	0.059	0.194	0.49
3		0.0035	0.00069	13.9	0.029	0.191	0.48
4		0.0014	0.00011	35.1	0.011	0.188	0.47
5	0.20	0.0140	0.00110	8.9	0.045	0.223	0.56
6		0.0070	0.00028	18.1	0.022	0.221	0.55
7		0.0035	0.00007	36.3	0.011	0.220	0.55
8		0.0014	0.00001	89.6	0.005	0.215	0.54
9	0.20	0.0140	0.11000	0.9	0.423	0.216	0.54
10		0.0070	0.02750	2.1	0.188	0.197	0.49
11		0.0035	0.00687	4.5	0.089	0.186	0.47
12		0.0014	0.00110	11.7	0.034	0.179	0.45
13	0.20	0	0.0100	3.8	0.105	0.182	0.45
14			0.0010	12.6	0.032	0.172	0.42
15			0.0001	40.5	0.009	0.168	0.42
16	0.10	0.0140	0.01100	2.3	0.086	0.117	0.59
17		0.0070	0.00275	4.9	0.041	0.112	0.56
18		0.0035	0.00069	10.0	0.020	0.110	0.55
19		0.0014	0.00011	25.2	0.008	0.108	0.54
20	0.40	0.0140	0.01100	4.4	0.181	0.325	0.41
21		0.0070	0.00275	9.1	0.088	0.317	0.40
22		0.0035	0.00069	18.5	0.043	0.313	0.39
23		0.0014	0.00011	46.7	0.017	0.311	0.39
24	0.10	any	0			0.167	0.83
25	0.20	any	0			0.286	0.71
26	0.40	any	0			0.444	0.56

We have seen that the data for homozygous effects of spontaneously occurring mutants having individually small effects on viability is much closer to a quadratic (Fig. 1 B) than to a sharp threshold model (Fig. 1 D). It is possible that with more intense competition the shape would be altered in the direction of D. On the other hand, Temin et al. (1969) found no substantial difference in the interaction of viability of second and third chromosome homozygotes between crowded and uncrowded cultures. It is also possible that the heterozygous effects on total fitness, which must include effects on fertility as well as viability, approach the shape of D. However, I know of no data in support of this and it is much in need of study.

The load for drastic mutants, mostly full lethals, is of course more nearly the theoretical value of twice the mutation rate, for the average number of mutant genes per fly is about one or two and this doesn't give much opportunity for interactions. However, according to Mukai's results, lethal mutations occur with a frequency probably less than 1/20 that of mutations with minor effects.

Until data on heterozygous effect, including fertility, can be obtained, I shall assume that Mukai's results are indicative of the heterozygous effects. This would then lead to the (quite tentative) conclusion that the mutation load for mutations causing effects on viability, at least in Drosophila, is about half the value expected if the genes were independent.

The Mutation Load in an Asexual Population

Epistasis has much less effect on the mutation load in an asexual population, provided the classes with one or more mutants are all less fit than the class with none. This is to be expected for the following reason: The only time when two or more mutants are eliminated simultaneously by a single genetic extinction is if they originated simultaneously, since there is no way in an asexual population for mutant genes to be reassorted. We do not have to consider mutations that occur in the descendants of a mutant individual. These may change the rate at which the original mutation is eliminated, but this has no effect on the load caused by the original mutant. There is still, on the average, one extinction per mutant whenever this occurs. Thus the mutation load is simpler in asexual than sexual species, being very close to the total mutation rate per zygote.

This is easily shown. Assume that genotype A is sufficiently more fit than the others to be the common type. Designate all the others collectively by B.

	Prevailing type	All others
Genotype(s)	A	B
Frequency	P_A	P_B
Fitness	w_A	w_B

In this notation w_B is the average fitness of all the other genotypes. U is the total rate of mutation from A to other types. More strictly, this counts as one mutation the simultaneous occurrence of more than one. Assume, as before, that the reverse mutation is negligible.

The frequency of genotype A next generation will be

$$P'_A = \frac{P_A(1 - U)\,w_A}{\overline{w}}. \tag{18}$$

At equilibrium, $P'_A = P_A$, and therefore $\bar{w} = (1 - U) w_A$. Thus the mutation load is

$$L = \frac{w_A - \bar{w}}{w_A} = U . \tag{19}$$

Since U is the rate of occurrence of one or more mutations per zygote, it is less than the total rate. If the mutants arise independently and the total rate per gamete is $\sum u$, then the load is

$$L = 1 - e^{-2\Sigma u} \tag{20}$$

as given by Kimura and Maruyama (1966).

To summarize, regardless of the manner of interaction of mutants as long as any number of mutants causes some detriment to fitness (large relative to the individual gene mutation rate) the load in an asexual population is the total mutation rate, U, where this is taken to mean the total rate at which zygotes with one or more new mutants appear.

These remarks do not apply to a threshold model where several mutants can accumulate in an individual and cause no change at all in fitness, and thus be eliminated in groups as the number of mutant genes becomes large enough to pass the truncation level.

Nearly Neutral Mutants in a Finite Population

In any real population there will be deviations from the equilibrium values for many reasons. The conditions determining the equilibrium are never completely stable, so the population is not able to keep up with the changes. Usually, although not necessarily, the population fitness is maximized at or near the equilibrium gene frequencies. Therefore the effect of departures from equilibrium is usually to increase the load.

In a finite population there will be random drift in gene frequencies, the effect of which is to carry the frequencies away from equilibrium values and thereby increase the load. The effect of random drift on the mutation load has been studied by Kimura et al. (1963), and I shall not repeat the results here. The most striking finding is that mutations giving rise to small selective differences can create a larger mutation load than those with a larger effect. This is true over a rather wide range of population sizes. Somehow, the greater amount of drift of such mutants more than compensates for their lesser deleterious effects.

These results apply to mutant genes whose deleterious effects, although small (e.g., 10^{-3}), are large relative to mutation rates. If the mutant effect becomes so small as to be comparable to the mutation rates, the load is reduced. I shall not attempt a mathematical treatment, but one can get a qualitative idea from the following considerations.

A mutant with a very mild deleterious effect may persist in the population for a long time, long enough that there is an appreciable chance of its mutating again before being eliminated. If the chance of mutating, either back to the original or, more likely, to a third allele, is u, then after n generations the proportion of descendant genes that have mutated will be $1 - (1 - u)^n$. If this happens, the mutant escapes extinction by selection, and the genetic death should be charged against the subsequent mutation. As the selective disadvantage gets to be less and less, more mutants will mutate before being eliminated. Thus the load gets less as s gets less, approaching zero as s becomes zero. The exact manner of approach will depend on whether the mutant is recessive or dominant, on its epistatic relations, and on the interaction of selection, mutation, and random drift. Random drift will be important because such nearly neutral mutations will be strongly influenced by random processes, even in large populations.

Molecular biology has shown dramatically the wide range of mutational possibility at a single gene with its several hundred nucleotides. The possibility that many nucleotide substitutions may cause inconsequential changes has become increasingly apparent. Some, or perhaps many, such changes may alter the fitness by an amount that is of the same order as the mutation rate or the reciprocal of the effective population number, or less. The fate of such mutants is determined largely by random processes. The possibility that such genes may be important in evolution has been discussed by several authors (Kimura, 1968; King and Jukes, 1969; Crow, 1969).

The possibility of such mutants seemed real enough several years ago that Kimura and I (1964) undertook to study the problem. The range of mutational possibility is so great that it is a reasonable model to assume that each mutant allele is a type not pre-existing in the population. On this model the average proportion of loci that are heterozygous is

$$H = \frac{4N_e u}{4N_e u + 1}, \tag{21}$$

where u is the total mutation rate to all neutral alleles and N_e is the inbreeding effective population number (Wright, 1931, 1949, 1951; Kimura and Crow, 1963; Crow and Morton, 1955).

In the derivation of the formula we assumed a population in which self-fertilization is possible. Actually the result is not appreciably different when there are separate sexes. This is easily demonstrated as follows:

Let F_t be the probability that two homologous genes in an individual are identical in the sense of being derived from the same gene in a common ancestor. Let G_t be the probability that two genes, one from each of two different individuals, are identical. Then, in the absence of mutation,

$$F_t = G_{t-1},$$
$$G_t = \frac{1}{N}\left(\frac{1}{2} + \frac{1}{2}F_{t-1}\right) + \left(1 - \frac{1}{N}\right)G_{t-1}.$$
(22)

The first equation follows directly from the fact that each individual gets a gene from each of his parents which, with random mating, is equivalent to a random sample of genes from two individuals, which by definition is G for the previous generation. The second equation comes from the fact that with random mating two genes in different individuals came from the same individual in the previous generation with probability $1/N$ and from different individuals with probability $1 - 1/N$. In the first case the probability of their being identical is 1 if they are descended from the same allele and F_{t-1} if they came from different alleles in the same individual. These two contingencies are equally likely, so each has a probability of $1/2$. In the second case the probability is G_{t-1}.

If mutation is considered, two identical alleles will still be identical in the next generation only if neither of them has mutated in the meantime. So the equations for identity in successive generations become

$$F_t = G_{t-1}(1-u)^2,$$
$$G_t = \left[\frac{1}{2N}(1 + F_{t-1}) + \left(1 - \frac{1}{N}\right)G_{t-1}\right](1-u)^2.$$
(23)

Putting these together,

$$F_{t+1} = \frac{1}{2N}(1 + F_{t-1})(1-u)^4 + \left(1 - \frac{1}{N}\right)F_t(1-u)^2.$$
(24)

At equilibrium, $F_{t+1} = F_t = F_{t-1}$. Neglecting terms involving higher powers of u, we obtain

$$F = \frac{1-4u}{4N_e u + 1} \approx \frac{1}{4N_e u + 1},$$
(25)

and

$$H = \frac{4N_e u}{4N_e u + 1}$$
(26)

as when self-fertilization is permitted.

If the numbers in the two sexes are different

$$\frac{1}{N_e} = \frac{1}{4N_m} + \frac{1}{4N_f}$$
(27)

where N_m and N_f are the number of males and females. For nonrandom variability in number of progeny the effective number can be defined in terms of the mean and variance of the number of progeny per parent (Kimura and Crow, 1963).

I have assumed that the number of possible kinds of mutations at a locus is so large that every new mutant is a type not currently represented in the population. If there are only k allelic states, Eq. (26) is modified to become

$$H = \frac{4N_e u}{4N_e \left(\dfrac{k}{k-1} \right) + 1} . \tag{28}$$

The Segregation Load

A Single Locus with Two Alleles

This question was first raised and answered by Haldane in his original paper (1937). The decreased fitness, in comparison with the heterozygote in situations where the heterozygote is favored, arises from the fact that in a Mendelian population there will always be inferior homozygotes arising by segregation.

To consider the question quantitatively, consider the following simple model.

Genotype	AA	AA'	$A'A'$
Fitness	$w(1-s)$	w	$w(1-t)$
Frequency	p^2	$2pq$	q^2

Assume that s and t are large enough, relative to the mutation rates, that mutation can be ignored. Following the procedure used above for the mutation load, we write the frequency of allele A one generation later in terms of the present genotypic frequencies and the selection coefficients. Thus, writing p' for the frequency next generation,

$$p' = \frac{p^2 w(1-s) + pqw}{\bar{w}}$$

where $\bar{w} = p^2 w(1-s) + 2pqw + q^2 w(1-t)$. Equating p to p' to specify an equilibrium and solving leads to

$$p = \frac{t}{s+t}, \qquad q = \frac{s}{s+t},$$
$$\bar{w} = w \left(1 - \frac{st}{s+t} \right), \tag{29}$$

and from the definition (Eq. (1)), the load is

$$L = \frac{st}{s+t}. \tag{30}$$

Notice that the formula can also be written as

$$L = sp^2 + tq^2 \tag{31}$$

$$= sp = tq \tag{32}$$

where p and q are the equilibrium values of the allele frequencies.

This example illustrates two principles of the segregation load. The first, illustrated by Eq. (31) is that the allele with the greatest selective disadvantage makes the least contribution to the segregation load. For example, if $s = 0.01$ and $t = 1$, which means that $A'A'$ is either lethal or sterile, $sp^2 = 100/101^2 = 0.01$ and $tq^2 = 1/101^2 = 0.0001$. So the more drastic allele contributes only one percent of the total segregation load.

The second principle, first noticed by N. E. Morton, is illustrated by Eq. (32). This is that one can compute the segregation load from information on only one allele. In the next section, it will be shown that this is also true for multiple alleles.

Multiple Alleles with Independent Loci

If there are multiple alleles, there are numerous possibilities for stable equilibria. The rules for determining the stability of such a system were given by Kimura (1956a) and Mandel (1959). For a recent review, see Li (1967). It is not necessary that every heterozygote be superior to any homozygote for there to be an equilibrium, nor does the superiority of the heterozygote for a particular pair over the two homozygotes insure that this pair will persist in a polyallelic system. There are also systems that are stable, but which become unstable when one allele is removed. In the following discussion, I shall assume that the conditions for a stable equilibrium have been met.

Assume that the genotypes and fitnesses are as follows.

Genotype	$A_i A_i$	$A_i A_j$	
Fitness	$w_{ii} = w(1 - s_{ii})$	$w_{ij} = w(1 - s_{ij})$	(33)
Frequency	p_i^2	$2 p_i p_j$	

From the general equation for gene frequency change (e.g., Wright, 1949)

$$\Delta p = \frac{p_i(w_i - \bar{w})}{\bar{w}} \tag{34}$$

where w_i is the average fitness of all genotypes which have gene A_i, weighted by the number of A_i genes carried (1/2 for heterozygotes, 1 for homozygotes) and $\overline{w}$ is the general mean. Thus

$$w_i = \left(\sum_j p_i p_j w_{ij}\right)/p_i = \sum_j p_j w_{ij}, \tag{35}$$

$$\overline{w} = \sum_i \sum_j p_i p_j w_{ij}. \tag{36}$$

At equilibrium $\Delta p = 0$, and therefore

$$w_i = \overline{w}. \tag{37}$$

So we are led to the very reasonable conclusion that, at equilibrium, each allele has the same average fitness as any other, or as the average of all. If this were not true, then the allele with a higher average fitness would tend to increase and there would be no equilibrium.

The segregation load is

$$L = \frac{w - \overline{w}}{w} = \frac{w - w_i}{w} = s_i, \tag{38}$$

where

$$w_i = w(1 - s_i).$$

This can also be written as

$$L = \sum_j p_j s_{ij} \tag{39}$$

by using Eq. (35). A more useful form is

$$L \geqq p_i s_{ii}. \tag{40}$$

This follows from Eq. (39) *a fortiori*, since the right side of Eq. (39) contains several non-negative terms of which $p_i s_{ii}$ is only one.

If all heterozygotes have equal fitnesses and all homozygotes are inferior, but not necessarily equal to each other, then the inequality in Eq. (40) becomes an equality. This is because with equal heterozygotes $s_{ij} = 0$ for all combinations where $i \neq j$ and therefore $p_i s_{ii}$ is the only non-zero term in Eq. (39). Writing (40) as

$$p_i = L/s_{ii}$$

and noting that $\sum p_i = 1$, we obtain

$$L \sum (1/s_{ii}) = 1,$$

and

$$L = \frac{\tilde{s}}{n}, \tag{41}$$

where $\bar{s}$ is the harmonic mean of the homozygous disadvantages, s_{ii}, s_{jj}, etc., and n is the number of alleles. This shows that for comparable fitness coefficients, the segregation load is inversely proportional to the number of alleles maintained in the population.

Eq. (40) extends the principle mentioned in the last section to multiple alleles. Regardless of the individual fitnesses, if the population is at equilibrium under selective balance, the segregation load (or at least a minimum estimate thereof) can be gotten from information on a single allele and its homozygous effect on fitness. The minimum estimate of the load is the product of the frequency of that allele and its homozygous selective disadvantage relative to the best genotype.

An example of a fairly common, yet harmful recessive gene, is the one causing cystic fibrosis. Suppose that the incidence of the disease is 0.0004. The frequency of the recessive allele is then 0.02. There is good evidence that this is caused by a single allele, or a group of noncomplementing alleles (Crow, 1965). Since the recessive homozygote rarely reproduces, $s_{ii} = 1$. So, if this allele is maintained at equilibrium by heterozygous superiority, regardless of the number of alleles, the segregation load is 0.02 or greater. In contrast, if this allele is maintained by a high mutation rate — and a rate of 0.0004 is required to maintain this frequency if the gene is fully recessive — the load is only 0.0004.

Eq. (40) was derived for a population mating at random. The same inequality holds if the population has come to equilibrium with consanguineous matings leading to a specified departure from multinomial frequencies (see Crow, 1961, p. 143).

A group of independent loci will have a collective segregation load that is roughly the sum of the individual loads until the number gets large. Suppose there were 100 loci, independent in inheritance and in their effect on fitness, and each with a load as large as that just mentioned. The total load would then be 100×0.02, or 2. This means that, with independence, the average fitness of the population relative to the best heterozygote is $(1 - 0.02)^{100}$, or roughly $e^{-2} = 0.14$. With 200 loci, $e^{-4} = 0.02$.

In general, the average fitness is

$$\bar{w} = w e^{-\Sigma l}, \quad \text{or} \quad L = 1 - e^{-\Sigma l}$$

where l is the load for an individual locus.

This can quickly get to be a very large fraction if the number of polymorphic loci is large. That this creates problems in accounting for large numbers of segregating loci in a population has been discussed repeatedly (e.g. Greenberg and Crow, 1960; Kimura and Crow, 1964; Lewontin and Hubby, 1966). I shall now consider briefly some ways in which the segregation load may be less.

Ways in Which the Segregation Load may be Lessened

There is increasing evidence for the existence of numerous poly-morphic loci in natural populations. For example, Lewontin and Hubby (1966) estimated from a group of isozymes where electrophoretic differences could be detected that some 12% of loci are heterozygous in the average *Drosophila pseudoobscura*, or about 30% of loci were polymorphic in the population. There are several reasons why these are more likely to be underestimates than overestimates of the true number.

If each of these loci is maintained independently by heterozygote advantage, then the segregation load appears to be enormous. If the number of genes is 10^4 or more and the data of Lewontin and Hubby are representative, then the typical fly is heterozygous for a thousand or more loci, and the population is polymorphic for several thousand loci. The value of e^{-1000l} is small even when l is nearly 0.

The load is increased if the population size is small, for this decreases the heterozygosity. The equilibrium between selection for heterozygosis and random loss of alleles has been studied by Kimura and Crow (1964) for a special case in which all heterozygotes are alike and all homozygotes have their fitness reduced by a fraction s. The equilibrium solutions were obtained by a diffusion approximation. All new mutants were assumed to enter into this mutually heterotic system.

For example, if the effective population number (N_e) is 10,000, the mutation rate (u) 10^{-5}, and $s = 0.01$, the probability of heterozygosity is 0.88 and the segregation load is 0.0012. This is equivalent to maintaining about eight alleles at equal frequencies, as can be confirmed from Eq. (41). If s is larger, more alleles are maintained, but the load is larger. If s is increased to 0.1, 95% of the loci are heterozygous, but the load increases to about 0.0045. In general, as can be seen from the graphs in the paper just mentioned, $\log L$ decreases approximately linearly with $\log N_e$.

There are several ways in which a large amount of polymorphism can be maintained without a large segregation load.

One possibility is that the selective differences are very small. If these are much less than the reciprocal of the effective population size, the allele frequencies will be largely determined by random drift and mutation. The considerations discussed earlier in the section entitled "Nearly neutral mutations in a finite population" become applicable. If the product $4N_e u \gg 1$, the locus will be polymorphic for neutral mutations. Of course, with neutral genes there is no load. Whether any large number of such mutants exist is an open question, but the evidence for them is increasing (e.g. King and Jukes, 1969).

A second possibility is that some polymorphic loci are maintained by frequency-dependent selection. If each allele is favored when rare, but not when common, there is a stable equilibrium when each is present in intermediate frequencies. The selective differences are minimized at or close to the equilibrium point, so that a population at equilibrium can be polymorphic with very little load. However, in any real population there will be drift away from the equilibrium point because of random processes, and the population is returned to the equilibrium only by selection. So there is necessarily a load created by this. As far as I know, no general treatment of this problem has been published, so I am unable to give any general comparison between the magnitude of load created by heterotic versus frequency-dependent selection in finite populations. For a discussion of frequency dependence, see Kojima and Yarbrough (1967).

I have been discussing multiple polymorphisms as if they were independent in inheritance and in the action of selection. Neither is strictly true, of course.

Nonindependence in inheritance can reduce the segregation load if a group of heterotic genes can be held together by linkage, at least under some circumstances. An extreme example is the case where there are several loci, each occupied by two alleles, and with both homozygotes lethal at every locus. Under this model the segregation load can be 1/2 at each locus, and for n loci will be $1 - 0.5^n$. However, if all the loci were linked into two chromosomes, complementary at each locus, the load would be reduced to 1/2.

In general, the effect of linkage will depend on the nature of the epistasis. If fitness is measured in terms of w, closer linkage will enhance the average fitness if the double heterozygote has a fitness beyond what would be expected from additive effects of the two single heterozygotes, as in the example above (Haldane, 1957b).

Sved, Reed and Bodmer (1967), King (1967), and Milkman (1967) have all suggested that a threshold or truncation selection model can greatly decrease the segregation load. The model assumes that beyond a certain level of heterozygosity additional heterozygous loci make no increased contribution to the average fitness. In an extreme form, all individuals beyond a number x of heterozygous loci have fitness one and those below this number have fitness zero. An alternative model with much the same consequences is that a certain fraction, p, of the individuals are selected and the rest are culled. Those individuals that are selected are the ones with the largest number of heterozygous loci.

It is doubtful that natural selection acts by counting the number of heterozygous loci and then sharply divides the population into two groups based on the number of such loci. But the animal breeder practices truncation selection with respect to phenotypes, and it may be argued that natural selection approximates this pattern sufficiently well to alter substantially the number of polymorphisms maintained by a certain amount of selection. In terms of the drawings in Fig. 1, the shape of the curve relating fitness as ordinate to homozygosity as abscissa may be somewhere between C and D. The question is one that requires empirical answers; clearly not enough is known about gene interactions to judge the realism of such models at present.

The introduction of linkage into threshold models leads to mathematical difficulties, but computer simulations have shown that it is possible to devise systems in which a great amount of polymorphism is maintained. Wills, Crenshaw and Vitale (1969) have studied one such model. They assume truncation selection of a certain intensity (e.g. 10% selective elimination, where eliminated individuals are those with the smallest number of heterozygous loci). With close linkage the population tends to retain particular chromosomes – generally those that are mutually complementary – and a moderate amount of selection can retain a large number of polymorphisms. Similar development of a high degree of organization along the chromosome has been found by Lewontin (personal communication). See also Sved (1968 b).

Which of these mechanisms are the more important in determining natural polymorphisms and whether additional mechanisms play a role are among the most intriguing questions of population genetics. The combined efforts of experimental studies, natural population censuses, and computer simulation studies will probably be required for an adequate answer.

The Load Due to Meiotic Drive

As one more example of a genetic load, consider the effect of meiotic drive (Sandler and Novitski, 1957). Typical examples are the t-alleles in mice and the SD factor in Drosophila. I shall consider only a simple case.

Assume that the homozygote is lethal or sterile (which is often true). Let h be the selection against heterozygotes. Assume further that the ratio of a to A genes contributed by males is $K : 1 - K$, but that the contribution from heterozygous females is the regular $1:1$ of Mendelian heredity. (Meiotic drive is typically found in one sex only.)

The genotypes and fitnesses will be designated as

Genotype	AA	Aa	aa	
Fitness	1	$1-h$	0	
Frequency	$p_f p_m$	$p_f q_m + p_m q_f$	$q_f q_m$	(42)

where p_f and p_m are the frequency of allele A in the gametes of females and males; $q_m = 1 - p_m$ and $q_f = 1 - p_f$. With meiotic drive we cannot make the simplifying assumption of Hardy-Weinberg zygote ratios.

The allele frequencies next generation are given by

$$q'_m = \frac{K(1-h)(p_m q_f + p_f q_m)}{1 - h(p_m q_f + p_f q_m) - q_m q_f} , \tag{43}$$

$$q'_f = \frac{q'_m}{2K} . \tag{43}$$

To specify the equilibrium conditions, let $q'_m = q_m = \hat{q}_m$ and $q'_f = q_f = \hat{q}_f = \hat{q}_m/2K$. This leads immediately to the quadratic

$$\hat{q}_m^2(1-2h) + \hat{q}_m[h(1+4K) - 2K] + K[2K - 1 - h(1+2K)] = 0 \tag{44}$$

or

$$A\hat{q}_m^2 + B\hat{q}_m + C = 0 \tag{44'}$$

where

$$A = 1 - 2h,$$
$$B = h(1+4K) - 2K,$$
$$C = K[2K - 1 - h(1+2K)].$$

The relevant solution is

$$\hat{q}_m = \frac{-B - \sqrt{B^2 - 4AC}}{2A} . \tag{45}$$

The load is

$$L = h(\hat{p}_m \hat{q}_f + \hat{p}_f \hat{q}_m) + \hat{q}_m \hat{q}_f = \frac{h\hat{q}_m}{2K}(1 + 2K - 2\hat{q}_m) + \frac{\hat{q}_m^2}{2K} . \tag{46}$$

The value of the meiotic drive load for several representative values of K and h are given in Table 4.

When $h = 0$,

$$\hat{q}_m = K - \sqrt{K(1-K)} , \tag{47}$$

$$L = \frac{\hat{q}_m^2}{2K} = \frac{1}{2} - \sqrt{K(1-K)} . \tag{48}$$

This system has the interesting and unusual property that the load is decreased when the lethal gene is partially dominant. This property is brought out in Table 4, which also gives an idea of the range of values for K and h that maintain the polymorphism.

Table 4. *The load due to meiotic drive. The gene favored by segregation distortion is assumed to be lethal when homozygous and to have a disadvantage of h relative to the normal homozygote. K is the proportion of gametes with the driven gene in one sex; the other sex is assumed to have normal Mendelian segregation*

K	h							
	0.00	0.01	0.02	0.05	0.10	0.20	0.30	0.50
0.5	0	0	0	0	0	0	0	0
0.6	0.010	0.010	0.010	0.007	0	0	0	0
0.7	0.042	0.042	0.041	0.039	0.029	0	0	0
0.8	0.100	0.100	0.100	0.097	0.087	0.033	0	0
0.9	0.200	0.200	0.200	0.197	0.187	0.129	0	0
0.95	0.282	0.282	0.282	0.279	0.267	0.205	0.031	0
0.98	0.360	0.360	0.359	0.356	0.342	0.270	0.083	0
0.99	0.401	0.400	0.400	0.396	0.379	0.298	0.103	0
1.00	0.500	0.495	0.490	0.472	0.438	0.333	0.125	0

The Cost of Natural Selection

In his classical papers on natural selection Haldane (1924–1932) discussed the number of generations required to change the gene or genotype frequency by a specified amount. Out of this comes the generalization (Haldane, 1932, 1966, p. 180):

"The number of generations required for a given change in the population is inversely proportional to the intensity of selection. This is true for all systems of slow selection."

For example, the number of generations (t) required to change the frequency of a dominant gene with selective advantage s from a frequency of p_O to a frequency p_t is

$$t = \frac{1}{s}\left[\log_e\frac{p_t(1 - p_O)}{p_O(1 - p_t)} + \frac{1}{1 - p_t} - \frac{1}{1 - p_O}\right] \tag{49}$$

and for other situations the time is in the form

$$t = \frac{1}{s}\,f(p) \tag{50}$$

where $f(p)$ is not a function of s.

From this we see that with slow selection the selective intensity times the time required for a change is a constant. For slower selection the time is spread out, but in some sense the total selection for the whole time involved is constant. Haldane made this idea more precise in two papers (1957a, 1960).

Haldane's Original Formulation — Haploid Model

I shall illustrate the procedure first with a haploid example and introduce the complications of diploidy later. Assume that the generations are discrete and that the frequencies and fitnesses are as follows:

Genotype	A	A'
Frequency	p	$q = 1 - p$
Fitness	w_A	$w_A(1 - s)$

$$\bar{w} = w_A(1 - sq).   \tag{51}$$

Assume that A, the favored gene, is initially rare, either because it is a new mutation or, more likely, because there has been a change in the environment such that a previously deleterious gene is now favorable.

In the inital generation (generation 0) the frequency of A is p_0. This genotype will be represented next generation by an average of w_A descendants. The average individual in the population will leave $\bar{w}$ descendants. The cost this generation, as defined by Haldane, is

$$\frac{w_A - \bar{w}}{w_A} = sq .   \tag{52}$$

The cost, as thus defined, is the same as the load except that here we are dealing with a changing rather than with an equilibrium population.

Haldane asked what is the value of the cost, summed over all the generations involved in the substitution. If the change is slow, we can replace summation with integration and write the total cost as

$$C = \int_0^\infty sq\,dt .   \tag{53}$$

The standard equation for gene frequency change (e.g. Wright, 1949) is

$$\Delta p = \frac{p(w_A - \bar{w})}{\bar{w}} ,   \tag{54}$$

which in this case is

$$\Delta p = \frac{pqs}{1 - sq} \approx \frac{dp}{dt} .   \tag{55}$$

Also, if selection is weak, $1 - sq$ is very nearly one. So with this simplification, substituting Eq. (55) into Eq. (53) and making the appropriate changes in the limits of integration to correspond to the change in the variable from t to p,

$$C = \int_{p_O}^{1} \frac{sq}{pqs} \, dp = -\log_e p_O . \tag{56}$$

The point of particular interest is that the s cancels, and the total cost is thus independent of s (to the limit of accuracy of the approximations used).

If $p_O = 10^{-4}$, $-\log_e p_O = 9.2$. With diploidy and various degrees of dominance the value is larger. In Haldane's words (1957, p. 520):

"The unit process of evolution, the substitution of one allele by another, if carried out by natural selection based on juvenile deaths, usually involves a number of deaths equal to about 10 or 20 times the number in a generation, always exceeding this number, and perhaps rarely being 100 times this number. To allow for occasional high values I take 30 as a mean. If natural selection acts by diminished fertility, the effect is equivalent."

Different Formulations of the Cost

The summation over a number of generations is more nearly exact if we change Haldane's definition slightly. Instead of measuring the cost per generation as the decrease in mean fitness relative to the favored genotype, we measure it relative to the population average. The quantity to be summed, then, is not $(w_A - \bar{w})/w_A$, but $(w_A - \bar{w})/\bar{w}$.

To be concrete, consider a population whose total size is determined mainly, if not entirely, by factors other than the genes whose frequency is changing by natural selection. The population may be growing or decreasing, although over any long period of time the value of $\bar{w}$ cannot be greatly different from one. The discussion is simpler if we think of the population as being of constant size. As the environment changes, previously deleterious genes become favorable, and vice versa. The process of bringing the newly favorable genes to high frequency requires differential mortality and fertility.

The cost is the excess in survival and fertility that the favored genotype must have in order to carry out the gene substitution at a specified rate, while the entire population stays roughly constant. Still considering the haploid model of the previous section, the cost is defined as

$$C = \sum \frac{w_A - \bar{w}}{\bar{w}} = \sum \frac{sq}{1 - sq} . \tag{57}$$

Some numerical values for this sum have been given earlier (Crow, 1968 b). The value is almost independent of s as long as this is less than about 0.1.

With slow gene frequency change we can replace summation by integration, as before, writing

$$C = \int_{0}^{\infty} \frac{sq}{1-sq} \, dt . \tag{58}$$

Substitution from Eq. (55) as before leads this time to the cancellation of $1 - sq$ as well as s. Thus

$$C = \int_{p_O}^{1} \frac{sq(1-sq)}{spq(1-sq)} \, dp$$

$$= \int_{p_O}^{1} \frac{dp}{p} = -\log_e p_O . \tag{59}$$

Notice that it is not necessary for s to be constant during the process, for the cancellation occurs each generation. As long as s is small enough for the continuous model to be applicable, the formula gives a good approximation whether s is constant or not.

If survival and fertility are measured in Malthusian parameters (Fisher, 1930, 1958) and we treat the population as changing continuously, then

$$\frac{dp}{dt} = p(m_A - \bar{m}) \tag{60}$$

exactly. The instantaneous cost in these units is $m_A - \bar{m}$, so the integral becomes

$$\int_{0}^{\infty} (m_A - \bar{m}) \, dt = \int_{p_O}^{1} \frac{dp}{p} = -\log_e p_O . \tag{61}$$

This is exact, provided the units of measurement are appropriate. The correspondence between this and the previous is, roughly, $w = e^m$.

An alternative formulation has been given by Joseph Felsenstein (personal communication). This brings out more naturally the geometric nature of the gene frequency change. Suppose that survival and fertility are measured in Malthusian parameters. If we are to substitute a gene in k generations, then the descendants of the original number must have increased in frequency from p_0 to one during that time. If the overall population size is not changing, then the average fitness, $\bar{m}$, is zero, or $e^{\bar{m}} = 1$. For the favored type to increase in k generations implies that m, the fitness of the favored type, must be such that

$$N p_0 e^{mk} = N ,\qquad (62)$$

where N is the population number, or

$$m = -n \log_e p_0 ,\qquad (63)$$

where $n = 1/k$, the number of gene substitutions per generation. In this formulation m is the cost per generation; the total cost over the k generation is $-\log_e p_0$ as before. If the population number is not constant, the cost per generation is $m - \bar{m}$.

The discrete-generation analogy to Eq. (62) is

$$N p_0 (1 + s)^k = N ,\qquad (64)$$

from which

$$s = \frac{1 - p_0^n}{p_0^n} .\qquad (65)$$

This gives the cost per generation of substituting genes at a rate of n per generation. This is more exact than Eq. (59), since it does not depend on s being small. As an extreme example, consider the case where the effect of the change in environment is such as to render the currently prevailing genotype lethal. Such a situation might be approached when a new insecticide is applied and only a minority of resistant genotypes survive. In this case the population must expand the initial $N p_0$ individuals into N in a single generation. The ratio of deaths to survivors is $N q_0 / N p_0$ $= (1 - p_0)/p_0$ which agrees with Eq. (65) for $n = 1$. In order to keep the size constant, the reproductive excess must be $(1 - p_0)/p_0$; that is to say that number of progeny of the resistant group must be $p_0 + (1 - p_0)/p_0 = 1/p_0$.

Returning to Eq. (59), the upper integration limit should not be exactly one because the gene frequency is kept from reaching this value by reverse mutation, but that makes only a trivial difference in the cost. Furthermore, since the cost depends on the logarithm of the initial frequency, it changes rather slowly with changes in the initial frequency, as the following examples show:

p_0	$C = -\ln p_0$
10^{-6}	14
10^{-4}	9
10^{-2}	4.6
10^{-1}	2.3

Most of the cost is during the early generations of the gene substitution while the favorable gene is rare and the ratio of eliminations to noneliminations is high. This means that genes that are initially common, either because of a high mutation rate or because they were only mildly disadvantageous previously, are the least costly to substitute. Moreover, it is probably just such genes with very slight effects that are most important in evolution, not only because they are initially less rare, but because a gene that is only mildly deleterious in the old environment has the best chance of being beneficial in the new.

The Cost With Diploidy and Dominance

The previous discussion has concerned haploids in order to isolate the problem from algebraic complications. I now consider briefly the diploid case. As long as s is not large, the various formulations are nearly equivalent; I shall proceed in the manner of Eq. (55) and (59).

Assume that, at the stage where enumeration occurs (e.g. immediately after fertilization), the genotypes are in Hardy-Weinberg ratios.

Genotype	AA	AA'	$A'A'$	
Frequency	p^2	$2pq$	q^2	(66)
Fitness	w	$w(1-hs)$	$w(1-s)$	

The change in frequency of gene A in one generation (Wright, 1949) is

$$\Delta p = \frac{p(w_A - \overline{w})}{\overline{w}} \approx \frac{dp}{dt} \qquad (67)$$

where w_A is the average fitness of individuals carrying gene A, weighted by the number of A genes carried, and w is the average fitness of the population.

$$\overline{w} = w[1 - s(1-p)(1-p+2hp)], \qquad (68)$$

$$w_A = w[1 - hs(1-p)]. \qquad (69)$$

By analogy with Eq. (58), the cost is

$$C = \int_0^\infty \frac{w - \overline{w}}{\overline{w}} dt . \qquad (70)$$

Substituting from Eq. (67), (68), and (69) into (70), changing the integration limits and integrating, we arrive at

$$C = \frac{-1}{1-h} \left[\log_e p_0 + h \log_e \frac{h}{1 - h - (1-2h)p_0} \right] \qquad (71)$$

when $h \neq 1$, and

$$C = -\log_e p_O + \frac{1-p_O}{p_O} \tag{72}$$

when $h = 1$. Note that again the equations are happily free of any dependence on s.

Here are some approximate representative values.

h	p_O	C	
0	10^{-4}	9	A dominant
0.5	10^{-4}	18	
0.9	10^{-3}	50	partial dominance
0.99	10^{-2}	70	
1.00	10^{-2}	100	A recessive.

A quantity analogous to the cost of a gene substitution is

$$C' = \sum \frac{w_A - \overline{w}}{\overline{w}} \approx \int_O^\infty \frac{w_A - \overline{w}}{\overline{w}} \, dt \tag{73}$$

which is easily calculated. Substituting from Eq. (67), we obtain

$$C' = \int_{p_O}^1 \frac{dp}{p} = -\log_e p_O \tag{74}$$

which is the same as for a haploid organism.

C' will be less than C because $w_A < w$; thus this is a lower limit. Furthermore, C' is completely general as regards dominance, epistasis, linkage, and mating system. Therefore it is useful in setting a lower limit on the cost of a gene substitution. Notice that in Eq. (71) when $h = 0$, $C = -\log_e p_O$. So the least cost is with a fully dominant mutant.

Thus, in a diploid system, the cost of substituting a moderately rare mutant (e.g. one previously maintained by a mutation rate of 10^{-5} to 10^{-4}) is from 10 to 100. That is to say the total number of eliminated individuals is 10 to 100 times the number of adults in any single generation.

Alternatively stated, if the average rate of gene substitution is n per generation, then the substituted genotype must have a fitness that exceeds that of the average of the population by nC. Taking C to be 30, as Haldane suggests, a rate of evolution in which the substitution rate, n, is 1/300 (one substitution, on the average, every 300 generations) would require a reproductive and survival excess of 10% of the favored genotype beyond that required to maintain the population size, or its characteristic growth rate. This assumes, of course, that the entire excess is used for selection of this trait and no allowance is made for selection on other loci, elimination of harmful mutants, and random effects.

In this discussion I have assumed that the environment changes in discrete steps so that a mutant changes from harmful to beneficial in a single generation. If a more realistic model, one in which the environment changes gradually so that the mutant changes slowly from harmful to beneficial, there is only a small difference in the cost and the general conclusion remains essentially unchanged (Kimura, 1967). For example, if s changes from -0.01 to 0 in 10,000 generations, the total cost is reduced to about 2/3 of what it would be if the change were instantaneous.

Is the Haldane Principle the Limiting Factor in Determining Evolutionary Rates?

It is clear that Haldane regarded this principle as at least a partial explanation for the observed slowness of evolution of morphological traits. He said, "I am convinced that quantitative arguments of the kind here put forward should play a part in all future discussions of evolution". Needless to say, he did not regard his formulation as more than a beginning, for he also said, "I am aware that my conclusions will probably need drastic revision".

The evolution of mammalian structural proteins has been at a rate averaging about 1.6×10^{-9} substitutions per codon per year (King and Jukes, 1969). I shall regard this as roughly 10^{-6} per cistron per generation. If the genes are partially dominant and the mutant phenotype had an initial frequency of about 10^{-4}, the cost is 10–50 per substitution. The cost per generation is then 10–50×10^{-6}. With 10^4 loci evolving at this rate the amount of excess fitness of the favored type would need to be 10 to 50%. This involves the necessary assumption that, if several genes are evolving simultaneously, selection acts on each independently.

On the other hand if the number of genes is much larger—as one might naively assume by dividing the amount of DNA in a mammalian nucleus by the size of a cistron – the cost of substituting this many independent loci becomes enormous. This has been used by Kimura (1968) as an argument for the belief that much evolutionary change in DNA composition is nonadaptive.

It is apparent from Eq. (59), (71), and (72) that the cost is determined very strongly by the initial frequency of the gene. One way in which the cost may be reduced is if the genes start at higher frequencies. Here are some representative values of the total cost required for a specified gene frequency change. These are for a gene with no dominance ($h = 1/2$ in Eq. (71)).

Gene Frequency Change		Total Cost
From	To	
0.0001	0.9999	18.42
0.001	0.999	13.81
0.01	0.99	9.19
0.1	0.9	4.39
0.2	0.8	2.77
0.3	0.7	1.69
0.4	0.6	0.81

So, if the same or equivalent phenotypic change can be accomplished by changing five genes from 0.4 to 0.6 as from changing one from 0.0001 to 0.9999, the cost is 5 (0.81) = 4.05 as compared to 18.42. If there are independent genes with equivalent effects, natural selection will change most rapidly those with intermediate frequencies, so there is a gain in time as well as in total cost. Sewall Wright has repeatedly suggested that adaptive changes may depend more on shifting balances among genes with moderate frequencies than by substitution of rare genotypes. If small frequency changes at several loci change the average phenotype, or fitness, by the same amount as one large frequency change, as might well be the case, Wright's model requires less reproductive excess.

The remarkable property that the cost does not depend on s when s is small implies that all gene substitutions involving the same dominance and the same initial frequency have the same cost. This suggests that if two mutants were linked they could both be substituted for half the cost that would be required if they were selected independently. A difficulty, however, is that the double mutant is more rare. In simple cases these exactly compensate. If the single mutant has frequency p_o and the double mutant p_o^2, then the cost for the linked double mutant is the same as for the two independent mutants, since $-\log_e p_o^2 = -2\log_e p_o$.

It would also appear that, since the cost is independent of s, twice as much cost would be involved in substituting two genes, each contributing one gram to the weight of a horse would be twice as costly as a single gene contributing two grams. But this too would be mitigated by the likelihood that the gene with greater effect is the more rare initially; it would have had a larger deleterious effect in the earlier environment before an increase in size was favored.

Probably of greater significance in application of the Haldane cost principle is uncertainty about the importance of gene interaction. The Haldane argument assumes that the genes are independent – that is, that the fitnesses are multiplicative if measured by w or additive if measured in the Malthusian parameter, m. Sved (1968a) and Maynard Smith (1968) have called attention to the great difference that ensues when there is truncation selection. The arguments are essentially the same as for the segregation load. If the genes contribute roughly additively to some underlying variable and selection acts to truncate the distribution, then the effect is the equivalent of strong epistasis. The consequences are greatly different. The number of loci at which selection can act simultaneously with the same total intensity of selection is many times greater.

Whether selection acts mainly on independent genes or in a threshold manner is not clear at present – to me, at least. Clearly some traits are quite independent of one another, as for example two different genes for specific resistance to differently-acting poisons or diseases. On the other hand, if the environment changes in such a way as to make increased size adaptive, presumably a number of genes would be selected simultaneously for the same trait. Selection is not likely to act in a strict threshold manner, but it might approach this nearly enough to reduce the cost considerably below what would be expected with independence.

Clearly there is a limitation somewhere, but our present ignorance as to the nature of fitness interactions of evolutionary important genes limits the utility of the principle at present.

Population Variability as a Limiting Factor

It is usually easier experimentally to measure the amount of variability in fitness than it is to measure the cost or the load. From the correlation between relatives the additive portion of the variability can also be measured, at least in principle.

One measure of variability is the coefficient of variation. With the haploid model Eq. (51) the variance in fitness is $pq(ws)^2$, so the coefficient of variation is $ws\sqrt{pq}/\bar{w}$, where $\bar{w}$ is the mean fitness $= w(1 - sq)$. The cumulative variability, summed over the entire period of time during which the A gene is being substituted for A' is

$$CV = \sum \frac{ws\sqrt{pq}}{\bar{w}} \, \Delta t, \tag{75}$$

where Δt is the duration of one generation.

The coefficient of variation is used instead of the standard deviation to allow for fluctuations in population size. I am assuming as before that the population size is determined mainly by environmental and genetic factors other than those genes which are in the process of substitution. Of course, if $\bar{w}$ is regarded as a measure of absolute fitness, its long time average value cannot differ greatly from one.

The change in gene frequency in one generation is

$$\Delta p = \frac{wspq}{\bar{w}} \, \Delta t \, . \tag{76}$$

Substituting from Eq. (76) into Eq. (75) gives

$$CV = \sum (pq)^{-1/2} \Delta p \, , \tag{77}$$

or, if the change in gene frequency can be regarded as essentially continuous over the (unknown) n generations required for the gene frequency change,

$$CV = \int_{p_0}^{p_n} (pq)^{-1/2} dp = \sin^{-1}(2p_n - 1) - \sin^{-1}(2p_0 - 1) \tag{78}$$

when $p_0 = 0$ and $p_n = 1$, $CV = \pi = 3.14$. The value, in contrast to Haldane's cost, is not strongly dependent on p_0. For example, if $p_0 = 1 - p_n = 0.001$, the value is 3.02 whereas with $p_0 = 1 - p_n = 0.01$ it is 2.74.

Notice that ws and $\bar{w}$ cancel within each generation, so there is no necessity to assume that either is constant.

For a diploid population we use the same model as before.

Genotype	AA	AA'	$A'A'$
Frequency	p^2	$2pq$	q^2
Fitness	w	$w(1-hs)$	$w(1-s)$

Consider first that the favored gene is recessive, that is, $h = 1$. The variance in fitness is $p^2(1-p^2)w^2s^2$ so the cumulative coefficient of variation is

$$CV = \sum ws[p^2(1-p^2)]^{1/2} \Delta t/\bar{w}, \tag{79}$$

and the gene frequency change is $\Delta p = wsp^2q\Delta t/\bar{w}$. Substituting this into Eq. (79) and treating the process as continuous, as before, gives

$$CV = \int_{p_0}^{p_n} [(1+p)/(1-p)]^{1/2} p \, dp \, . \tag{80}$$

This can be readily integrated by using the transformation $x^2 = (1+p)/(1-p)$, yielding

$$C = 2[\tan^{-1}x_n - \tan^{-1}x_0] + \log_e \left[\frac{(x_n - 1)(x_0 + 1)}{(x_n + 1)(x_0 - 1)}\right]. \tag{81}$$

When $p_0 = 0.01$, $p_n = 1$, $x_0 = 1.01$, $x_n = \infty$, the value of C is 6.9. When $p_0 = 0.001$, it is 9.2.

If there is no dominance ($h = 1/2$), the situation is almost the same as with a haploid. The variance is $p(1-p)s^2w^2/2$, $\Delta p = swp(1-p)/2\bar{w}$, and

$$\begin{aligned} C &= \sqrt{2}[\sin^{-1}(2p_n - 1) - \sin^{-1}(2p_0 - 1)] \\ &= \sqrt{2}\pi = 4.44, \quad \text{when} \quad p_0 = 1 - p_n = 0. \end{aligned} \tag{82}$$

The situation is symmetrical, so the formula for a dominant gene is the same as for a recessive with p and q exchanged.

The cumulative coefficient of variation during the process of gene substitution is therefore independent of the intensity of selection. Its value ranges from about 4 for an initially rare gene with no dominance to about 10 for a recessive.

Unfortunately, I fear that this measure, although simple and elegant and having the desirable property of being independent of s, is nearly useless. The reason is the elementary fact that the coefficient of variation is not additive. Thus this principle cannot be used if several genes are evolving simultaneously as in a Mendelian population. It may be of use in thinking of evolution in an asexual population where substitutions occur in tandem.

So I ruefully give up the useful property of having s drop out in order to have additivity. The natural measure of variability of fitness is $V_w/\bar{w}^2$, where V_w is the variance of the w's. I have called this the index of opportunity for selection (Crow, 1958, 1962). If fitness is measured in Malthusian parameters, then the denominator term is not needed. The cumulative variance cost

$$VC = \sum(V_w/\bar{w}^2)\Delta t = \sum \frac{w^2 s^2 pq}{\bar{w}^2}\Delta t. \tag{83}$$

Substituting for Δt from Eq. (76) and treating the process as continuous, as before, yields the variance cost

$$VC = \int_{p_0}^{p_l} \frac{ws}{\bar{w}}\,dp = s'(p_l - p_0) \tag{84}$$

which is equal to s' when $p_O = 1 - p_l = 0$. In this formulation $s' = ws/\bar{w}$ is the increase in fitness brought about by one gene substitution and is measured as a fraction of the average fitness at the time. I have treated $\bar{w}$ as if it were a constant on the assumption that the change in fitness resulting from one gene substitution is very small relative to $\bar{w}$. In fact, I suspect that s' is as likely to be constant as s is.

The simply stated result is: The total variance cost for selection for all the generations involved in a gene substitution is equal to the relative increase in fitness brought about by that substitution.

For the diploid model, in the absence of dominance ($h = 1/2$) the result is the same as for the haploid case. With a recessive gene ($h = 1$) the variance is $p^2(1 - p^2)w^2s^2$ and $\Delta p = wsp^2(1 - p)/\bar{w}$ so the total is

$$VC = \int_{p_O}^{p_l} \left[\frac{p^2(1 - p^2)w^2s^2}{\bar{w}^2} \right] \left[\frac{\bar{w}}{wsp^2(1 - p)} \right] dp$$

$$= \int_{p_O}^{p_l} s'(1 + p)\, dp = s'[p_l(2 + p_l) - p_O(2 + p_O)]/2 \qquad (85)$$

$$= 3s'/2 \quad \text{when} \quad p_O = 1 - p_l = 0,$$

and

$$s' = ws/\bar{w}.$$

The situation is symmetrical, so the variance cost for substitution of a dominant gene is the same as for a recessive.

The solution can be stated much more generally. The rate of change in fitness is given very nearly by

$$\frac{\Delta\bar{w}}{\bar{w}} = \frac{V_g}{\bar{w}^2} \Delta t \qquad (86)$$

where V_g is the *genic* or *additive* variance in fitness. This is Wright's discrete generation analog of Fisher's "Fundamental Theorem of Natural Selection".

Substituting Eq. (86) into Eq. (83) and considering only the additive component of V_w (that is, using V_g instead), I obtain

$$VC = \int_{w_O}^{w_n} \frac{1}{\bar{w}}\, d\bar{w} = \log(w_n/w_O) \approx s' \qquad (87)$$

where, as before, s' is the *relative* increase in fitness accomplished by one gene substitution.

This can be stated more simply with a continuous model with fitness measured in Malthusian parameters.

Letting $\bar{m}$ be the average fitness of the population, measured in Malthusian parameters, and V_g the genic variance in fitness the Fundamental Theorem states that

$$d\bar{m} = V_g dt .$$ (88)

Integrating both sides of Eq. (88)

$$\Delta m = m_l - m_0 = \int_{t_0}^{t_l} V_g dt$$ (89)

where Δm is the increase in fitness during the time interval. If V_g is approximately constant,

$$\Delta m = V_g(t_l - t_0) = nV_g$$ (90)

where n is the number of generations involved.

This gives a way of relating the change in fitness associated with a gene substitution to the genic variance and the rate of evolution. For example, if a gene substitution produces a fitness change of Δm, the average number of generations required for a fitness change equivalent to a gene substitution is

$$n = \frac{\Delta m}{V_g} .$$ (91)

Concluding Remarks

There is a large and growing literature on loads and costs, only part of which has been referred to here. I should like to mention a few additional subjects and give references to discussions of them.

Attempts to determine the relative magnitude of the mutation, segregation, and other loads, as expressed in natural populations, have not met with appreciable success in any organism. Arguments can be made for a large mutational component in the hidden load that is brought out by inbreeding, but this too is not clear at present. Some pertinent references are: Dobzhansky, 1957; Crow, 1958, 1963; Levene, 1963; Haldane and Jayakar, 1965; Nei, 1965, 1968b.

The theory of genetic loads, combined with segregation analysis, has been particularly useful in analyzing the inheritance of human characteristics. This field has been developed mainly by Morton (Morton, 1965; Morton and Chung, 1959; Dewey, et al., 1965; Chung, et al., 1959).

Finally, for a general review of population genetics, including some perceptive remarks on genetic loads, see Lewontin (1967).

Summary

The word "load" was introduced by H. J. Muller in an attempt to assess the impact of mutation on human welfare. Muller independently rediscovered the powerful principle, earlier formulated by Haldane, that the impact of mutation on fitness of the population is equal approximately to the mutation rate and is independent of the effect of the individual mutant on fitness. In Muller's thinking the load was clearly a burden, which could be measured only abstractly in terms of fitness, but which was felt in terms of death, sterility, illness, pain, and frustration. The word has been extended by Dobzhansky to include the effects of deleterious homozygotes maintained in balanced polymorphism. Although it might have been better to choose a word with fewer emotional overtones, it has now become too widely used to change, I think. But it should be emphasized that the load is not necessarily bad; it is often the *sine qua non* for evolution.

The mutation load, defined as the amount of fitness decrease relative to a population that has reached equilibrium with a mutation rate zero, is equal to the total gametic mutation rate if the mutants are completely recessive, but multiplied by a factor of two if the mutants have any appreciable heterozygous effect, as is usually the case. Data from Drosophila on the amount of epistasis of viability mutants indicate that the synergistic interaction of deleterious mutants is such as to reduce the mutation load by roughly one-half. So the overall mutation load is estimated as being somewhere near the total gametic mutation rate.

The principle holds for mutants whose harmful effects are large relative to the mutation rate and becomes less as the mutations become more nearly neutral. For completely neutral mutations the load becomes zero, of course, and the probability of a given locus being heterozygous in a particular individual if there are many potential alleles is $4N_e u/(4N_e u + 1)$, where u is the mutation rate and N_e is the effective population number.

The segregation load, defined as the reduction in average fitness compared with a population without segregation (i.e. asexual), is ordinarily much larger per locus if the selection coefficient is large relative to the mutation rate. The most useful formula is $L \geq ps$, where L is the segregation load, p is the frequency of one allele, and s is its homozygous effect on fitness. This permits the segregation load to be estimated from knowledge of only one allele regardless of the number, known or unknown, that are maintained in the population. Ways in which the segregation load may be reduced, especially by truncation selection and linkage, are discussed.

The genetic load due to meiotic drive is shown to differ from the mutation load in that selection against the heterozygote decreases rather than increases the load of a recessive mutant.

Other loads due to recombination, heterogeneous environment, maternal-fetal incompatibility, and random drift are briefly mentioned.

The Haldane principle of the cost of natural selection is developed in several ways, first as originally formulated by Haldane, and then with several modifications. All agree that with slow selection the total amount of selection that is needed for a gene substitution is $-\log_e p_0$, where p_0 is the initial frequency of the mutant. This is for a haploid population; formulae for the larger amount with diploidy and dominance are also given.

Whether the Haldane cost principle is the main limitation on the rate of evolution is not known, but it is important in being the first quantitative formulation of the problem of how much differential mortality and fertility is required for natural selection at a specified rate. An alternative approach, based on variances rather than means, is also presented.

References

Brues, Alice M.: The cost of evolution vs. the cost of not evolving. Evolution **18**, 379—383 (1964).

— Genetic load and its varieties. Science **164**, 1130—1136 (1969).

Chung, C. S.: Relative genetic load due to lethal and detrimental genes in irradiated populations of *Drosophila melanogaster*. Genetics **47**, 1489—1504 (1962).

—, O. W. Robinson, and N. E. Morton: A note of deaf mutism. Ann. Hum. Genet. **23**, 357—366 (1959).

Crow, J. F.: Some possibilities for measuring selection intensities in man. Human Biol. **30**, 1—13 (1958).

— Population genetics. Amer. J. Hum. Genet. **13**, 137—150 (1961).

— Population genetics: Selection. Methodology in human genetics. Ed. by W. Burdette. San Francisco: Holden-Day 1962.

— The concept of genetic load: A reply. Amer. J. Hum. Genet. **15**, 310—315 (1963).

— Problems of ascertainment in the analysis of family data. Genetics and the epidemiology of chronic diseases. Publ. **1163**, U.S. Public Health Service, pp. 23—44 (1965).

— Some analyses of hidden variability in Drosophila populations. Population biology and evolution, pp. 71—86. Ed. by R. Lewontin. Syracuse: University Press 1968a.

— The cost of evolution and genetic loads. In: Haldane and modern biology, pp. 165—178. Ed. by K. Dronamraju. Baltimore: Johns Hopkins Press 1968b.

— Molecular genetics and population genetics. Proc. Int. Cong. Genet. **3**, 105—113 (1969).

—, and N. E. Morton: Measurement of gene frequency drift in small populations. Evolution **9**, 202—214 (1955).

—, — The genetic load due to mother-child incompatibility. Amer. Natur. **94**, 413—419 (1960).

Dewey, W. J., I. Barrai, N. E. Morton, and M. P. Mi. Recessive genes for severe mental defect. Amer. J. Hum. Genet. **17**, 237—256 (1965).

Dobzhansky, Th.: A review of some fundamental concepts and problems of population genetics. Cold Spr. Harb. Symp. Quant. Biol. **20**, 1—15 (1955).

— Genetic loads in natural populations. Science **126**, 191—194 (1957).

Feller, W.: On the influence of natural selection on population size. Proc. Natl. Acad. Sci. (Wash.) **55**, 733—738 (1966).

— On fitness and the cost of natural selection. Genet. Res. **9**, 1—15 (1967).

Fisher, R. A.: The Genetical Theory of Natural Selection. Oxford: Clarendon Press 1930; Rev. ed.: New York: Dover Press 1958.

Greenberg, Rayla, and J. F. Crow: A comparison of the effect of lethal and detrimental chromosomes from Drosophila populations. Genetics **45**, 1153—1168 (1960).

Haldane, J. B. S.: A mathematical theory of natural and artificial selection. Part I. Trans. Cambridge Phil. Soc. **23**, 19—41; Part II. Biol. Proc. Cambridge Phil. Soc. **1**, 158—163; Part III. Proc. Cambridge Phil. Soc. **23**, 363—372; Part IV. Proc. Cambridge Phil. Soc. **23**, 607—615; Part V. Proc. Cambridge Phil. Soc. **23**, 838—844; Part VI. Proc. Cambridge Phil. Soc. **26**, 220—230; Part VII. Proc. Cambridge Phil. Soc. **27**, 131—136; Part VIII. Proc. Cambridge Phil. Soc. **27**, 137—142; Part IX. Proc. Cambridge Phil. Soc. **28**, 244—248 (1924—1932).

— The causes of evolution. New York: Harper and Brothers 1932; Reprinted, Ithaca, N. Y.: Cornell University Press 1966.

— The effect of variation on fitness. Amer. Natur. **71**, 337—349 (1937).

— The cost of natural selection. J. Genet. **55**, 511—524 (1957a).

— The conditions for co-adaptation in polymorphism for inversions. J. Genet. **55**, 218—225 (1957b).

— More precise expressions for the cost of natural selection. J. Genet. **57**, 351—360 (1960).

—, and S. D. Jayakar. The nature of human genetic loads. J. Genet. **59**, 53—59 (1965).

Kimura, M.: Rules for testing the stability of a selective polymorphism. Proc. Natl. Acad. Sci. (Wash.) **42**, 336—340 (1956a).

— A model of a genetic system which leads to closer linkage by natural selection. Evolution **10**, 278—287 (1956b).

— Optimum mutation rate and degree of dominance as determined by the principle of minimum genetic load. J. Genet. **57**, 21—34 (1960).

— Some calculations on the mutation load. Jap. J. Genet. (Suppl.) **36**, 179—190 (1961).

— On the evolutionary adjustment of spontaneous mutation rates. Genet. Res. **9**, 23—34 (1967).

— Evolutionary rate at the molecular level. Nature (Lond.) **217**, 624—626 (1968).

—, and J. F. Crow: The measurement of effective population number. Evolution **17**, 279—288 (1963).

—, — The number of alleles that can be maintained in a finite population. Genetics **49**, 725—738 (1964).

—, and T. Maruyama: The mutational load with epistatic gene interactions in fitness. Genetics **54**, 1337—1351 (1966).

— —, and J. F. Crow: The mutation load in small populations. Genetics **48**, 1303—1312 (1963).

King, J. L.: The gene interaction component of the genetic load. Genetics **53**, 403—413 (1966).

— Continuously distributed factors affecting fitness. Genetics **55**, 483—492 (1967).

—, and T. H. Jukes: Non-Darwinian evolution. Science **164**, 788—798 (1969).

Kojima, K., and K. M. Yarbrough: Frequency-dependent selection at the esterase-6 locus in *Drosophila melanogaster*. Proc. Natl. Acad. Sci. (Wash.) **57**, 645—649 (1967).

Levene, H.: Genetic equilibrium when more than one ecological niche is available. Amer. Natur. **87**, 311—313 (1953).

— Inbred genetic loads and the determination of population structure. Proc. Natl. Acad. Sci. (Wash.) **50**, 587—592 (1963).

Lewontin, R. C.: Population genetics. Ann. Rev. Genet. **1**, 37—70 (1967).

—, and J. L. Hubby: A molecular approach to the study of genic heterozygosity in natural populations. II. Amount of variation and degree of heterozygosity in natural populations of *Drosophila pseudoobscura*. Genetics **54**, 595—609 (1966).

Li, C. C.: Decrease of population fitness upon inbreeding. Proc. Natl. Acad. Sci. (Wash.) **49**, 439—445 (1963a).

— Genetic aspects of consanguinity. Amer. J. Med. **34**, 702—714.

— The way the load ratio works. Amer. J. Hum. Genet. **15**, 316—321 (1963c).

— Genetic equilibrium under selection. Biometrics **23**, 397—484 (1967).

Mandel, S. P. H.: The stability of a multiple allelic system. Heredity **13**, 289—302 (1959).

Maynard Smith, J.: "Haldane's Dilemma" and the rate of evolution. Nature (Lond.) **219**, 1114—1116 (1968).

Milkman, R. D.: Heterosis as a major cause of heterozygosity in nature. Genetics **55**, 493—495 (1967).

Morton, N. E.: The mutational load due to detrimental genes in man. Amer. J. Hum. Genet. **12**, 348—364 (1960).

— Models and evidence in human population genetics. Proc. XI Int. Cong. Genet., pp. 935—951, 1965.

— and C. S. Chung: Formal genetics of muscular dystrophy. Amer. J. Hum. Genet. **11**, 360—379 (1959).

— —, and L. D. Friedman: Relation between homozygous viability and average dominance in *Drosophila melanogaster*. Genetics **60**, 601—614 (1968).

—, J. F. Crow, and H. J. Muller: An estimate of the mutational damage in man from data on consanguineous marriages. Proc. Natl. Acad. Sci. (Wash.) **42**, 855—863 (1956).

Mukai, T.: The genetic structure of natural populations of *Drosophila melanogaster*. VII. Synergistic interaction of spontaneous mutant polygenes controlling viability. Genetics **61**, 149—161 (1969).

Muller, H. J.: Our load of mutations. Amer. J. Hum. Gent. **2**, 111—176 (1950).

Nei, M.: Effect of linkage on the genetic load manifested under inbreeding. Genetics **51**, 679—688 (1965).

— Evolutionary change of linkage intensity. Nature (Lond.) **218**, 1160—1161 (1968a).

— On the genetic load manifested under inbreeding. (personal communication) (1968b). (1968b).

Prout, T.: Sufficient conditions for multiple niche polymorphism. Amer. Natur. **102**, 493—496 (1968).

Sandler, L., and E. Novitski: Meiotic drive as an evolutionary force. Amer. Natur. **91**, 105—110 (1957).

Sanghvi, L. D.: The concept of genetic load: A critique. Amer. J. Hum. Genet. **15**, 298—309 (1963).

Sved, J. A.: Possible rates of gene substitution in evolution. Amer. Natur. **102**, 283—293 (1968a).

— The stability of linked systems of loci with a small population size. Genetics **59**, 543—563 (1968b).

—, T. E. Reed, and W. F. Bodmer: The number of balanced polymorphisms that can be maintained in a natural population. Genetics **55**, 469—481 (1967).

Temin, Rayla G.: Homozygous viability and fertility loads in *Drosophila melanogaster.* Genetics **53**, 27—46 (1966).

—, Helen U. Meyer, P. S. Dawson, and J. F. Crow: The influence of epistasis on homozygous viability depression in *Drosophila melanogaster.* Genetics **61**, 497—519 (1969).

Van Valen, L.: Haldane's dilemma, evolutionary rates and heterosis. Amer. Natur. **97**, 185—190 (1963).

Wallace, B.: Topics in Population Genetics. New York: W. W. Norton Comp. (1968).

Wills, C.: The mutational load in two natural populations of *Drosophila pseudocbscura.* Genetics **53**, 281—294 (1966).

—, J. Crenshaw, and J. Vitale: A computer model allowing maintenance of large amounts genetic variability in Mendelian populations. I. Assumptions and results for large populations. Genetics (in press) (1969).

Wright, S.: Evolution in Mendelian populations. Genetics **16**, 97 —159 (1931).

— Adaptation and selection. In: Genetics, paleontology, and evolution, pp. 365—389. Ed. by G. L. Jepson, G. G. Simpson, and E. Mayr. Princeton, N. J.: Princeton Univ. Press 1949.

— The genetical structure of populations. Ann. Eugen. **15**, 323—354 (1951).

Stochastic Processes in Population Genetics, with Special Reference to Distribution of Gene Frequencies and Probability of Gene Fixation*

M. KIMURA

1. Introduction

The fundamental quantity which is used in population genetics to describe the genetic composition of a Mendelian population (i.e. reproductive community) is the gene frequency, or the proportion of a given gene in the population. Thus, a population is characterized by a set of gene frequencies.

Adequacy of using gene frequencies for describing the genetic composition of a Mendelian population stems from the fact that each gene is a self-reproducing entity, and therefore, its frequency changes almost continuously with time as long as the population is reasonably large. It is sometimes remarked, however, that genotypic frequencies would give a more realistic description of the genetic composition of a population. Shortcomings of such a measure will become evident if we consider a natural population as an aggregate of genotypes. First, each genotype does not necessarily reproduce its own kind under sexual reproduction. In other words, each genotype does not have continuity as does each gene. Secondly, the number of possible genotypes greatly exceeds the total number of individuals in a population. As a result, each individual is quite likely to represent a unique genotype in the entire history of the species (except for monozygotic twins). Thus, as an aggregate of genotypes, a population is an enormously complicated system. For example, with 100 segregating loci each with a pair of alleles, the possible number of genotypes is 3^{100} or roughly 5.2×10^{47}. This may be compared with 6.6×10^{10} an estimated cumulative total of the number of individuals in the hominid line from its inception a million years ago down to the invention of agriculture (cf. Deevey, 1960). Actually, for man, the number of possible genotypes is about $10^{4 \times 10^9}$ rather than 10^{30}, because a haploid set of human chromosomes comprises some 4×10^9 nzcleotide

* Contribution No. 697 from the National Institute of Genetics, Misima, Shizuoka-ken, Japan.

pairs (Muller, 1958), each of which may take one of the four states, $A - T$, $T - A$, $G - C$ and $C - G$.

In the mathematical theory of population genetics founded by the great trio, Fisher, Haldane, and Wright, one of the central problems is to investigate the change of gene frequencies by mutation, natural selection, migration and random sampling of gametes. The sampling of gametes introduces an element of chance into the change of gene frequencies so that the process must be treated as a stochastic process. (Here the term "stochastic process" refers to the mathematical formulation of a chance event that proceeds with time.) For the historical development of this subject together with the works that follow them, the readers may refer to Kimura (1964).

A mathematical approach which has emerged through those studies and has proved most powerful is called "diffusion models". In this approach, the process of change in gene frequency is treated as a continuous stochastic process. This is essentially an approximation that takes advantage of the fact, as mentioned earlier, that for any reasonably large population the gene frequencies change almost continuously with time.

Consideration of stochastic processes in population genetics leads us to the problems of gene frequency distribution at equilibrium and the probability of gene fixation. These are of special importance for understanding the genetic structure of Mendelian populations. Gene frequency distributions have been studied extensively by Wright since 1931 and much of what we know about them is due to his studies. This problem, once considered highly academic, is now becoming increasingly important to treat concrete situations. Thus, using Wright's concept of the gene frequency distribution together with the mathematical work of Miller (1962), Robertson (1962) has shown that the genetic composition of laboratory populations consisting of a few hundred individuals may be very different from the wild populations of much larger size. Kimura, Maruyama, and Crow (1963) have shown, using Wright's distribution formula, that the mutational load in a small population is considerably larger than in a large population. Furthermore, for a wide range of population sizes a mutant that is slightly harmful is more damaging to the fitness of the population than a mutant with much greater harmful effect. Also, Nei and Imaizumi (1966) extended these works to investigate the relationship between genetic variation and population size for three different cases: reversible mutation, balance between mutation and selection, and overdominance. Kimura and Crow (1964) studied the number of alleles maintained in a finite population. They investigated the cases of selectively neutral isoalleles and mutually heterotic alleles, assuming that each mutant represents an allelic state not preexisting in the population. The last assumption was based on the consideration that a gene is composed of hundreds or thousands of nucleotide pairs each of which is capable of base substitutions and therefore the number of possible allelic states must be astronomical. Their investigation has shown that overdominance is rather inefficient as a factor for maintaining a large number of alleles in a finite population. In other words, to maintain a large number of alleles, a high rate of production of new mutation is required. A fuller treatment of the problem for the case of neutral mutations was recently presented by Kimura (1968b).

In the above two papers, a single locus was considered in which an infinite sequence of alleles may be produced by mutation. On the other hand, Kimura (1969) studied the number of heterozygous nucleotide sites per individual and related quantities that describe the statistical properties of the mutant frequency distribution attained under steady flux of molecular mutations in a finite population. It was assumed that a very large number of nucleotide sites are available for mutation and that whenever a mutant appears it represents a mutation at a different site. This study extended the work of Wright (1938, 1942, and 1945) on the distribution of gene frequencies under irreversible mutation. It is significant that the concept of gene frequency distribution is applicable not only to individual gene but even such small genetic units as nucleotide pairs.

The second problem, the probability by which a mutant gene becomes established (fixed) in a population, is of special importance for the study of evolutionary genetics. Pioneering works on this problem by Fisher (1922, 1930) and Haldane (1927) were followed by more general treatment by Kimura (1957) who introduced the method of the Kolmogorov backward equation to solve the problem. That this problem too has bearing to more practical situations is evidenced by the outstanding work of Robertson (1960) who developed a theory of limit in artificial selection based on the concept of gene fixation in a small population.

In a finite population, a mutant gene becomes either fixed or lost from the population within a finite length of time. The formulation for predicting the number of generations until fixation (excluding the cases of loss) has been obtained by Kimura and Ohta (1969).

2. The Diffusion Equation Method

The diffusion models in population genetics are founded on the assumption that the process of change in gene frequencies may be treated with sufficient accuracy by the use of the diffusion equations, particularly, the Kolmogorov forward and backward equations (Kimura, 1964).

Let us consider a locus in which a pair of alleles A_1 and A_2 are segregating. We will denote by $\phi(p, x; t)$ the transition probability density that the frequency (i.e., relative proportion in the population) of A_1 lies between x and $x + dx$ at time t, given that it is p at the start (i.e., at time $t = 0$). Here, time t may most conveniently be measured with one generation as unit length of time. Then ϕ satisfies the following Kolmogorov forward equation

$$
\frac{\partial \phi(p, x; t)}{\partial t} = \frac{1}{2} \frac{\partial^2}{\partial x^2} \left\{ V_{\delta x} \phi(p, x; t) \right\}
$$
$$
- \frac{\partial}{\partial x} \left\{ M_{\delta x} \phi(p, x; t) \right\}, \tag{1}
$$

where $M_{\delta x}$ and $V_{\delta x}$ are respectively the mean and the variance of δx, the amount of change in gene frequency x per generation. This equation is also called the Fokker-Planck equation. Here, both $M_{\delta x}$ and $V_{\delta x}$ may depend on x and t. In the above formulation, we considered the process of change in the forward direction, assuming that x is a random variable while p is fixed. In this context, we will often omit letter p and write $\phi(x, t)$ for $\phi(p, x; t)$ in the following treatments.

In applying the above equation to obtain the gene frequency distribution at statistical equilibrium, it is often useful to note that

$$
P(x, t) = - \frac{1}{2} \frac{\partial}{\partial x} \left\{ V_{\delta x} \phi(x, t) \right\} + M_{\delta x} \phi(x, t), \tag{2}
$$

which constitutes the right hand side of Eq. (1) as $- \partial P(x, t)/\partial x$, represents the probability flux, that is, the net probability mass that flows across the point x per generation.

Next, let us consider the process retrospectively and regard x as fixed and p (initial frequency) a variable. Then ϕ satisfies the following Kolmogorov backward equation

$$
\frac{\partial \phi(p, x; t)}{\partial t} = \frac{1}{2} V_{\delta p} \frac{\partial^2}{\partial p^2} \phi(p, x; t)
$$
$$
+ M_{\delta p} \frac{\partial}{\partial p} \phi(p, x; t). \tag{3}
$$

This equation is actually a time homogeneous form of the Kolmogorov backward equation and is valid only when both $M_{\delta x}$ and $V_{\delta x}$ do not depend on time parameter t. Except for such a restriction the equation is quite general.

For the derivation of the above two diffusion Eq. (1) and (3), the reader may refer to Kimura (1964) and also Crow and Kimura (1970).

One of the very important uses of the above backward equation is its application to the problem of gene fixation. Namely, if we take $x = 1$, then ϕ in Eq. (3) may be interpreted as the probability that gene A_1 becomes fixed (established) in the population by the t-th generation, given that it is p at the initial generation ($t = 0$). Thus, if we denote this probability by $u(p, t)$, we have

$$\frac{\partial u(p, t)}{\partial t} = \frac{V_{\delta p}}{2} \frac{\partial^2 u(p, t)}{\partial p^2} + M_{\delta p} \frac{\partial u(p, t)}{\partial p}. \tag{4}$$

This type of equation may be extended to treat the probability of joint fixation of mutant genes in two or more loci. For example, suppose p is the frequency of mutant gene A_1 in the first locus and q is the frequency of mutant gene B_1 in the second locus. Then the probability $u(p, q; t)$ that both A_1 and B_1 become fixed in the population by the t th generation satisfies the equation

$$\begin{aligned}
\frac{\partial u(p, q; t)}{\partial t} &= \frac{1}{2} V_{\delta p} \frac{\partial^2 u(p, q; t)}{\partial p^2} + W_{\delta p \delta q} \frac{\partial^2 u(p, q; t)}{\partial p \partial q} \\
&+ \frac{1}{2} V_{\delta q} \frac{\partial^2 u(p, q; t)}{\partial q^2} + M_{\delta p} \frac{\partial u(p, q; t)}{\partial p} \\
&+ M_{\delta q} \frac{\partial u(p, q; t)}{\partial q},
\end{aligned} \tag{5}$$

where δp and δq respectively stand for the amount of change of p and q in one generation and, furthermore, $M_{\delta p}$ and $M_{\delta q}$ denote their means, $V_{\delta p}$ and $V_{\delta q}$ their variances, and $W_{\delta p \delta q}$ their covariance.

3. Random Genetic Drift Due to Small Population Number

As a concrete example of stochastic processes in population genetics, let us study the process of random genetic drift in the narrow sense. Assume that mutation, migration, and selection are absent, and the change of gene frequencies from generation to generation is caused by random sampling of gametes in reproduction.

Consider a pair of alleles A_1 and A_2 segregating with respective frequencies x and $1 - x$ in a random mating population consisting of N monoecious individuals. We assume that the mode of reproduction is such that N male and N female gametes are drawn as a random sample from the population to form the next generation. Then, the mean and the variance in the change of gene frequency are

$$M_{\delta x} = 0 \tag{6}$$

and

$$V_{\delta x} = x(1 - x)/(2N). \tag{7}$$

The first relation Eq. (6) is an expression of the fact that the mean gene frequency does not change with generations because of no mutation, no migration, and no selection (i.e., no systematic evolutionary pressures).

The second relation is derived by the consideration that the binomial variance of the number (rather than proportion) of A_1 carrying gametes among random samples of gametes of size $2N$ is $2Nx(1-x)$. If mating is not random, or the distribution of number of offspring does not follow a Poisson distribution, the "variance" effective number N_e (cf. Kimura and Crow, 1963) may be substituted for N in Eq. (7).

In the following treatment, for simplicity's sake, we will assume that the effective number N_e is equal to the actual number. Thus with the mean and variance given by Eq. (6) and (7), the forward Eq. (1) becomes

$$\frac{\partial \phi(p, x; t)}{\partial t} = \frac{1}{4N} \frac{\partial^2}{\partial x^2} \{x(1-x)\, \phi(p, x; t)\}, \quad (0 < x < 1). \quad (8)$$

In this equation $\phi(p, x; t)$ stands for the transition probability density that the frequency of A_1 becomes x in the t-th generation given that it is p at the initial (0-th) generation.

The solution of this equation was obtained by Kimura (1955a). It is expressed in terms of Gegenbauer polynomials as follows,

$$\phi(p, x; t) = \sum_{i=1}^{\infty} \frac{(2i+1)(1-r^2)}{i(i+1)}\, T_{i-1}^1(r)\, T_{i-1}^1(z) \cdot e^{-i(i+1)t/(4N)}, \quad (9)$$

where $r = 1 - 2p$, $z = 1 - 2x$, and $T_{i-1}^1(\cdot)$ stands for a Gegenbauer polynomial such that $T_0^1(z) = 1$, $T_1^1(z) = 3z$, $T_2^1(z) = \frac{3}{2}(5z^2 - 1)$ etc. The solution may also be expressed in terms of the hypergeometric function, that is,

$$\phi(p, x; t) = \sum_{i=1}^{\infty} p(1-p) i(i+1)(2i+1) F(1-i, i+2, 2, p)$$
$$\cdot F(1-i, i+2, 2, x)\, e^{-i(i+1)t/(4N)}, \quad (0 < x < 1), \quad (10)$$

where $F(\cdot, \cdot, \cdot, \cdot)$ denotes the hypergeometric function such that

$$F(1-i, i+2, 2, x) = 1 + \frac{(1-i)(i+2)}{1 \cdot 2} x$$
$$+ \frac{(1-i)(2-i)(i+2)(i+3)}{1 \cdot 2 \cdot 2 \cdot 3} x^2 + \cdots . \quad (11)$$

Though the right hand side of Eq. (9) or (10) is an infinite series, for a large value of t, only the first few terms are of any significance. Thus, when t is very large, we obtain the asymptotic formula

$$\phi \sim 6p(1-p)\, e^{-t/(2N)}, \quad (t \to \infty), \quad (12)$$

which shows that after a large number of generations the probability distribution of unfixed classes becomes flat and decays at the rate of $1/(2N)$ per generation.

Figs. 1a and 1b illustrate the process of random genetic drift based on the Eq. (10). They show that as time goes on the gene frequency tends to deviate from the initial frequency and the probability that both alleles coexist within a population gradually decreases due to chance fixation of one of the alleles. Eventually this probability decreases at the rate of $1/(2N)$ in each generation. As seen from Fig. 1a, if the initial frequency (p) is 0.5, the distribution curve becomes practically flat by the $2N$-th generation, after which Eq. (12) holds as a good approximation. On the other hand, as seen from Fig. 1b, if the initial frequency is 0.1, it takes about $4N$ generations before the distribution curve becomes practically flat.

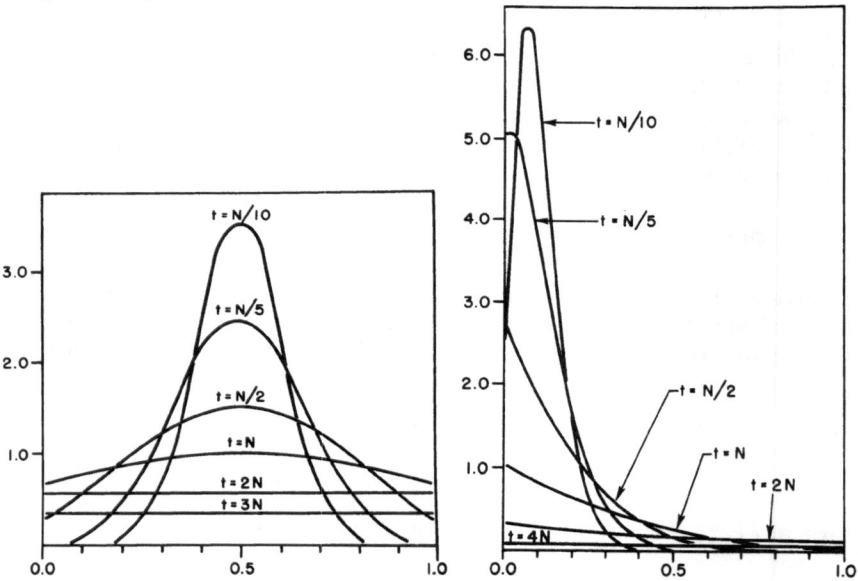

Figs. 1a and b. The process of random genetic drift due to random sampling of gametes in finite populations. The initial frequency (p) is 0.5 in Fig. 1a (left) and it is 0.1 in Fig. 1b (right) (from Kimura, 1955a, with a slight modification)

The probability that both alleles coexist in the population may be obtained by integration $\phi(p, x, t)$ in Eq. (9) with respect to x over (0,1). Thus, denoting this probability at the t-th generation by Ω_t, we have

$$\Omega_t = \int_0^1 \phi(p, x; t) \, dx = \sum_{j=0}^{\infty} [P_{2j}(r) - P_{2j+2}(r)] \, e^{-(2j+1)(2j+2)t/(4N)}$$
$$= 6p(1-p) \, e^{-t/(2N)} + \cdots, \tag{13}$$

where $r = 1 - 2p$ and $P(\cdot)$ represent the Legendre polynomials; $P_0(r) = 1$, $P_1(r) = r$, $P_2(r) = \frac{1}{2}(3r^2 - 1)$, etc. The average frequency of heterozygotes can also be calculated by using Eq. (9) as follows

$$H_t = \int_0^1 2x(1-x)\,\phi(p, x; t)\,dx = 2p(1-p)\,e^{-t/(2N)} = H_0 e^{-t/(2N)}\,. \quad (14)$$

showing that the heterozygosity decreases exactly at the rate of $1/(2N)$ per generation rather than asymptotically as in Ω_t of Eq. (13). The variance of the frequency of A_1 among populations at the t-th generation is

$$V_t = p(1-p) - \tfrac{1}{2}H_t = p(1-p)\,\{1 - e^{-t/(2N)}\}\,. \quad (15)$$

Finally, the probability of the gene A_1 being fixed in the population by the t-th generation is

$$f(p, 1; t) = p + \sum_{i=1}^{\infty}(2i+1)p(1-p)(-1)^i F(1-i, i+2, 2, p)\,e^{-i(i+1)t/(4N)}\,,$$

$$\quad (16)$$

which was first obtained by the study of the moments of the distribution (Kimura, 1955a). Random genetic drift with multiple alleles at a single locus was studied by Kimura (1955b, 1956) and Karlin and McGregor (1964).

Linkage disequilibrium due to random genetic drift when two loci are considered simultaneously was first studied by Hill and Robertson (1968) followed by Ohta and Kimura (1969).

4. Stationary Gene Frequency Distribution

In natural populations, it is expected that there is a balance between systematic evolutionary pressures such as recurrent mutation and stochastic factors such as random sampling of gametes. This will lead to a stable distribution of gene frequencies. For a single locus with a pair of alleles A_1 and A_2, a general formula was derived by Wright (1938) to describe such a distribution. The formula is

$$\phi(x) = \frac{C}{V_{\delta x}}\exp\left\{2\int \frac{M_{\delta x}}{V_{\delta x}}\,dx\right\}, \quad (17)$$

where C is a constant, $M_{\delta x}$ and $V_{\delta x}$ are the mean and the variance of the change of x (frequency of A_1) per generation, and $\exp\{\cdot\}$ stands for the exponential function. It is important to note that the distribution $\phi(x)$ is independent of the initial frequency p. It represents the probability density that the frequency of A_1 in a population is x, but we can also interprete $\phi(x)\,dx$ as the proportion of populations having gene frequency within the range x to $x + dx$ among a hypothetical aggregate consisting of the infinite number of populations satisfying the same conditions. Constant C in Eq. (17) is determined such that

$$\int_0^1 \phi(x)\,dx = 1 . \tag{18}$$

Wright's formula Eq. (17) may most easily be derived by noting that at stationary state the probability flux given in Eq. (2) is zero so that

$$-\frac{1}{2}\frac{d}{dx}\{V_{\delta x}\phi(x)\} + M_{\delta x}\phi(x) = 0 . \tag{19}$$

Let $V_{\delta x}\phi(x) = f$, then

$$\frac{1}{f}\frac{df}{dx} = \frac{2M_{\delta x}}{V_{\delta x}}$$

or

$$\frac{d}{dx}(\log f) = 2M_{\delta x}/V_{\delta x} .$$

Integrating both sides of this equation, we get

$$\log f = \text{Constant} + \int 2\frac{M_{\delta x}}{V_{\delta x}}dx$$

or

$$f = C e^{2\int (M_{\delta x}/V_{\delta x})dx} .$$

This is equivalent to Eq. (17) if we note $f = V_{\delta x}\phi(x)$.

As an application of Wright's formula, let us consider a case of reversible mutation between a pair of alleles A_1 and A_2, assuming that they are selectively neutral. Let u be the mutation rate from A_1 to A_2 and let v be the mutation rate in the reverse direction. Then, if x is the frequency of A_1,

$$M_{\delta x} = -ux + v(1 - x) . \tag{20}$$

In a population of effective size N_e,

$$V_{\delta x} = x(1 - x)/(2N_e) . \tag{21}$$

Substituting these expressions for $M_{\delta x}$ and $V_{\delta x}$, we obtain, from Eq. (17) the following formula for the distribution of x;

$$\phi(x) = C x^{4N_e v - 1}(1 - x)^{4N_e u - 1} \quad \text{(Wright, 1931).} \tag{22}$$

Constant C as determined by Eq. (18) is

$$C = 1/B(4N_e v, 4N_e u) = \frac{\Gamma(4N_e(u+v))}{\Gamma(4N_e u)\,\Gamma(4N_e v)}, \tag{23}$$

where $B(\cdot,\cdot)$ stands for the Beta function.

The mean and the variance of this distribution are

$$\bar{x} = \int_0^1 x\phi(x)\,dx = v/(u+v) \tag{24}$$

and

$$\sigma_x^2 = \int_0^1 x^2\phi(x)\,dx - \bar{x}^2 = \frac{\bar{x}(1-\bar{x})}{4N_e(u+v)+1}. \tag{25}$$

The distribution curve is unimordal and concentrated around the mean ($\bar{x}$) if both $4N_e u$ and $4N_e v$ are much larger than unity, but it is U-shaped if they are less than unity. When we consider each nucleotide site rather than conventional genetic locus, the rate of mutation due to nucleotide substitution is of the order of 10^{-9} per generation in mammals (Kimura, 1968a, b). So $4N_e u$ and $4N_e v$ are almost always much less than unity for each site. Thus, U-shaped distribution has a realistic meaning. For such a distribution, an allele (or a mutant form in a nucleotide site) is either fixed in the population or lost from it most of the time. The probability that allele A_1 is temporally fixed (established) in the population may be obtained from

$$f(1) = \int_{1-1/(2N)}^1 \phi(x)\,dx, \tag{26}$$

and the probability that A_1 is temporally lost from the population

$$f(0) = \int_0^{1/(2N)} \phi(x)\,dx, \tag{27}$$

where N is the actual number of individuals in the population (Kimura, 1968 b). Substituting Eq. (22) for $\phi(x)$ in Eq. (26) and (27), and assuming that u, v and $1/(2N)$ are all much smaller than unity, we obtain

$$f(1) = C \int_{1-1/(2N)}^1 x^{4N_e v-1}(1-x)^{4N_e u-1}\,dx \approx C/\{4N_e u(2N)^{4N_e u}\} \tag{28}$$

and

$$f(0) = C \int_0^{1/(2N)} x^{4N_e v-1}(1-x)^{4N_e u-1}\,dx \approx C/\{4N_e v(2N)^{4N_e v}\}. \tag{29}$$

For $N_e = N$, these results agree with the frequencies of terminal classes given by Wright (1931) who obtained his results by a different method.

5. Number of Alleles Maintained in a Finite Population

In this section we will study the number of alleles maintained per locus in an equilibrium population of finite size in which mutational production of new alleles is balanced by extinction of alleles caused by random sampling of gametes.

Selectively Neutral Isoalleles

First we will investigate the case of selectively neutral isoalleles. Consider a particular locus and assume that there are K possible allelic states $A_1, A_2, ..., A_K$. Furthermore, to simplify the treatment, we assume that mutation rate u per gene is equal for all the alleles and that each allele mutates with rate $u/(K-1)$ to one of the remaining $(K-1)$ alleles in each generation. Let x_i be the frequency of A_i in the present generation. Assuming that alleles are selectively neutral, the frequency of A_i in the next generation is

$$x'_i = (1-u) x_i + u_1(1-x_i) + \xi_i, \tag{30}$$

where $u_1 = u/(K-1)$, and ξ_i represents the part of change due to random sampling of gametes. If we assume that sampling of gametes takes place after mutation, then ξ_i has the mean and the variance,

$$E(\xi_i) = 0 \tag{31}$$

and

$$E(\xi_i^2) = X_i(1-X_i)/(2N_e), \tag{32}$$

where $X_i = (1-u) x_i + u_1(1-x_i)$ is the frequency of A_i after mutation but before the sampling. E stands for an operator of "taking expectation". We note here that Eq. (30) may also be expressed as

$$x'_i = X_i + \xi_i. \tag{30'}$$

At equilibrium in which the gene frequency distribution does not change with generations, we have $E(x'_i) = E(x_i) \equiv \mu'_1$. Thus, taking expectation of both sides of Eq. (30), we obtain

$$\mu'_1 = (1-u) \mu'_1 + u_1(1-\mu'_1)$$

or

$$\mu'_1 = 1/K, \tag{33}$$

where μ'_1 is the first moment of x_i at equilibrium. Similarly, the second moment at equilibrium, i.e. $E(x'^2_i) = E(x_i^2) \equiv \mu'_2$, may be obtained by squaring both sides of Eq. (30) or (30′) and then taking expectation. Namely,

$$E(x_i'^2) = E\{(X_i + \xi_i)^2\} = E\left\{X_i^2 + \frac{X_i(1 - X_i)}{2N_e}\right\}$$

$$= \frac{1}{2N_e} E\{(1 - u) x_i + u_1(1 - x_i)\} + \left(1 - \frac{1}{2N_e}\right)$$

$$\cdot E\{[(1 - u) x_i + u_1(1 - x_i)]^2\},$$

and this yields

$$\mu_2' = \frac{1}{K}\left\{\frac{1}{2N_e} + \left(1 - \frac{1}{2N_e}\right)\frac{u}{K-1}\left(2 - \frac{K}{K-1}u\right)\right\}\Bigg/ \left\{1 - \left(1 - \frac{1}{2N_e}\right)\left(1 - \frac{Ku}{K-1}\right)^2\right\}. \tag{34}$$

The average homozygosity or the expectation of the sum of squares of allelic frequencies is

$$\bar{H}_0 = E\left\{\sum_{i=1}^{K} x_i^2\right\} = K\mu_2'. \tag{35}$$

Following Kimura and Crow (1964), we define the *effective number of alleles* (n_e) by the reciprocal of the average homozygosity, i.e.

$$n_e = 1/(K\mu_2'). \tag{36}$$

Thus, if both u and $1/(2N_e)$ are much smaller than unity, we have

$$n_e = \left(1 + \frac{4N_e K u}{K-1}\right)\Bigg/\left(1 + \frac{4N_e u}{K-1}\right), \tag{37}$$

approximately.

On the other hand, if the number of allelic states is so large that each mutant form represents an allelic state not pre-existing in the population, $K = \infty$ in Eq. (34), and we obtain

$$n_e = 2N_e - (2N_e - 1)(1 - u)^2, \tag{38}$$

where N_e is the variance effective number of the population. We should note here that Eq. (38) is exact and no restrictions are placed on effective population number N_e and mutation rate u. However, since u is in general much smaller than unity, Eq. (38) is simplified to give

$$n_e = 4N_e u + 1. \tag{39}$$

This agrees with the formula given by Kimura and Crow (1964), and above derivation shows that it is valid under rather mild restrictions,

$$4N_e u + 1 \ll K$$

and

$$u \ll 1 .$$

It is sometimes remarked that formula like Eq. (39) is valid only for u up to $1/N_e$, but no such restriction is required. Formula (39) may also be obtained by the frequency distribution given by Kimura and Crow (1964),

$$\Phi(x) = 4M(1 - x)^{4M-1}x^{-1} , \tag{40}$$

where $M = N_e u$ in which u is the mutation rate to new not pre-existing alleles. This distribution has a meaning such that $\Phi(x)\,dx$ represents the *expected number* of alleles whose frequencies lie in the range x to $x + dx$ within the population rather than representing the probability that a particular allele lies in the frequency range x to $x + dx$ as in Eq. (22). Using this distribution, the expectation of the sum of squares of allelic frequencies is

$$\bar{H}_0 = \int_0^1 x^2 \Phi(x)\,dx = 1/(4M + 1) .$$

Thus, we obtain

$$n_e = 1/\bar{H}_0 = 4M + 1 = 4N_e u + 1 , \tag{41}$$

which agrees with Eq. (39). The above distribution Eq. (40) also enables us to obtain the *average number of alleles* maintained in the equilibrium population. This is the expectation of the number of alleles actually contained in the population. Thus, if we denote by n_a the average number of alleles,

$$n_a = \int_{1/(2N)}^1 \Phi(x)\,dx = 4N_e u \int_{1/(2N)}^1 (1 - x)^{4N_e u - 1}x^{-1}dx . \tag{42}$$

Since the effective number of alleles (n_e) is equal to the reciprocal of the sum of squares of allelic frequencies, it is generally less than the actual number of alleles, so that $n_e \leq n_a$. Only when the frequencies of all the alleles are equal do these two numbers agree with each other.

Fig. 2 illustrates a result of Monte Carlo experiment to check the validity of the above treatment. In this experiment, the population consists of 50 males and 50 females, of which only 25 males and 25 females actually participate in breeding, i.e. $N = 100$, $N_e = 50$. Mutation to a new, not pre-existing allele is induced in each gamete with probability 0.005 so that $u = 0.005$ and $4N_e u = 1.0$. The initial population contained 200 different alleles, but the majority of them were lost by generation 20. The balance between mutation and random extinction of alleles was reached well before generation 100. The figure depicts the course of fluctuation of the average and the effective numbers of alleles in the population at the intervals of 40 generations from generation 120 through generation 2080.

Neutral alleles random mutation
N = 100, Ne = 50, u = 0.005
●——— Average (actual) number
○------ Effective number

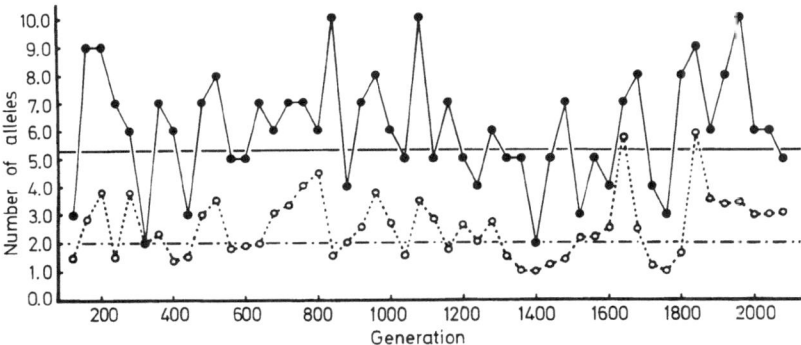

Fig. 2. A result of Monte Carlo experiment on the number of neutral isoalleles. $N = 100$, $N_e = 50$, $u = 0.005$. The average and the effective numbers of alleles are plotted respectively by solid and open circles. Horizontal lines, solid and broken, represent corresponding theoretical values (from Kimura, 1968b, with a slight modification)

The actual computer outputs were at intervals of 20 generations and gave 100 observed values over generations 120–2100. They gave

$$n_a = 6.05, \quad n_e = 2.07 \quad \text{(observed values)}.$$

On the other hand, the corresponding values derived from Eq. (42) and (41) are

$$n_a = 5.30, \quad n_e = 2.00 \quad \text{(theoretical values)}.$$

Despite the smallness of the population number assumed in the experiment, the agreement between the observed and theoretical values is fairly good.

Mutually Heterotic Alleles

Next, we will consider the case in which alleles are overdominant. We assume that each mutant represents an allelic state not pre-existing in the population. In addition, we assume that all the heterozygotes have equal fitness and all the homozygotes also have equal fitness which is lower by s as compared with that of the heterozygotes. Even with this simplification, the problem is much more difficult than in the neutral case. However, a theory was developed by Kimura and Crow (1964) which enables us to obtain the effective number of alleles (n_e) maintained in a population of effective size N_e.

Let u be the mutation rate per gene and let s be the selection coefficient against homozygotes. We will denote by f the expectation of the sum of squares of allelic frequencies so that

$$n_e = 1/f. \tag{43}$$

Furthermore, let

$$r = 2M/\sqrt{S} \tag{44}$$

in which $M = N_e u$ and $S = N_e s$.

It was shown by Kimura and Crow (1964) that for a given set of values of M and S, the value of f may be obtained first by solving for X the equation

$$e^{-X^2/2} = r \int_{-X}^{2\sqrt{S}-X} e^{-\lambda^2/2} d\lambda, \tag{45}$$

and then calculating f from

$$f = (r + X)/(2\sqrt{S}). \tag{46}$$

Actually, Eq. (45) contains some approximations, but it is valid if $s \gg u$ and also $f \ll 1$. In addition, if S is large, Eq. (45) may be replaced with good approximation by

$$\frac{e^{-X^2/2}}{\int_{-X}^{\infty} e^{-\lambda^2/2} d\lambda} = r. \tag{47}$$

Note that the left hand side of the above equation is equal to the ratio between the height of the normal distribution (mean 0 and variance unity) at X, and the area of the distribution from $-X$ to ∞. For example, if $N_e = 10^5$, $s = 10^{-3}$, $u = 10^{-5}$, we have $S = 10^2$, $M = 1$, and $r = 0.2$. Using tables of the normal distribution and its integral, we find that the value of X which satisfies Eq. (47) is $X = 1.28$ approximately. Therefore, from

Eq. (46) we get $f = 0.074$ so that $n_e = 13.5$. Namely, the effective number of alleles maintained in the population is about 13.5.

In order to check validity of the above theoretical treatments, Monte Carlo experiments were performed. Table 1 lists a result of an experiment performed by using the IBM 7090 computer, simulating a diploid population consisting of 50 breeding individuals. In the experiment, one new mutation was fed into the population in each generation, and the

Table 1. *Result of Monte Carlo experiment on the number of overdominant isoalleles maintained in a finite population, assuming $u = 0.01$, $s = 0.1$, and $N = 50$. In the table, $\bar{f}$ stands for the sum of squares of allelic frequencies in the population. The observed effective number $n_e = 1/\bar{f} = 4.7$*

Generation	f
60	0.2070
80	0.3198
100	0.2308
120	0.1944
140	0.1914
160	0.2354
180	0.2090
200	0.3150
220	0.2066
240	0.1700
260	0.1884
280	0.2870
300	0.1616
320	0.1496
340	0.1258
360	0.1560
380	0.2382
400	0.2400
average	$\bar{f} = 0.2126$

relative fitnesses of the heterozygote and the homozygote were assumed to be 1.0 and 0.9 respectively. Namely, $N = N_e = 50$, $u = 0.01$, $s = 0.1$. Also, the population started from a composition containing 100 different alleles. Balance between random extinction of alleles and mutational production of new alleles was established well before generation 60. As shown in the table, the average of f values (the sums of squares of allelic frequencies) from generation 60 through generation 400 examined at 20 generation intervals (18 computer outputs) turned out to be $\bar{f} = 0.2126$. This gives the observed effective number, $n_e = 4.7$. On the other hand, the corresponding theoretical value obtained by applying Eq. (45) is $n_e = 4.19$. Thus, agreement between the observed and the expected values on the number of alleles are reasonably good despite the smallness of the population number.

Recently, the theory of Kimura and Crow (1964) on the number of overdominant alleles was applied by Kerr (1967) who investigated the number of sex-determining alleles in bee populations in Brazil. The alleles concerned are haplo-viable but homozygous lethal so that the theory in this section is applicable by putting $s = 1$ while regarding males as "gametes". In the bee population of Piracicaba, the effective number (N_e) of the population was estimated to be 428, and the observed number of alleles was $11.0 \sim 12.4$. On the other hand, the theoretical number of alleles derived by using Eq. (45) was 11.2 if one assumes $u = 10^{-5}$ and it was 10.9 if $u = 10^{-6}$. Thus, the agreement between the observed and expected numbers of alleles appears satisfactory.

6. Number of Heterozygous Nucleotide Sites

In the foregoing section, we studied the number of alleles maintained per locus in a finite population assuming that each mutant is an allele not pre-existing in the population.

In this section, we will consider a different model and will investigate the average number of heterozygous nucleotide sites per individual in a diploid population consisting of N individuals and having the variance effective number N_e. The main assumption of the model is that the total number of sites per individual is so large and the mutation rate per site is so low that whenever a mutant appears it represents a mutation at a previously homoallelic site. Throughout this section, "site" refers to a single nucleotide pair, though the theory is appropriate to a small group of nucleotides such as a codon which consists of three nucleotides.

Let us assume that in the entire population in each generation mutants appear on the average in v_m sites.

Consider a particular site in which a mutant has appeared. Let $\phi(p, x; t)$ be the probability density that the frequency of the mutant in the population becomes x after t generations, given that it is p at $t = 0$. Then, $\phi(p, x; t)$ satisfies partial differential Eq. (3). Here we assume that the process is time homogeneous so that neither $M_{\delta p}$ nor $V_{\delta p}$ depends on time parameter t. In a typical situation which is of particular interest in population genetics, we have

$$M_{\delta p} = sp(1 - p) \{h + (1 - 2h) p\} \tag{48}$$

and

$$V_{\delta p} = p(1 - p)/(2N_e). \tag{49}$$

Namely, the mutant form in the site has a selective advantage s in homozygotes and sh in heterozygotes over the pre-existing form, and the factor causing random fluctuation in the mutant frequency is the random sampling of gametes.

In the following treatment, we assume that the same condition applies to different sites so that p, s, and sh are the same among the sites. Actually, the mutation rate per site is so low that each mutant gene is likely to be represented only once at the moment of its occurrence, that is, $p = 1/(2N)$ and the assumption that p is the same among different sites is a realistic one. On the other hand, selection coefficients s and sh both may vary from site to site. For such a case, however, we may use their means $\bar{s}$ and $\overline{sh}$ in the following treatment.

Now, since each mutant gene that appears in a finite population reaches either fixation or loss within a finite number of generations, under continued appearance of new mutations over many generations, a balance will be reached between appearance of mutants and their random extinction or fixation. In such a state of statistical equilibrium a stable frequency distribution will be realized among mutant forms at different sites if we consider only the segregating sites. Since v_m new mutations appear in the population in each generation, $v_m \phi(p, x; t)\,dx$ represents the contribution made by mutants which appeared t generations earlier with initial frequency p to the present frequency class in which the mutant frequencies are in the range $x \sim x + dx$. Under random mating, the frequency of heterozygotes is $2x(1 - x)$ for a site having mutant frequency x. Thus, considering all the contributions made by mutations in the past (from $t = 0$ to $t = \infty$), the expected number of heterozygous nucleotide sites per individual is

$$H(p) = v_m \int_0^\infty \left[\int_0^1 2x(1 - x)\, \phi(p, x; t)\, dx \right] dt . \tag{50}$$

In order to obtain an equation for $H(p)$, we multiply each term of Eq. (3) by $v_m 2x(1 - x)$, and then integrate each of the resulting terms first with respect to x over the interval $(0,1)$, and then with respect to t over $(0, \infty)$. This leads to

$$\int_0^\infty \frac{\partial}{\partial t} \left\{ v_m \int_0^1 2x(1 - x)\, \phi(p, x; t)\, dx \right\} dt$$
$$= \frac{1}{2} V_{\delta p} \frac{\partial^2}{\partial p^2} H(p) + M_{\delta p} \frac{\partial}{\partial p} H(p) . \tag{51}$$

The left hand side of this equation is simplified if we apply the conditions

$$\phi(p, x; \infty) = 0 \tag{52}$$

and

$$\phi(p, x; 0) = \delta(x - p) , \tag{53}$$

where $\delta(\cdot)$ is the Dirac delta function.

Namely,

$$\int_0^\infty \frac{\partial}{\partial t}\left\{ v_m \int_0^1 2x(1-x)\,\phi(p,x;t)\,dx \right\} dt$$

$$= v_m \int_0^1 2x(1-x)\,\phi(p,x;\infty)\,dx - v_m \int_0^1 2x(1-x)\,\phi(p,x;0)\,dx$$

$$= -2v_m p(1-p)\,.$$

The first condition Eq. (52) follows from the fact that the mutant form either becomes fixed or lost within a finite number of generations. The second condition Eq. (53) represents that the initial frequency of the mutant is p.

Then, Eq. (51) yields the ordinary differential equation

$$\frac{1}{2} V_{\delta p} \frac{d^2}{dp^2} H(p) + M_{\delta p} \frac{d}{dp} H(p) + 2v_m p(1-p) = 0\,. \tag{54}$$

Furthermore, since "mutations" at $p=0$ and $p=1$ do not contribute to the heterozygous sites, we have the boundary conditions

$$H(0) = H(1) = 0\,. \tag{55}$$

The solution of Eq. (54) which satisfies boundary conditions Eq. (55) is

$$H(p) = \{1 - u(p)\} \int_0^p \psi_H(\xi)\,u(\xi)\,d\xi$$
$$+ u(p) \int_p^1 \psi_H(\xi)\,\{1 - u(\xi)\}\,d\xi\,, \tag{56}$$

where

$$\psi_H(\xi) = 4v_m \xi(1-\xi) \int_0^1 G(x)\,dx / \{V_{\delta\xi} G(\xi)\} \tag{57}$$

and

$$u(p) = \int_0^p G(x)\,dx \Big/ \int_0^1 G(x)\,dx\,, \tag{58}$$

in which

$$G(x) = e^{-2\int_0^x (M_{\delta\xi}/V_{\delta\xi})\,d\xi}\,. \tag{59}$$

As will be shown in the next section, $u(p)$ is the probability of ultimate fixation of a mutant in the population.

In what follows we will consider a special but important case in which the mutant form has a selective advantage s in homozygotes and $s/2$ in heterozygotes (i.e. no dominance in fitness) so that $M_{\delta p} = sp(1-p)/2$ from Eq. (48), and the sole factor causing random fluctuation in the mutant frequency is random sampling of gametes so that $V_{\delta p} = p(1-p)/(2N_e)$. In this case, we have

$$G(x) = e^{-2Sx},$$

$$\psi_H(\xi) = 4N_e v_m e^{2S\xi}(1 - e^{-2S})/S$$

and

$$u(p) = (1 - e^{-2Sp})/(1 - e^{-2S}),$$

where $S = N_e s$. Then, we obtain, from Eq. (56)

$$H(p) = \frac{4N_e v_m}{S} \left(\frac{1 - e^{-2Sp}}{1 - e^{-2S}} - p \right). \tag{60}$$

If the mutant form in each site is represented only once at the moment of its occurrence, $p = 1/(2N)$ in the above formula. This yields the following simple approximation formulas:

If the mutant is sufficiently advantageous such that $2N_e s \gg 1$, we have

$$H(1/2N) \approx 4v_m(N_e/N), \tag{61}$$

but, if it is sufficiently disadvantageous such that $2N_e s' \gg 1$ in which $s' = -s$,

$$H(1/2N) \approx 2v_m/(Ns'). \tag{62}$$

On the other hand, if it is almost neutral such that $|2N_e s| \ll 1$, we have

$$H(1/2N) \approx 2v_m(N_e/N). \tag{63}$$

These results seem to show that mutations having a definite advantage or disadvantage cannot contribute greatly to the heterozygosity of an individual because of the rare occurrence of advantageous mutations and rapid elimination of deleterious ones.

There is some reason to assume that the majority of molecular mutations due to base substitution is almost neutral and that they occur at the rate of about two per gamete per generation (i.e. $v_m/(2N) = 2$) in mammals (Kimura, 1968a). Also, the variance effective number is likely to be hundreds of thousands or less in most mammalian populations. So, if we take $N_e = 10^5$ we have, from Eq. (63),

$$H(1/2N) \approx 8 \times 10^5.$$

Namely, the total number of heterozygous nucleotide sites per individual is some eight hundred thousands. However, since the total number of nucleotide sites per individual is estimated to be four billion (4×10^9) in mammals, the fraction of heterozygous sites is about 2×10^{-4}. If an average cistron consists of one thousand sites, it is heterozygous at one or more sites with probability $1 - e^{-0,2} \approx 0.18$.

7. Probability of Gene Fixation

In this section we will investigate the probability by which a mutant gene becomes fixed (established) in a finite population. Consider a pair of alleles A_1 and A_2, and let $u(p, t)$ be the probability that the mutant gene A_1 becomes fixed by the t-th generation, given that it is p at the 0-th generation. Then, as shown in Section 2, $u(p, t)$ satisfies the partial differential equation

$$\frac{\partial u(p, t)}{\partial t} = \frac{V_{\delta p}}{2} \frac{\partial^2 u(p, t)}{\partial p^2} + M_{\delta p} \frac{\partial u(p, t)}{\partial p}, \tag{64}$$

where $M_{\delta p}$ and $V_{\delta p}$ are respectively the mean and the variance of δp, the amount of change in the frequency (p) of A_1 in one generation. Here we assume that both $M_{\delta p}$ and $V_{\delta p}$ are independent of time parameter t, namely, the process is time homogeneous.

The probability of fixation $u(p, t)$ will then be obtained by solving Eq. (64) with the boundary conditions

$$u(0, t) = 0, \quad u(1, t) = 1. \tag{65}$$

These boundary conditions follow from the fact that the probability of fixation is zero if A_1 is absent at the start $(p = 0)$, but it is unity if it is initially fixed $(p = 1)$.

In the simplest case of random drift studied in Section 3 in which $M_{\delta p} = 0$ and $V_{\delta p} = p(1-p)/(2N_e)$, the solution of Eq. (64) is given by

$$u(p, t) = p + \sum_{i=1}^{\infty} (2i+1) p(1-p)(-1)^i F(1-i, i+2, 2, p) e^{-\lambda_i t}, \tag{66}$$

where $\lambda_i = i(i+1)/(4N_e)$ and N_e is the "variance" effective number of the population (see also Eq. (16)). In a more general case, however, the exact solution of Eq. (64) is rather difficult to obtain.

Next, let us consider the probability by which the mutant gene ultimately becomes fixed in the population. We will denote such a probability by $u(p)$, that is,

$$u(p) = \lim_{t \to \infty} u(p, t). \tag{67}$$

Since $\partial u(p)/\partial t = 0$ for this quantity, $u(p)$ satisfies the ordinary differential equation

$$\frac{V_{\delta p}}{2} \frac{d^2 u(p)}{dp^2} + M_{\delta p} \frac{du(p)}{dp} = 0 \tag{68}$$

with boundary conditions

$$u(0) = 0, \quad u(1) = 1. \tag{69}$$

In a special but important case in which the mutant gene (A_1) has selective advantage s in homozygotes and sh in heterozygotes over A_2 and the sole factor causing random fluctuation is random sampling of gametes, we have

$$M_{\delta p} = sp(1 - p)\{h + (1 - 2h)p\} \tag{70}$$

and

$$V_{\delta p} = p(1 - p)/(2N_e). \tag{71}$$

Thus, Eq. (68) becomes

$$\frac{p(1 - p)}{4N_e} \frac{d^2 u(p)}{dp^2} + sp(1 - p)\{h + (1 - 2h)p\} \frac{du(p)}{dp} = 0. \tag{72}$$

This equation was used by the author to obtain the formula for the probability of fixation

$$u(p) = \int_0^p e^{-2cDx(1-x)-2cx}dx \Big/ \int_0^1 e^{-2cDx(1-x)-2cx}dx, \tag{73}$$

(Kimura, 1957), where $c = N_e s$, $D = 2h - 1$ and p is the initial frequency of A_1. If there is no dominance, that is, if the mutant heterozygote $A_1 A_2$ has fitness midway between the fitnesses of the two homozygotes $A_1 A_1$ and $A_2 A_2$, then $h = 1/2$ or $D = 0$ in the above formula, and we have

$$u(p) = (1 - e^{-2cp})/(1 - e^{-2c}), \tag{74}$$

where $c = N_e s$ in which s is the selective advantage of $A_1 A_1$ over $A_2 A_2$. In this case, it may be convenient to designate the selective advantage of A_1 over A_2 by s_1 so that $s_1 = s/2$. Then (74) becomes

$$u(p) = \frac{1 - e^{-4N_e s_1 p}}{1 - e^{-4N_e s_1}}. \tag{75}$$

If $|2N_e s_1|$ is small, this yields

$$u(p) = p + 2N_e s_1 p(1 - p) \tag{76}$$

approximately. Thus, $u(p) - p$ is $2N_e$ times $s_1 p(1 - p)$. This is the basis of Robertson's principle concerning the limits in artificial selection which states that, for additive genes and under low selection intensity, the "total advance" is $2N_e$ times the gain in the first generation (cf. Robertson, 1960). On the other hand, if $4N_e s_1$ is much larger than unity but $4N_e s_1 p$ is small ($4N_e s_1 p \ll 1$), we have

$$u(p) = 4N_e s_1 p \tag{77}$$

approximately.

In the above treatments, we have assumed an arbitrary initial frequency p for the mutant gene A_1. In a special case in which the mutant gene is represented only once at the start in a population consisting of N diploid individuals, we may take $p = 1/(2N)$ to obtain the probability of its eventual fixation. Thus, if we denote the probability of eventual fixation of individual mutant gene by u, we have

$$u = u(1/(2N)). \tag{78}$$

However, if the number of males (N^*) is very different from that of females (N^{**}), an adjustment of taking $p = 1/(4N^*)$ or $p = 1/(4N^{**})$ may be required depending on whether the mutant gene occurred in a male or in a female (Moran, 1961, and Watterson, 1962).

Substituting $p = 1/(2N)$ in Eq. (75) and assuming that $|s_1|$ is small, we have

$$u = \frac{2s_1(N_e/N)}{1 - e^{-4N_e s_1}} \tag{79}$$

as a good approximation. In addition, if s_1 is positive and $4N_e s_1$ is much larger than unity, the above reduces to

$$u = 2s_1(N_e/N), \quad \text{(Kimura, 1964),} \tag{80}$$

where N_e/N is the ratio between the effective and actual numbers of the population. Thus, in a special case in which $N_e = N$, we obtain $u = 2s_1$, a well known result that the probability of ultimate survival of an advantageous mutant gene is approximately twice the selection coefficient (Haldane, 1927). In natural populations, however, the effective population number (N_e) may usually be much smaller than the actual number (N). According to Crow (1954) and also Crow and Morton (1955), estimated values of N_e/N based on the distribution of progeny number are $0.48 \sim 0.9$ for Drosophila, 0.75 for *Lymnaea* and $0.69 \sim 0.95$ for man. In general, if the distribution of the progeny number is known, we may compute the ratio by

$$\frac{N_e}{N} \approx \frac{\bar{k}}{1 + V_k/\bar{k}} \tag{81}$$

(cf. Kimura and Crow, 1963), where $\bar{k}$ and V_k are respectively the mean and the variance in number of progeny per parent. We should note here that in computing N_e/N from observational data, adjustment has to be made so that $\bar{k} = 2$ (cf. Crow, 1954). Kojima and Kelleher (1962) studied the ultimate survival probability of a single mutant gene using the method of branching process assuming that the progeny distribution follows a negative binomial distribution. They found that if the variance of the family size distribution is about twice the mean of the distribution ($V_k/\bar{k} \approx 2$ in our terminology), the ultimate survival probability is about 2/3 the value obtained under the assumption of Poisson progeny distribution (that is, when $V_k/\bar{k} = 1$). This may immediately be understood from Eq. (80) and (81), since the ratio N_e/N is 2/3 for $V_k/\bar{k} = 2$ and $\bar{k} = 2$, while it is unity for the Poisson distribution with $\bar{k} = 2$.

In natural populations, fluctuation in population number is rather common. In this case, as shown by Wright (1939), the effective number over a period is given by the harmonic mean of the effective numbers of the successive generations during that period. In particular, if the fluctuation is cyclic with a period of τ generations, then

$$N_e = \tau \bigg/ \sum_{t=1}^{\tau} \{1/N_e^{(t)}\}, \tag{82}$$

where $N_e^{(t)}$ is the effective number in the t-th generation within the cycle.

Eq. (75) or (77) may also be adapted to such a situation by noting that the initial frequency of a single mutant gene is $p_i = 1/(2N^{(i)})$ if it appeared in the i-th generation of a cycle, where $N^{(i)}$ is the actual population number in that generation. Thus, Eq. (80) still holds for such a cyclic fluctuation if we take for N_e the harmonic mean of the effective population number as given by Eq. (82), and for N the actual population number of the generation when the mutant gene appeared. This agrees with the result of Ewens (1967) who used the method of branching processes.

Incidentally, these examples show the usefulness of Eq. (80) and indirectly the great power of our diffusion models.

Going back to Eq. (73), let us consider the probability of fixation of a completely recessive mutant gene. For such a gene, $h = 0$ or $D = -1$ in Eq. (73), and we have

$$u(p) = \int_0^p e^{-2cx^2} dx \bigg/ \int_0^1 e^{-2cx^2} dx, \tag{83}$$

where $c = N_e s$. If s is positive and small but c is large, the above formula
leads approximately to

$$u = u(1/2N) = \frac{\sqrt{2N_e s}}{N\sqrt{\pi}}, \tag{84}$$

which gives the probability of ultimate fixation of an individual mutant
gene with small selective advantage in the homozygous state. If $N = N_e$,
this formula reduces to

$$u = \sqrt{2s/(\pi N)} = 1.128\sqrt{s/(2N)}. \tag{85}$$

Treating the same problem, Haldane (1927) estimated the chance of fixation to be of the order of $\sqrt{s/N}$ using the method of branching processes,
while Wright (1942) estimated it to be of the order of $\sqrt{s/(2N)}$ by his
method of integral equations. It is interesting to note that the value given
in Eq. (85) lies between those of Haldane and Wright. Also, Wright
obtained numerically the formula $1.1\ (s/2N)^{1/2}$ as the average chance
of fixation for values of s ranging from $4/2N$ to $64/2N$. It is remarkable
that Wright's empirical formula is very close to Eq. (85).

   Although Eq. (73) is fairly general for practical purposes, a still more
general formula is available. Namely, by solving equation Eq. (68) with
boundary conditions (69), we obtain

$$u(p) = \int_0^p G(x)\,dx \bigg/ \int_0^1 G(x)\,dx, \quad \text{(Kimura, 1962),} \tag{86}$$

where

$$G(x) = e^{-2\int (M_{\delta x}/V_{\delta x})dx} \tag{87}$$

in which $M_{\delta x}$ and $V_{\delta x}$ are the mean and the variance of the change in
gene frequency per generation. This formula is quite general and can
cover the cases of frequency dependent selection and random fluctuation
of gene frequency due to random fluctuation of selection intensity.

   The formula is also useful for obtaining the probability of fixation
in terms of a discrete generation model, especially when the selection
coefficient is not small. In such a case, it may be convenient to express
the fitnesses in Wright's selective values such that the selective values
of A_1A_1, A_1A_2, and A_2A_2 are respectively w_{11}, w_{12}, and w_{22}. Under
random mating and constant fitness of individual genotypes, if x is the
frequency of A_1 in the population, the expected amount of change in x
in one generation is given by Wright's formula (Wright, 1937),

$$M_{\delta x} = \frac{x(1-x)}{2\bar{w}}\frac{d\bar{w}}{dx} = \frac{x(1-x)}{2}\frac{d}{dx}\log\bar{w}, \tag{88}$$

where $\bar{w} = w_{11}x^2 + 2w_{12}x(1-x) + w_{22}(1-x)^2$ is the mean selective value of the population. For a population of effective size N_e, in which the variance in the amount of change in x in one generation is

$$V_{\delta x} = x(1-x)/(2N_e), \tag{89}$$

we obtain

$$-2 \int \frac{M_{\delta x}}{V_{\delta x}} dx = -2N_e \int \frac{d}{dx}(\log \bar{w}) dx = -2N_e \log \bar{w}.$$

Therefore Eq. (87) becomes

$$G(x) = (\bar{w})^{-2N_e}.$$

Thus, from Eq. (86), we obtain

$$u(p) = \int_0^p (\bar{w})^{-2N_e} dx \Big/ \int_0^1 (\bar{w})^{-2N_e} dx. \tag{90}$$

Note here that if we denote the selective advantage of A_1A_1 and A_1A_2 over A_2A_2 by s and sh respectively so that $w_{11} = 1 + s$, $w_{12} = 1 + sh$ and $w_{22} = 1$, we have

$$\bar{w} = 1 + s(2h-1)x(1-x) + sx \tag{91}$$

in the above formula. When $|s|$ is small,

$$(\bar{w})^{-2N_e} \approx e^{-2N_e\{s(2h-1)x(1-x)+sx\}}$$

approximately, and Eq. (90) agrees with (73). Namely, when selection coefficients are small, the two models, continuous and discrete on the generation time, give essentially the same result.

However, when selection coefficients are fairly large and if the discrete generation model is used, Eq. (90) may be more appropriate than Eq. (73). For example, in the case of genic selection, if we designate the selective values of A_1A_1, A_1A_2, and A_2A_2 respectively as $1+2s_1$, $1+s_1$, and 1, we have $h = 1/2$ and $s = 2s_1$ in Eq. (91), and therefore (90) reduces to

$$u(p) = \frac{1 - (1+2s_1p)^{-(2N_e-1)}}{1 - (1+2s_1)^{-(2N_e-1)}}. \tag{92}$$

This may be compared with Eq. (75) the corresponding formula for the continuous generation time model in which s_1 is the selective advantage of A_1 over A_2 measured in malthusian parameters.

The method of the Kolmogorov backward equation may also be used to obtain the probability of joint fixation of mutant genes at two or more loci. As an example, let us consider a haploid population of effective size N_e in which two loci are segregating independently with alleles A_1 and A_2 at the first locus and alleles B_1 and B_2 at the second locus. Let us suppose also that mutant genes A_1 and B_1 are individually neutral but become advantageous in combination, such that the selective values of the four genotypes A_1B_1, A_2B_1, A_1B_2, and A_2B_2 are $1+\varepsilon$, 1, 1, and 1 respectively. If we denote by p the frequency of A_1 in the first locus and by q the frequency of B_1 in the second locus, then, assuming independent assortment, the expected frequencies of A_1B_1, A_2B_1, A_1B_2, and A_2B_2 are pq, $(1-p)q$, $p(1-q)$, and $(1-p)(1-q)$ respectively. Thus the expected amount of change in p in one generation is

$$M_{\delta p} = \frac{(1+\varepsilon)pq + p(1-q)}{1+\varepsilon pq} - p = \frac{\varepsilon p(1-p)q}{1+\varepsilon pq}$$

or

$$M_{\delta p} = \varepsilon qp(1-p)/(1+\varepsilon pq). \tag{93}$$

Similarly, the corresponding change in q is

$$M_{\delta q} = \varepsilon pq(1-q)/(1+\varepsilon pq). \tag{94}$$

The variances of δp and δq and their covariance are

$$V_{\delta p} = p(1-p)/N_e, \qquad V_{\delta q} = q(1-q)/N_e, \tag{95}$$

and $W_{\delta p \delta q} = 0$.

Let $u(p,q)$ be the probability of joint fixation of mutant genes A_1 and B_1 at $t = \infty$, given that their initial frequencies are p and q, then, as seen from Eq. (5), $u(p,q)$ satisfies the partial differential equation

$$\frac{p(1-p)}{2N_e} \frac{\partial^2 u}{\partial p^2} + \frac{q(1-q)}{2N_e} \frac{\partial^2 u}{\partial q^2} + \frac{\varepsilon qp(1-p)}{1+\varepsilon pq} \frac{\partial u}{\partial p} + \frac{\varepsilon pq(1-q)}{1+\varepsilon pq} \frac{\partial u}{\partial q} = 0. \tag{96}$$

The boundary conditions required are: (i) $u(p,1)$ agrees with the formula for the fixation probability in the first locus derived by assuming that the mutant gene in the second locus is fixed in the population, and (ii) $u(1,q)$ agrees with that in the second locus derived by assuming that the mutant gene in the first locus is fixed. Namely,

$$u(p,1) = \frac{1-(1+\varepsilon p)^{-2N_e+1}}{1-(1+\varepsilon)^{-2N_e+1}} \tag{97}$$

and

$$u(1,q) = \frac{1-(1+\varepsilon q)^{-2N_e+1}}{1-(1+\varepsilon)^{-2N_e+1}}. \tag{98}$$

It turns out that the pertinent solution of Eq. (96) is

$$u(p, q) = \frac{1 - (1 + \varepsilon pq)^{-2N_e + 1}}{1 - (1 + \varepsilon)^{-2N_e + 1}} \,. \tag{99}$$

If $|\varepsilon| \ll 1$ and $N_e \gg 1$, this solution is approximated by

$$u(p, q) = \frac{1 - e^{-2N_e \varepsilon pq}}{1 - e^{-2N_e \varepsilon}} \,. \tag{100}$$

The validity of this formula was checked by Monte Carlo experiments by Ohta (1968). When $2N_e\varepsilon$ is much larger than unity, but initial frequency of both mutants are low so that $2N_e\varepsilon pq$ is much less than unity, we have

$$u(p, q) \approx 2N_e\varepsilon pq \tag{101}$$

approximately. Since, for neutral mutations the corresponding probability of joint fixation is pq, Eq. (100) means that the epistasis of the kind as considered here increases the probability of fixation by the factor $2N_e\varepsilon$. For the general type of epistasis the problem is much more difficult, and a general formula for the probability of joint fixation at multiple loci announced by Kimura (1964) as formula (10.17) in the References 18 unfortunately was incorrect.

8. Average Number of Generations Until Fixation

In the previous section, the probability by which a mutant gene becomes fixed (established) in a finite population was studied. Here, we will investigate the average number of generations involved for such a process of gene fixation.

As before, let $u(p, t)$ be the probability of fixation by the t-th generation and assume that it satisfies the differential Eq. (64). If we define a new quantity $T_1(p)$ by the relation

$$T_1(p) = \int_0^\infty t \, \frac{\partial u(p, t)}{\partial t} \, dt \,, \tag{102}$$

then

$$\bar{t}_1(p) = T_1(p)/u(p) \tag{103}$$

represents the average number of generations until fixation of A_1 having initial frequency p. Note that in calculating the average number, we exclude the cases in which A_1 is lost from the population.

In order to obtain $\bar{t}_1(p)$, we will derive the equation for $T_1(p)$ as follows: Differentiate each term of Eq. (64) with respect to t, multiply each of the resulting terms by t, and then integrate them with respect to t from 0 to ∞. This gives

$$\int_0^\infty t\, \frac{\partial^2 u(p,t)}{\partial t^2}\, dt = \frac{1}{2}\, V_{\delta p}\, \frac{\partial^2}{\partial p^2}\, T_1(p) + M_{\delta p}\, \frac{\partial}{\partial p}\, T_1(p). \qquad (104)$$

Then if we assume that $t\partial u(p,t)/\partial t$ vanishes at $t = \infty$, the left hand side of Eq. (104) is simplified:

$$\int_0^\infty t\, \frac{\partial^2 u(p,t)}{\partial t^2}\, dt = \left[t\, \frac{\partial u(p,t)}{\partial t} \right]_0^\infty - \int_0^\infty \frac{\partial u(p,t)}{\partial t}\, dt = -u(p,\infty) = -u(p),$$

where $u(p)$ is the probability of ultimate fixation of A_1 given in Eq. (86) Thus, Eq. (104) reduces to the ordinary differential equation

$$\frac{d^2}{dp^2}\, T_1(p) + \frac{2M_{\delta p}}{V_{\delta p}}\, \frac{d}{dp}\, T_1(p) + \frac{2u(p)}{V_{\delta p}} = 0. \qquad (105)$$

The solution of this equation which satisfies the boundary conditions

$$\lim_{p \to 0} \bar{t}_1(p) = \text{finite} \qquad (106)$$

and

$$\bar{t}_1(1) = 0 \qquad (107)$$

turns out to be as follows;

$$T_1(p) = u(p) \int_p^1 \psi(\xi)\, u(\xi)\, \{1 - u(\xi)\}\, d\xi + \{1 - u(p)\} \int_0^p \psi(\xi)\, u^2(\xi)\, d\xi, \qquad (108)$$

where

$$\psi(x) = 2 \int_0^1 G(x)\, dx / \{V_{\delta x} G(x)\}, \qquad (109)$$

in which

$$G(x) = \exp\left\{ -\int_0^x \frac{2M_{\delta\xi}}{V_{\delta\xi}}\, d\xi \right\}.$$

Of the two boundary conditions, the first, i.e. Eq. (106), specifies that in a finite population a single mutant gene which appeared in the population reaches fixation within a finite length of time. The second states that in a population in which A_1 is already fixed, the time until fixation is zero.

Thus, the average number of generations until fixation of A_1 in the population *excluding the cases of its loss* is

$$\bar{t}_1(p) = \int_p^1 \psi(\xi)\, u(\xi)\, \{1 - u(\xi)\}\, d\xi + \frac{1 - u(p)}{u(p)} \int_0^p \psi(\xi)\, u^2(\xi)\, d\xi , \quad (110)$$

where p is the initial frequency of A_1, and $\psi(\cdot)$ and $u(\cdot)$ are given respectively by Eq. (109) and (86)

Using a similar procedure, we can obtain the average number of generations until loss of mutant gene A_1 from the population, excluding the cases of its ultimate fixation. This is given by

$$\bar{t}_0(p) = \frac{u(p)}{1 - u(p)} \int_p^1 \psi(\xi)\, \{1 - u(\xi)\}^2\, d\xi + \int_0^p \psi(\xi)\, \{1 - u(\xi)\}\, u(\xi)\, d\xi . \quad (111)$$

As an example, let us consider a neutral mutant gene which appeared in a finite population of effective size N_e. In this case, $M_{\delta p} = 0$ and $V_{\delta p} = p(1 - p)/(2N_e)$ so that the number of generations until fixation, from Eq. (110), is

$$\bar{t}_1(p) = -\frac{1}{p} \{4N_e(1 - p)\log_e(1 - p)\} . \quad (112)$$

Similarly, the number of generations until loss, from Eq. (111) is

$$\bar{t}_0(p) = -4N_e\left(\frac{p}{1 - p}\right)\log_e p . \quad (113)$$

If the mutant gene is represented only once at the moment of occurrence, $p = 1/(2N)$ in which N is the actual number of individuals in the population. Then, assuming N large, we obtain from Eq. (112),

$$\bar{t}_1(1/2N) = 4N_e , \quad (114)$$

approximately. Namely, a single mutant gene, if it is neutral, takes about $4N_e$ generations until fixation if we disregard the cases in which it is eventually lost from the population. On the other hand, from Eq. (113), we obtain

$$\bar{t}_0(1/2N) = 2\left(\frac{N_e}{N}\right)\log_e(2N) \quad (115)$$

approximately. As an example, if we take $N = 10^4$ and $N_e/N = 0.8$, the average numbers of generations until loss turns out to be about 16 generations.

For more detailed studies on the subject, including Monte Carlo experiments, readers may refer to Kimura and Ohta (1969).

References

Crow, J. F.: Breeding structure of populations. II. Effective population number. In: Statistics and mathematics in biology, pp. 543–556. Ames, Iowa: Iowa State College Press 1954.

—, and N. Morton: Measurement of gene frequency drift in small populations. Evolution 9, 202–214 (1955).

—, and M. Kimura: An introduction to population genetics theory. New York: Harper and Row 1970.

Deevey, E. S. Jr.: The human population. Sci. Am. 203, 195–204 (1960).

Ewens, W. J.: The probability of a new mutant in a fluctuating environment. Heredity 22, 438–443 (1967).

Fisher, R. A.: On the dominance ratio. Proc. Roy. Soc. Edinburgh 42, 321–341 (1922).

— The genetical theory of natural selection. Oxford: Clarendon Press 1930.

Haldane, J. B. S.: A mathematical theory of natural and artificial selection. Part V: Selection and mutation. Proc. Cambridge Phil. Soc. 23, 838–844 (1927).

Hill, W. G., and A. Robertson: Linkage disequilibrium in finite populations. Theor. Appl. Genet. 38, 226–231 (1968).

Karlin, S., and J. McGregor: On some stochastic models in genetics. In: Gurland, J. (Ed.): Stochastic models in medicine and biology, pp. 245–271. Madison: University of Wisconsin Press 1964.

Kerr, W. E.: Multiple alleles and genetic load in bees. J. Apicult. Res. 6, 61–64 (1967).

Kimura, M.: Solution of a process of random genetic drift with a continuous model. Proc. Natl. Acad. Sci. U.S. 41, 144–150 (1955a).

— Random genetic drift in multi-allelic locus. Evolution 9, 419–435 (1955b).

— Random genetic drift in a tri-allelic locus: Exact solution with a continuous model. Biometrics 12, 57–66 (1956).

— Some problems of stochastic processes in genetics. Ann. Math. Stat. 28, 882–901 (1957).

— On the probability of fixation of mutant genes in a population. Genetics 47, 713–719 (1962).

—, and J. F. Crow: The measurement of effective population number. Evolution 17, 279–288 (1963).

—, T. Maruyama, and J. F. Crow: The mutation load in small populations. Genetics 48, 1303–1312 (1963).

— Diffusion models in population genetics. J. Appl. Prob. 1, 177–232 (1964).

—, and J. F. Crow: The number of alleles that can be maintained in a finite population. Genetics 49, 725–738 (1964).

— Evolutionary rate at the molecular level. Nature (Lond.) 217, 624–626 (1968a).

— Genetic variability maintained in a finite population due to mutational production of neutral and nearly neutral isoalleles. Genet. Res. 11, 247–269 (1968b).

— The number of heterozygous nucleotide sites maintained in a finite population due to steady flux of mutations. Genetics 61, 893–903 (1969).

—, and T. Ohta: The average number of generations until fixation of a mutant gene in a finite population. Genetics 61, 763–771 (1969).

Kojima, K., and T. M. Kelleher: Survival of mutant genes. Amer. Natur. 96, 329–346 (1962).

Miller, G. F.: The evolution of eigenvalues of a differential equation arising in a problem in genetics. Proc. Cambridge Phil. Soc. 58, 588–593 (1962).

Moran, P. A. P.: The survival of a mutant under general conditions. Proc. Cambridge Phil. Soc. **57**, 304–314 (1961).

Muller, H. J.: Evolution by mutation. Bull. Am. Math. Soc. **64**, 137–160 (1958).

Nei, M., and Y. Imaizumi: Effects of restricted population size and increase in mutation rate on the genetic variation of quantitative characters. Genetics **54**, 763–782 (1966).

Ohta, T.: Effect of initial linkage disequilibrium and epistasis on fixation probability in a small population, with two segregating loci. Theor. Appl. Genet. **38**, 243–248 (1968).

—, and M. Kimura: Linkage disequilibrium due to random genetic drift. Genet. Res. **13**, 41–55 (1969).

Robertson, A.: A theory of limits in artificial selection. Proc. Roy. Soc. B, **153**, 234–249 (1960).

— Selection for heterozygotes in small populations. Genetics **47**, 1291–1300 (1962).

Watterson, G. A.: Some theoretical aspects of diffusion theory in population genetics. Ann. Math. Stat. **33**, 939–957 (1962).

Wright, S.: Evolution in Mendelian populations. Genetics **16**, 97–159 (1931).

— The distribution of gene frequencies in populations. Proc. Natl. Acad. Sci. U.S. **23**, 307–320 (1937).

— The distribution of gene frequencies under irreversible mutation. Proc. Natl Acad. Sci. U.S. **24**, 253–259 (1938).

— Statistical genetics in relation to evolution. Actualités Scientifiques et Industrielles 802. Exposés de Biométrie et de Statistique Biologique. Paris: Hermann et Cie. 1939.

— Statistical genetics and evolution. Bull. Am. Math. Soc. **48**, 223–246 (1942).

— The differential equation of the distribution of gene frequencies. Proc. Natl Acad. Sci. U.S. **31**, 382–389 (1945).

Theory of Limits to Selection with Line Crossing

W. G. HILL

Introduction

Breeders of many species of animals and plants make breed or strain crosses and market the crossbred progeny in order to utilize heterosis. Several breeding schemes have been suggested for the improvement of such crossbreds without necessarily improving the parental strains. (The progeny will be called "crossbred" even if their parents are of different strains but from the same breed). There are essentially two types of programme. In the first, most of the emphasis is based on selection *between* lines. Typically these lines are developed by rapid inbreeding, and little or no selection is practised within them. Large scale testing is then undertaken to find those lines which produce the best cross. Other breeding programmes typically start with a pair of lines already known to give potentially useful crossbreds, and then selection is practised *within* these lines, with the aim of improving this cross. Of course, mixtures of these schemes are also used in practice.

In this paper the analysis will be almost entirely restricted to the alternative methods of practising selection within lines. This may be based only on performance in the individual pure strains, which will be referred to as pure line selection (PLS). Alternatively, selection can be carried out in one line for cross performance against a tester strain which may be highly inbred. This method was proposed by Hull (1945) and is commonly called recurrent selection to a tester (RST). Comstock, Robinson, and Harvey (1949) suggested the method of reciprocal recurrent selection (RRS) in which selection on cross performance is practised within both populations making the cross. However it is not necessary for us to assume that reciprocal crosses between the populations are actually made. In both the RST and RRS methods, individuals are crossed to the other strain, and those which have the best crossbred progeny or half-sibs are selected as parents of the next generation. These methods are thus intended to utilize non-additive genetic variation, particularly from overdominant genes.

Theoretical comparisons of the efficiency of the alternative selection and crossing schemes have been made by Comstock *et al.* (1949), Dickerson (1952), Crow (1953), and Cress (1966). However, as Bowman (1959) has pointed out in a review, these theoretical calculations are generally based on three suppositions: (1) no epistasis, (2) no more than two alleles per locus, and (3) linkage equilibrium. Bowman considers that if the literature regarding heterosis is taken into account, then comparisons based on these assumptions must be of limited value. In fact, these studies also make another important assumption, namely that the populations are of very large size (i.e. approaching infinitely large size) so that the maximum improvement possible in any scheme would be expected to be attained eventually. Robertson (1960), however, has drawn attention to the problems of limits to selection in populations of finite size. During the selection programme some favourable genes may be lost from the population by chance, so that the final advance depends on the effective population size as well as the initial frequencies and effects of the individual genes. Also, average rates of advance are likely to be influenced by population size. This is particularly important when genes with heterozygote superiority are initially at equilibrium so that no progress would initially be made in an RRS programme. Both Arthur (1964) and Cress (1967) have used Monte Carlo methods to study the effects of initial restriction of population size in causing drift from equilibrium gene frequency situations. Also Dickerson (1952) showed that RST with a highly inbred tester could be more efficient in terms of initial response than RRS since there is then no unstable equilibrium state. However, in Arthur's model the gene effects, selection intensity, and population size were sufficiently large that eventually the best possible limit was attained, and Cress investigated the rate of advance for only five generations.

Some long-term experimental comparisons of alternative cross breeding experiments have been made with *Drosophila*. Bell, Moore, and Warren (1955) compared several selection schemes for improvement of egg size and production in *D. melanogaster*. Egg size appeared to be largely controlled by additive genes, and conventional pure line family selection proved most efficient. For egg production, Bell *et al.* found that the response with RRS was superior to that with either PLS or RST. However the RRS line was not superior to the best single crosses between inbred lines developed (with much less effort) from the base populations. Rasmuson (1956) compared PLS and RRS for egg production, hatchability, and body weight in *D. melanogaster*. Only with egg production was more progress made with RRS and then by just 6–7%, with most of this superiority obtained after a few generations of selection. Kojima and Kelleher (1963) obtained substantial response to RRS for egg production in *D. pseudoobscura* but only for the first 10 or so generations of selection, after which the population reached a plateau at a level equivalent to that of the best 4% of all possible two-way crosses in the base population. The earlier experiments are reviewed in more detail by Bowman (1959) who concluded that no direct proof had been published at that time to indicate that the methods of RST and RRS were at all successful in what they were theoretically intended to achieve, namely to utilize non-additive variation. Large experiments with breeding schemes in poultry are being carried out in the United States but the results are as yet, unpublished.

Since the earlier theoretical studies have been primarily concerned with infinitely large populations, or have not considered the finite population model in any detail, it seemed worthwhile to study the expected rates of advance in a finite model with the alternative within-line selection schemes for producing crosses. The theory may help to throw some light on the experimental results and perhaps give some pointers towards breeding practice. Unfortunately there are so many possible parameters which can be studied that a very simple model has to be used, and several approximations will be made in the course of the analysis. In particular, we shall ignore epistasis and linkage and, again in common with Comstock *et al.* (1949) and Dickerson (1952), assume that there are only two alleles per locus. For simplicity, we shall usually also assume that the tester strain in an RST programme is already homozygous, or that it becomes homozygous immediately with inbreeding. With poultry, for example, this idealized situation cannot be attained, and the errors in this approximation will be investigated briefly.

Model. Response to a Single Cycle of Selection

Let us consider an autosomal locus with alternative alleles A_1 and A_2. The average genotypic values of A_1A_1 and A_2A_2 are assumed to be a_1 and a_2 units, respectively, poorer than the heterozygote for the quantitative trait under selection. We let q be the frequency of A_1 and let $\bar{q} = a_2/(a_1 + a_2)$. Since only differences in genotypic value are important, we can arbitrarily let the genotypic value of the heterozygote be $a = a_1 + a_2$. Thus we have:

Genotype	A_1A_1	A_1A_2	A_2A_2
Genotypic value	$(a_1+a_2)-a_1$	a_1+a_2	$(a_1+a_2)-a_2$
=	$a\bar{q}$	a	$a(1-\bar{q})$

The alternative types of gene action can be summarised as follows:

Overdominant $\qquad\qquad\qquad\qquad\quad 0 < \bar{q} < 1$

A_1 completely dominant over A_2 $\qquad \bar{q} = 1$

A_1 partially dominant over A_2 $\qquad 1 < \bar{q} < \infty$

A_1 recessive to A_2 $\qquad\qquad\qquad \bar{q} = 0$

Additive $\qquad\qquad\qquad\qquad\qquad \bar{q} \to \infty \quad$ but $(a\bar{q})$ finite.

This way of expressing the gene effects is most suitable for the case of overdominant gene action, with which we shall be particularly concerned, when $\bar{q}$ is the equilibrium gene frequency for large populations. Otherwise $\bar{q}$ has no obvious interpretation and merely becomes a convenient parameter. With additive gene action, where $\bar{q}$ becomes infinite, $\bar{q}$ always appears in expressions for changes in gene frequency as $a\bar{q}$, which is assumed to be finite. Although this definition of the model is less suitable for additive gene action, we shall rarely be concerned with additivity in a theory of selection for cross performance. In other definitions of gene effects (e.g. Comstock *et al.*, 1949; Dickerson, 1952) some terms become infinitely large when there is "pure" overdominance which is $\bar{q} = 0.5$ in this model.

1. Pure Line Selection

If truncation selection is practised on individual phenotypes in a large single population the change in gene frequency in one generation is, approximately,

$$\delta q = \frac{-i_m a}{\sigma} q(1 - q)(q - \bar{q}) \tag{1}$$

where i_m is the selection differential in standard units and σ is the phenotypic standard deviation. Formulae similar to Eq. (1) have been derived by various authors, notably Haldane (1931), Comstock *et al.* (1949), Crow (1953), and Griffing (1960). Eq. (1) holds only if gene effects are small such that terms in $(a/\sigma)^r$ can be ignored relative to a/σ for $r > 1$. Latter (1965) has examined the errors induced by this approximation for additive gene action in infinite populations. In populations of finite size δq represents the expected (i.e. mean) change in gene frequency, and can be predicted more accurately if i_m is computed from order statistics than from the normal integral. A more detailed discussion is given by Kojima (1961) and Hill (1969a). If progeny testing, for example, is practised in a pure line, the response becomes

$$\delta q = -\tfrac{1}{2} i a / \sigma_f \tag{2}$$

where σ_f is the standard deviation of progeny test means. More generally the response will be proportional to the average of i/σ_f for the two sexes, if, as is probable, they are not tested with exactly the same design. The relative responses with different schemes, such as individual selection or progeny testing are well known (e.g. Falconer, 1960), and we shall return to the problem of generation interval later. Of course Eq. (2) is subject to the same assumptions as Eq. (1) on population size and magnitude of gene effects. The formulae can be greatly simplified if we let $s = ia/\sigma_f$ so that

$$\delta q = -\tfrac{1}{2} sq(1 - q)(q - \bar{q}). \tag{3}$$

Thus s may be regarded as a selective value.

2. Selection on Cross Performance

In an RRS scheme those individuals with the highest average cross-bred progeny test are assumed to be chosen as parents of the next generation. Let us denote the populations X and Y, and for the allele A_1 let p and q be their respective frequencies and r and s their respective selective values (i.e. the mean over sexes of ia/σ_f). Predictions of changes in gene frequency in RRS programmes have been given by Comstock et al. (1949) and Dickerson (1952). They are similar to those for pure line selection and will be stated here without derivation. With small effects and progeny testing the changes in gene frequency will be:

$$\begin{aligned} \text{Population } X:\ &\delta p = -\tfrac{1}{2} rp(1 - p)(q - \bar{q}), \\ \text{Population } Y:\ &\delta q = -\tfrac{1}{2} sq(1 - q)(p - \bar{q}). \end{aligned} \tag{4}$$

In an RRS programme we might expect r and s to be equal, but this would not be the case if no reciprocal crosses were made.

If the tester strain, X, is assumed to be homozygous in an RST breeding programme, the response in population Y will depend on which allele is fixed in X. With a similar progeny testing scheme as in the RRS programme, changes in gene frequency will be as follows:

Allele fixed in population X Changes in gene frequency in Y

$$A_1(p = 1) \qquad\qquad \delta q = -\tfrac{1}{2} sq(1 - q)(1 - \bar{q}) \qquad (5a)$$

$$A_2(p = 0) \qquad\qquad \delta q = \tfrac{1}{2} sq(1 - q)\bar{q}. \qquad (5b)$$

In general, of course, Eq. (4) can be applied to RST also, even if the tester is not inbred.

When comparing the PLS, RRS, and RST schemes we may not assume that the selective value, s, is the same in each case, because the selection intensities or variances of progeny test means may not be the same. For example, σ_f may be less in an RST scheme if the tester is homozygous.

3. Changes in the Mean of the Quantitative Trait

If random mating is practised between individuals of the opposite strain, the mean, μ, of the crossbred progeny for the quantitative trait is

$$\mu = a[1 - \bar{q}(1 - \bar{q}) - (p - \bar{q})(q - \bar{q})] . \tag{6}$$

This mean is maximized with overdominance if $p = 1$ and $q = 0$ or vice versa, with complete dominance if $p = 1$ or $q = 1$, and with partial dominance or additivity if $p = q = 1$. The change in the mean with one cycle of selection is:

$$\delta\mu = -a[(q - \bar{q})\delta p + (p - \bar{q})\delta q + \delta p \delta q] .$$

Thus if the product term $\delta p \delta q$ is ignored, which should introduce little error if changes in gene frequency are small each generation, the responses to a single cycle of selection for the alternative schemes are as follows:

System	$\delta\mu$
PLS	$\dfrac{a}{2}(p - \bar{q})(q - \bar{q})[rp(1 - p) + sq(1 - q)]$
RRS	$\dfrac{a}{2}[rp(1 - p)(q - \bar{q})^2 + sq(1 - q)(p - \bar{q})^2]$
RST $p = 1$	$\dfrac{a}{2}sq(1 - q)(1 - \bar{q})^2$
$p = 0$	$\dfrac{a}{2}sq(1 - q)\bar{q}^2$

In the PLS system selection is carried out independently in the two populations, which are then crossed. The selective values are assumed to be r and s in populations X and Y, respectively.

Extension of the Theory to Finite Populations

Explanation of the theory for finite populations may be clarified if we first consider selection in a single population, for which the theory of limits was first discussed by Robertson (1960). Imagine that an identical selection programme is practised in a large number of replicate lines in which the frequency of the A_1 allele was originally q_0. With genetic sampling (drift) the gene frequencies will no longer remain the same in all lines, and we can thus discuss the distribution of gene frequency among these lines. Eventually all lines will reach fixation and a limit will be reached, although with overdominant genes fixation may occur very slowly, and accompany a decline in response (Robertson, 1962; Hill and Robertson, 1968). The chance of fixation of the allele A_1 is defined as the proportion of lines in which it is eventually fixed, for specified initial frequency.

Changes in the mean of a cross between two finite populations have to be studied in terms of the joint distribution of gene frequency and joint probabilities of fixation in the two lines. Thus we let $w(p_0, q_0)$, or simply w, be the probability that A_1 is fixed in both lines X and Y, given that their initial frequencies were p_0 and q_0 respectively. Similarly we define $u(p_0)$ and $v(q_0)$, or u and v, as the marginal probabilities of fixation of A_1 in X and Y respectively. Drift occurs independently in the two populations, but changes in gene frequency with selection are not independent in an RRS programme. With pure line selection we can assume that $w = uv$, but in a successful reciprocal recurrent selection programme for an overdominant gene we would hope that A_1 was not frequently fixed in both lines making the cross, so that $w < uv$. The probabilities of fixation in the various states can be summarized thus:

Allele fixed in X, Y	A_1, A_1	A_1, A_2	A_2, A_1	A_2, A_2
Probability	w	$u - w$	$v - w$	$1 - u - v + w$
Crossbred mean	$a\bar{q}$	a	a	$a(1 - \bar{q})$

Let us assume that the effective sizes of the populations X and Y are M and N respectively. With pure line selection in population Y, for example, the conditional probability that among the total of $2N$ alleles at some cycle $t + 1$ there are jA_1 alleles, given that there were iA_1 alleles at generation t, can be shown to be approximately

$$b_{ij} = \binom{2N}{j} (q + \delta q)^j (1 - q - \delta q)^{2N - j}, \qquad 0 \le i, j \le 2N \qquad (7)$$

where $q = i/2N$, the gene frequency at generation t, and $\delta q = -\frac{1}{2} sq(1 - q)$ $(q - q)$ from Eq. (3). The approximations associated with Eq. (7) for truncation selection are discussed by Hill (1969a), but in the context of individual selection rather than progeny testing. With PLS b_{ij} is independent of t if it is assumed that the selective values do not change, and similar transition probabilities can be specified for selection in X. Similarly, with RST, so long as the tester strain is homozygous, δq can be evaluated using Eq. (5a) or (5b) and the transition probabilities of the form of Eq. (7) become independent of t. But with RRS, although single generation responses can be expressed in this form, it is necessary to consider the joint transition probabilities in the two populations in order to describe long-term response. Since the genetic sampling occurs independently in the two populations, we have

$$d_{(h, i, j, k)} = \binom{2M}{j} (p + \delta p)^j (1 - p - \delta p)^{2M - j} \binom{2N}{k} (q + \delta q)^k (1 - q - \delta q)^{2N - k} \qquad (8)$$

where $p = h/2M$; $q = i/2N$; $0 \leqq h, j \leqq 2M$; $0 \leqq i, k \leqq 2N$;

$$\delta p = -\frac{r}{2} p(1-p)(q-\bar{q}); \qquad \delta q = -\frac{s}{2} q(1-q)(p-\bar{q})$$

and $d_{(h,i,j,k)}$ is the probability that, conditional on there being h and i A_1 alleles in populations X and Y respectively at some cycle t, there are j and k respectively in the succeeding cycle. Again, so long as r and s remain constant for all t, so does $d_{(h,i,j,k)}$. Of course transition probabilities for PLS can be written in the same form as Eq. (8), but they factor into terms of the form of Eq. (7).

The transition probabilities of Eq. (7) and Eq. (8) can be expressed in matrix form. In Eq. (8) a row of the matrix specifies both h and i and a column j and k so that the matrix is square of dimension $(2M+1)(2N+1)$. Standard techniques can then be used to obtain numerical results for distribution of gene frequency and the chances of fixation, which need not be discussed here. The type of method adopted is described elsewhere (Hill, 1969a). On an I.C.T. Atlas computer it was practicable to work with $N = M = 8$ with the transition matrix for RRS.

Diffusion Equation and Simple Approximations

Diffusion models have been widely applied to problems of selection at a single locus in a single finite population (Kimura, 1964). In particular, the chance of fixation has been derived for such models (Kimura, 1957) and has been applied to the theory of selection limits (Robertson, 1960). The diffusion equation is continuous in both time and gene frequency, but the chance of fixation computed from the equation has been shown to be a good predictor for a model of artificial selection in a finite population with discrete generations and values of gene frequency (Hill, 1969a).

1. Pure Line Selection

If the diffusion equation is used to approximate the gene frequency distribution for selection in a single finite population, say Y, and if time is measured on a scale proportional to the effective population size N the selection advance is a function of only Ns, $\bar{q}$, and the initial frequency, q_0 (Robertson, 1962). If the expected change in gene frequency is given by Eq. (3), the chance of fixation of A_1 in Y is given

$$v = \int_0^{q_0} e^{Ns(x-\bar{q})^2} dx \Big/ \int_0^1 e^{Ns(x-\bar{q})^2} dx \qquad (9)$$

(Kimura, 1957) and similarly u is given in terms of Nr, $\bar{q}$, and p_0 in population X. As we have mentioned, the joint chance of fixation of A_1 in the two populations is the product of the marginal probabilities, so that μ_L, the crossbred mean at the limit, is given by substitution in Eq. (6) as

$$\mu_L = a[1 - \bar{q}(1 - \bar{q}) - (u - \bar{q})(v - \bar{q})]. \tag{10}$$

If Ns and $Ns\bar{q}$ are small relative to unity, series expansion of Eq. (9) yields

$$v = q_0 + Nsq_0(1 - q_0)[\bar{q} - \tfrac{1}{3}(1 + q_0)] + 0[(Ns)^2]. \tag{11}$$

Robertson (1960) has expressed v in this form for $\bar{q} = 1$.

Under the diffusion approximations the selection limit, μ_L, will be a function of Mr, Ns, p_0, q_0, and $\bar{q}$. We shall be particularly interested in the case where the same breeding programme is practised in each line, so that $M = N$ and $Mr = Ns$. The time scale of the process will then be proportional to N.

2. Recurrent Selection to an Inbred Tester

If the tester strain X is homozygous at locus A, selection in line Y for cross performance is then equivalent to selection for an additive gene with selective value $-s(1 - \bar{q})/2$ or $s\bar{q}/2$ depending on whether A_1 or A_2, respectively, is fixed in X (Eq. (4)). Thus, from Kimura (1957), the chance of fixation of A_1 in Y is given by

$$v = [1 - e^{2Ns(p^* - \bar{q})q_0}]/[1 - e^{2Ns(p^* - \bar{q})}] \tag{12}$$

where p^* is the frequency of A_1 in the tester and takes values $p^* = 0$ or 1. If X were a cross between two homozygous lines, p^* could then take the value 0.5. The population mean at the limit is simply obtained from the definition of genotypic values.

If population X initially is segregating at locus A, but is inbred very rapidly without selection so that we can assume it is homozygous from the outset of the RST programme, the probability that X is fixed for A_1 is p_0 and the probability is $1 - p_0$ that X is fixed for A_2. The expected value of the selection limit becomes

$$\mu_L = a[1 - (1 - \bar{q})p_0v_1 - \bar{q}(1 - p_0)v_2] \tag{13}$$

where v_1, v_2 are the conditional chances of fixation of A_1 in Y given that A_1, A_2 are fixed in X and are given by substitution into Eq. (12) of $p^* = 1$ or $p^* = 0$, respectively.

If Ns and $Ns\bar{q}$ are of small order, Eq. (12) and (13) become

$$v = q_0 + Nsq_0(1 - q_0)(p^* - \bar{q}) + 0[(Ns)^2], \tag{14}$$

$$\mu_L - \mu_0 = Nsaq_0(1 - q_0)[p_0(1 - p_0) + (p_0 - \bar{q})^2]. \tag{15}$$

3. Reciprocal Recurrent Selection

No explicit formulae for the limit have been obtained from diffusion equations for RRS. However, a formula developed by Kimura (personal communication) and cited by Ohta (1968) for a special case of additive x additive epistatic interaction between independent loci in a single population can be modified to give the selection limit for RRS with complete dominance ($\bar{q} = 1$) when the same breeding programme is practised in both lines, i.e. $N = M$, $r = s$. The selection limit, μ_L, in the cross becomes

$$\mu_L = a[e^{2Ns} - e^{2Ns(1 - p_0)(1 - q_0)}]/(e^{2Ns} - 1). \qquad (16)$$

However, both forward and backward Kolmogorov diffusion equations can be set up to approximate selection with RRS for all values of $\bar{q}$ and $N \neq M$, $r \neq s$. The equations can be obtained by substitution into the multivariate formulae given by Kimura (1964), and we make no attempt at rigour here. Let $\phi(p, q, p_0, q_0, t)$ be the joint density of the gene frequency distribution in the two populations at time t, for initial frequencies p_0 and q_0. The mean changes in gene frequency ($M_{\delta p}$ and $M_{\delta q}$ of Kimura) are given by Eq. (4), the variances of changes ($V_{\delta p}$ and $V_{\delta q}$) are

$$V_{\delta p} = \frac{p(1 - p)}{2M}, \qquad V_{\delta q} = \frac{q(1 - q)}{2N}$$

and the covariance of change is zero since sampling of genes occurs independently in the two populations. The forward equation is

$$\begin{aligned}
\frac{\partial \phi}{\partial t} &= \frac{1}{2} \frac{\partial^2}{\partial p^2} \left[\frac{p(1 - p)}{2M} \phi \right] + \frac{1}{2} \frac{\partial^2}{\partial q^2} \left[\frac{q(1 - q)}{2N} \phi \right] \\
&+ \frac{\partial}{\partial p} \left[\frac{1}{2} rp(1 - p)(q - \bar{q}) \phi \right] + \frac{\partial}{\partial q} \left[\frac{1}{2} sq(1 - q)(p - \bar{q}) \phi \right].
\end{aligned} \qquad (17)$$

We are particularly interested in the model in which the same breeding programme is practised in both populations (with $M = N$ and $r = s$). Eq. (17) may be written

$$\begin{aligned}
\frac{\partial \phi}{\partial (t/N)} &= \frac{1}{4} \frac{\partial^2}{\partial p^2} [p(1 - p) \phi] + \frac{1}{4} \frac{\partial^2}{\partial q^2} [q(1 - q) \phi] \\
&+ \frac{1}{2} Ns \left\{ \frac{\partial}{\partial p} [p(1 - p)(q - \bar{q}) \phi] + \frac{\partial}{\partial q} [q(1 - q)(p - \bar{q}) \phi] \right\}.
\end{aligned} \qquad (18)$$

Therefore on a time scale proportional to N, and under the diffusion equation assumptions, changes in gene frequency will be a function of only Ns, $\bar{q}$, and the initial frequencies p_0 and q_0.

When r and s are very small, the value of μ_L can be obtained approximately for RRS for all values of $\bar{q}$ and $N \neq M$, $r \neq s$, and the derivation is given in the appendix. The main result is, from Eq. (A 3) of the appendix

$$\mu_L - \mu_0 = a \left\{ M r p_0 (1 - p_0)(q_0 - \bar{q})^2 + N s q_0 (1 - q_0)(p_0 - \bar{q})^2 \right.$$
$$\left. + \left[M r + N s - \frac{2MN(r+s)}{2M + 2N - 1} \right] p_0 (1 - p_0) q_0 (1 - q_0) \right\} \tag{19}$$

plus terms containing higher powers of r and s. With $M = N$ and $r = s$, Eq. (19) reduces to

$$\mu_L - \mu_0 = N s a \left[p_0 (1 - p_0)(q_0 - \bar{q})^2 + q_0 (1 - q_0)(p_0 - \bar{q})^2 \right.$$
$$\left. + \left(1 + \frac{1}{4N - 1} \right) p_0 (1 - p_0) q_0 (1 - q_0) \right]. \tag{20}$$

If N is large, the term $1/(N-1)$ in Eq. (20) may be ignored, and when $q = 1$ (complete dominance), Eq. (20) reduces to

$$\mu_L - \mu_0 = N s a (1 - p_0)(1 - q_0)(p_0 + q_0 - p_0 q_0) \tag{21}$$

which can also be obtained by series expansion of Eq. (16), taking only terms up to order $N s$.

Checks on the adequacy of the diffusion equation approximation for chances of fixation in single populations have been reported (Ewens, 1963; Hill, 1969a). These indicate that the diffusion results are certainly adequate for descriptive purposes and can therefore be used for PLS and RST in this study. In Table 1 diffusion equation results for RRS with complete dominance (Eq. (16)) are compared with those from the transition probability matrix (8) for $N = 4$ and $N = 8$. We see that the

Table 1. *Comparison of selection limits μ_L/a for reciprocal recurrent selection with complete dominance estimated from transition probability matrices with $N = 4(T4)$ and $N = 8(T8)$, from the diffusion approximation (DA), and by approximation ignoring terms of order greater than $Ns(A)$*

Ns	Initial frequencies											
	0.25, 0.25				0.25, 0.5				0.5, 0.5			
	$T4$	$T8$	DA	A	$T4$	$T8$	DA	A	$T4$	$T8$	DA	A
$\frac{1}{2}$	0.5589	0.5598	0.5606	0.5605	0.7336	0.7345	0.7352	0.7422	0.8329	0.8339	0.8347	0.8437
1	0.6702	0.6723	0.6744	0.6836	0.8223	0.8237	0.8252	0.8594	0.8956	0.8971	0.8985	0.9375
2	0.8329	0.8370	0.8416	—	0.9311	0.9329	0.9350	—	0.9650	0.9664	0.9679	—
4	0.9618	0.9655	0.9701	—	0.9918	0.9925	0.9936	—	0.9970	0.9974	0.9979	—
8	0.9977	0.9983	0.9991	—	0.9999	0.9999	1.0000	—	1.0000	1.0000	1.0000	—

agreement is generally very good, although for a given value of Ns the limit is consistently higher with larger N and with the diffusion approximation, which may be regarded as the limiting value as N becomes infinite. A similar bias has been observed in single population studies. Comparisons of alternative breeding methods have therefore always been made at the *same* population size, either using the diffusion equation or transition probability matrices with $N = 8$ for RRS, RST, and PLS. However, the results of Table 1 indicate that we can express the parameters in terms of Ns rather than N and s separately and thus draw inferences about a wide range of population sizes from results obtained at one population size, thereby reducing greatly the amount of computation required.

Also included in Table 1 is a check on the accuracy of the simple formula (21) including only linear terms in Ns. For $Ns = \frac{1}{2}$ there seems to be reasonable agreement, but the simple formula becomes strongly biased upwards if Ns is much larger. This range of validity is not surprising, since μ_L can be written as an expansion of exponential terms in Ns from Eq. (16).

Comparison of Selection Methods

We now have sufficient theory to enable us to compare the efficiencies of the alternative breeding schemes. Since the relative efficiencies differ considerably from one model of gene action to another, we shall consider these in turn, starting with complete dominance. All comparisons will be made in terms of the parameters Mr, Ns, and so on, although the cases with $Mr = Ns$ will be studied in greatest detail.

In practice it should be possible to attain higher values of s and Ns with pure line selection in particular, for recourse need not be made to progeny testing and the generation interval can be reduced. Using as a basis a single two year cycle of one generation with progeny testing, we have defined $s = \dfrac{ia}{\sigma_f}$ where σ_f is the standard deviation of progeny test means, and the response *per cycle* with PLS is $\delta q = -\dfrac{s}{2} q(1-q)(q-\bar{q})$. With mass selection (and, of course, PLS) the response *per generation* (Eq. (1)) is $\delta q = -\dfrac{i_m a}{\sigma} q(1-q)(q-\bar{q})$ where σ is the phenotypic standard deviation. Thus the approximate value of s over two generations is $\dfrac{4 i_m a}{\sigma}$, but the effective population size for the two generation period is $N/2$ since the sampling variance for one generation with $N/2$ is approximately the same as with two generations with population size N. An example may help to illustrate these comparisons.

(i) Progeny testing: two year cycle $\sigma_f = 8$, $i = 1$, $N = 32$.

(ii) Mass selection: one year cycle $\sigma = 30$, $i_m = 1.5$, population size $= 40$ each generation. Also let $a = 2$, with units, say, eggs per hen in poultry. With progeny testing we have $N = 32$, $s = 1/4$, $Ns = 8$ and in the same context with mass selection $N = 20$, $s = 2/5$, $Ns = 8$. We might thus predict the same total advance using these two schemes with pure line selection, since the limit is a function of Ns, but the rate of advance, inversely proportional to N, would be faster with mass selection. Other variations on selection programmes which affect their efficiency can be included in the same way. Comparisons of the different schemes should not necessarily be made for the same values of Ns; these Ns values can be modified to suit the individual's own prejudices. Unfortunately in the simple model which has been adopted, it is not possible to modify the values of s as selection proceeds, when, for example, some loci may become fixed and the phenotypic variance is reduced. Strictly, therefore, we are considering only single genes.

1. Complete Dominance ($\bar{q} = 1$)

With complete dominance the optimum selection limit is attained if either population is fixed for the favourable dominant allele A_1 and the optimum crossbred can therefore be reached with RST.

In Fig. 1 the selection limit, μ_L, is plotted for the three alternative breeding schemes PLS, RRS, and RST for a model of complete dominance and three different pairs of initial frequencies. The mean at the limit is plotted relative to the gene effect and so it can range from 0 to 1. The results were obtained from the diffusion equation and $Mr = Ns$. With RST it is assumed that selection is practised in one population and the tester is fixed instantaneously, with probability p_0, say, that it is fixed for A_1. When $p_0 \neq q_0$, two alternative sets of results are shown for RST, depending on the initial frequency in the population used as the tester.

For a given value of Ns it can be seen that the highest limit is always attained with RRS, but the superiority over PLS is never large. A doubling of s using PLS, which might be attained by practising mass selection thus reducing the generation interval and increasing the selection intensity, would reverse these rankings. The RST method is competitive so long as selection is practised in the correct population (an ideal which might be difficult to attain in a breeding programme!). We notice in Fig. 1 that the greater response occurs if the selected line has the higher initial frequency, in this case 0.5 as opposed to 0.1. It is coincidental in this

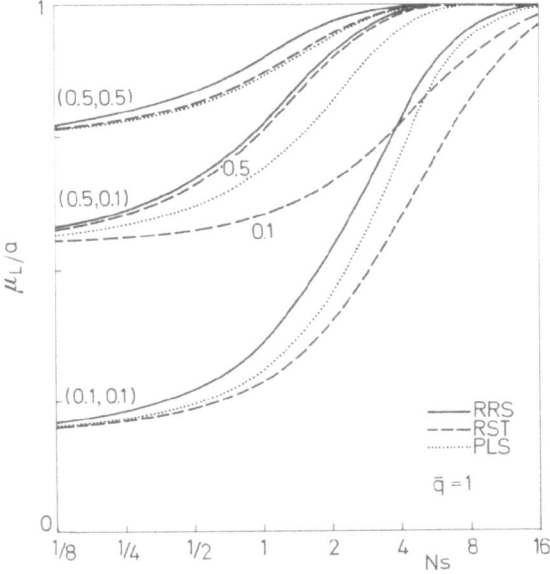

Fig. 1. The selection limit for the crossbred mean with complete dominance expressed as a proportion of the gene effect, a, for three pairs of initial frequencies and a range of Ns values. The initial frequency of the selected population with RST is also shown

example that the frequency is also nearer 0.5 where the variance of gene frequency is highest for we now show that it is always more efficient to select by RST in the line at higher initial frequency if there is complete dominance. From Eq. (12) and assuming selection in population Y, with initial frequency q_0, we find with RST that if the tester is fixed instantaneously

$$\mu_L = a[p_0 + (1 - p_0)(1 - e^{-2Nsq_0})/(1 - e^{-2Ns})] . \qquad (22)$$

With rearrangement and series expansion Eq. (22) gives

$$\mu_L/a = 1 - \frac{2Ns(1-p_0)(1-q_0)}{e^{2Ns}-1} \left\{ 1 + \frac{2Ns(1-q_0)}{2!} + \frac{[2Ns(1-q_0)]^2}{3!} + \cdots \right\}. \qquad (23)$$

For $s > 0$ a higher limit is therefore reached if $q_0 > p_0$ than if $p_0 > q_0$, so that selection should be practised in the population with higher initial frequency. The most plausible verbal interpretation seems to be that by selection in the population at high frequency we almost ensure fixation of the favourable allele. Since this is dominant, the optimum limit is reached in the cross.

In the RST system only one population has to be maintained at a large size (i.e. N), for drift is of no consequence in the tester. Thus if the total number of breeding animals is restricted by the facilities available, it may be possible to use a value of N almost twice as large in the single selected population with RST as compared with RRS or PLS, so that the RST system may be relatively more efficient than Fig. 1 indicates. Some of this advantage may be lost if the reproductive rate in the inbred tester is poor, so that family sizes are smaller and the progeny tests less efficient.

When Ns is small, the selection limits under the different systems can easily be compared. The results can be summarised as follows for complete dominance:

PLS: $\mu_L - \mu_0 = \frac{1}{3} N sa(1 - p_0)(1 - q_0)[p_0(2 - p_0) + q_0(2 - q_0)]$,

RRS: $\mu_L - \mu_0 = N sa(1 - p_0)(1 - q_0)(p_0 + q_0 - p_0 q_0)$,   (24)

RST: $\mu_L - \mu_0 = N sa(1 - p_0)(1 - q_0) q_0$.

For example if $p_0 = q_0$, the ratios of advance with the different systems are for small Ns, $\dfrac{\text{PLS}}{\text{RST}} = \dfrac{2}{3}(2 - q_0)$, $\dfrac{\text{RRS}}{\text{RST}} = 2 - q_0$, and $\dfrac{\text{PLS}}{\text{RRS}} = \dfrac{2}{3}$,

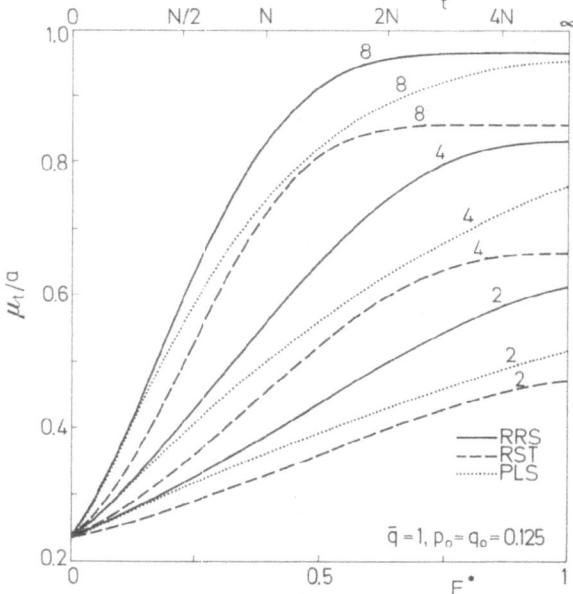

Fig. 2. Progress in the crossbreds from selection with complete dominance, initial frequencies 0.125 in each population and $Ns = 2$, 4 or 8. Time is expressed as $F^* = 1 - e^{-t/2N}$ and the crossbred mean as a proportion of the effect a

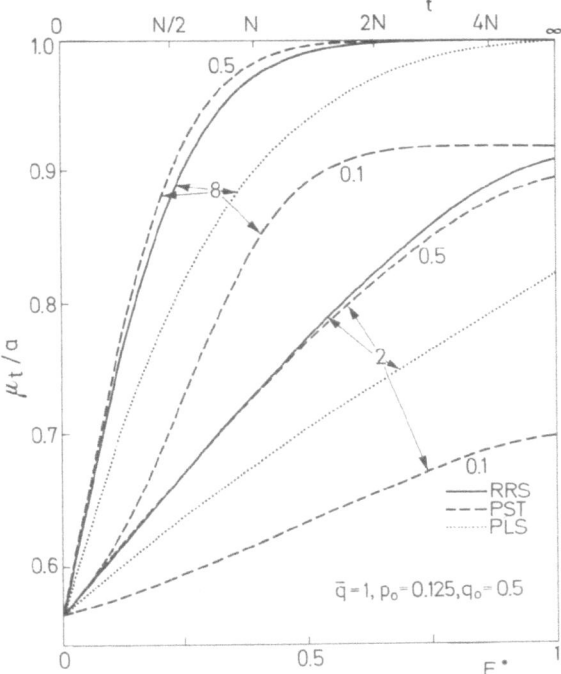

Fig. 3. Progress in the crossbreds from selection with complete dominance, initial frequencies 0.125 and 0.5 and $Ns = 2$ or 8. The initial frequency of the selected population with RST is shown. Time is expressed as $F^* = 1 - e^{t/2N}$ and the crossbred mean as a proportion of the effect a

again assuming that Ns is the same for each system. A doubling of N with RST would yield $\dfrac{RRS}{RST} = 1 - q_0/2$ which is less than 1. Again we note that RST is more efficient at high initial frequencies in the tester. The difference in advance (RRS − PLS) with $p_0 \neq q_0$ may be written RRS − PLS $= \frac{1}{3} Nsa(1 - p_0)(1 - q_0)[(p_0 + q_0) - 5p_0q_0]$ which is positive for $Ns > 0$ and $0 < p_0, q_0 < 1$.

The progress from selection before the limit is reached is shown in Figs. 2 and 3 for two pairs of inital frequencies, $p_0 = q_0 = 0.125$ (Fig. 2) and $p_0 = 0.125$, $q_0 = 0.5$ (Fig. 3), each with $Mr = Ns = 2$ and 8. In these graphs the expected value of μ_t/a at cycle t is plotted in successive cycles. However, the time scale (which ranges to $t \to \infty$) has been condensed by using the transformation $F^* = 1 - e^{-t/2N}$. With no selection F^* is approximately equal to the inbreeding coefficient, where N is the effective population size for a complete cycle of testing and selection. We see in the figures that although the pattern of response is not identical for the different methods there is essentially no change in ranking from the outset. The selection advance is seen to slow down earlier with RRS and RST than with PLS. Since RST is equivalent to selection for an additive gene, RST and PLS merely reflect the patterns of response for additive and dominant genes in single populations and these have been described elsewhere (Robertson, 1960; Hill, 1969b). Assuming instantaneous fixation of population X with RST, the initial rates of advance ($\delta\mu$ with $t=1$) are

$$\text{PLS: } \delta\mu = \frac{sa}{2}(1-p_0)(1-q_0)\left[p_0(1-p_0)+q_0(1-q_0)\right],$$

$$\text{RRS: } \delta\mu = \frac{sa}{2}(1-p_0)(1-q_0)\left[p_0(1-q_0)+q_0(1-p_0)\right], \qquad (25)$$

$$\text{RST: } \delta\mu = \frac{sa}{2}(1-p_0)(1-q_0)\,q_0 \,.$$

Eq. (25) can be compared with the approximate limit formulae (24). The total advance for RST is $2N$ times that in the first generation and rather more than $2N$ times for RRS. However Eq. (24) have very strong restrictions on the magnitude of Ns. Usually less than $2N$ times the initial advance is made with additive gene action (Robertson, 1960; Hill, 1969b).

2. Partial Dominance ($1 < \bar{q} < \infty$)

We shall consider the model of partial dominance in less detail than that of either complete dominance or overdominance. In Fig. 4 the crossbred mean at the limit is plotted for the case of $\bar{q}=1.5$. The genotypic values would then be $1.5\,a$, a and $-0.5\,a$ for A_1A_1, A_1A_2, and A_2A_2 respectively, but have been transformed in the graph to $1.0, 0.75,$ and $0.0,$ respectively. Thus if y is the ordinate, $y=0.25+\mu_L/(2a)$ with $0\le y\le 1$.

The initial frequencies chosen are similar to those given in Fig. 1 for complete dominance, but we notice some distinct differences in the results. Since the optimum crossbred mean is only attained when both populations are fixed for the favoured allele A_1, this limit cannot be reached with RST unless the tester is already fixed for A_1. Thus the RST system is relatively inefficient, particularly when Ns values are large. There is no simple rule about the gene frequencies in the population in which selection should be practised with RST. For $\bar{q}=1.5$, we have from Eq. (12) for RST.

$$\mu_L/a = \frac{3}{2}p_0 - \frac{1}{2} + \frac{1}{2}p_0\left[\frac{1-e^{-Nsq_0}}{1-e^{-Ns}}\right] + \frac{3}{2}(1-p_0)\left[\frac{1-e^{-3Nsq_0}}{1-e^{-3Ns}}\right]. \qquad (26)$$

As $Ns \to \infty$, $\mu_L/a \to 1 + p_0/2$, and the limit is maximized if selection is practised in the population at lower initial frequency, which is capable of making the greatest total advance. For very small Ns, $(\mu_L - \mu_0)/a \to Nsq_0(1 - q_0)$ $(9/4 - 2q_0)$ which turns out to be higher if q_0 (the initial frequency in the selected population) exceeds p_0, except when q_0 is very close to 0 or 1, when the ranking may be reversed.

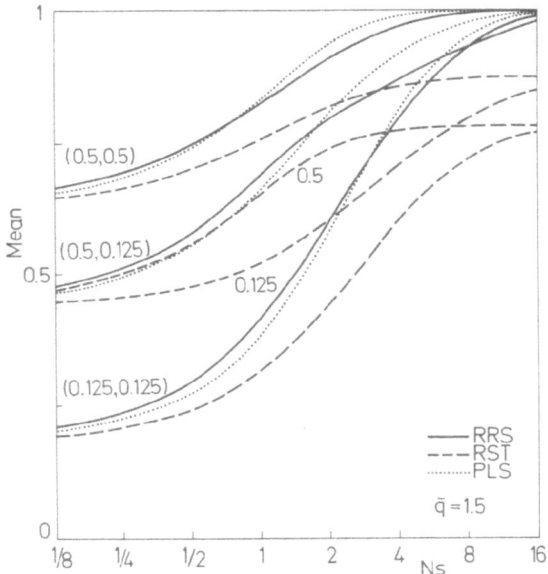

Fig. 4. As Fig. 1, but for partial dominance with $\bar{q} = 1.5$. The mean is expressed as $0.25 + \mu_L/2a$

At low Ns values RRS is more efficient than PLS, but the ranking changes as Ns becomes large. However, at no Ns value is the difference in gain large between these alternative schemes. Since higher Ns values are likely to be possible with PLS, this method is likely to be most efficient in practice if there is partially dominant gene action. We also note in Fig. 4 that with the highest Ns value (16) and RRS a higher limit is attained with $p_0, q_0 = 0.125$, 0.5 than with $p_0 = q_0 = 0.5$. Perhaps the explanation for this phenomenon is that the selection pressures are initially weaker in, for example, population X if Y has frequency $q_0 = 0.5$ rather than 0.125 and X becomes fixed for A_2 in the early generations more frequently when $q_0 = 0.5$.

The rates of advance with partial dominance are similar for RRS and PLS, and we shall not discuss them further.

3. Additivity ($\bar{q} \to \infty$)

If there is additive gene action the ranking of individuals on pure or cross performance will be the same, so that for given Ns both RRS and PLS have the same efficiency and rates of advance, but higher Ns values may be attainable with PLS. The RST system is not suitable for additive gene action, as Comstock *et al.* (1949) and Dickerson (1952) have pointed out, for with selection in only one population the advance can not exceed one-half of that possible with fixation of the favourable allele in both populations.

4. Overdominance ($0 < \bar{q} < 1$)

With overdominant gene action PLS is not a successful breeding system. If both homozygotes have the same genotypic value ($\bar{q} = 0.5$) PLS leads eventually to random fixation of either homozygote. Otherwise the more favourable homozygote is more frequently fixed (Robertson, 1962; Hill and Robertson, 1968), and, especially when $\bar{q}$ lies far from 0.5 and Ns is large, almost all lines will be fixed for the same homozygote and the line cross will not show heterosis (Robertson, 1962). However, with intermediate equilibrium frequencies and large Ns the progress to fixation is very slow, and we might expect to find populations selected on pure line performance having overdominant genes near their equilibrium frequency. This situation has received some attention, for the initial rate of advance with RRS will be zero if both populations are at equilibrium (Comstock *et al.*, 1949; Dickerson, 1952). Arthur (1964) demonstrated that the early response with RRS is greatly increased by a short period of inbreeding before selection is started, so that the genes drift from their equilibrium frequency. Dickerson (1952) showed that recurrent selection to an inbred tester gave greater initial response, for additive variance is immediately obtained. Much of the following discussion on the overdominance model will therefore center on the case of initial equilibrium; PLS will be ignored.

Initial Equilibrium. The mean of either line or the crossbred is initially $\mu_0 = a[1 - \bar{q}(1 - \bar{q})]$. If the maximum possible gain is made, the final crosses will all be heterozygotes with mean a, and this gain is $a\bar{q}(1 - \bar{q})$. It also turns out that at equilibrium $\sigma_d = a\bar{q}(1 - \bar{q})$, where σ_d^2 is the dominance variance at this locus. There is, of course, no additive variance at equilibrium. The initial rate of advance is $\delta\mu = 0$ with RRS and $\delta\mu = \dfrac{sa}{2} \bar{q}(1 - \bar{q})^3$ or $\delta\mu = \dfrac{sa}{2} \bar{q}^3(1 - \bar{q})$ with RST, according to whether A_1 or A_2, respectively, is fixed in the inbred tester. As before we shall assume that one population is fixed instantaneously, with probability $\bar{q}$ that A_1 is the allele fixed. Then the initial rate of advance with RST is on average $\delta\mu = \dfrac{sa}{2}\,\bar{q}^2(1-\bar{q})^2 = i\sigma_d^2/2\sigma_f$. The additive variance in the cross is now σ_d^2. Only after some drift has occurred is any additive variance generated with RRS and a response obtained.

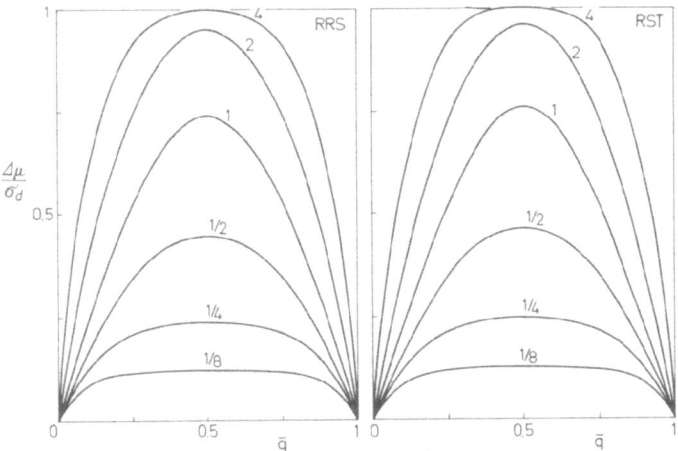

Fig. 5. Ratio of expected to maximum gain ($\Delta\mu/\sigma_d$) in the crossbreds for overdominance with initial equilibrium. Curves are plotted for different values of $N i\sigma_d/\sigma_t$

In Fig. 5 the selection limits with initial equilibrium for RRS and RST (assuming instantaneous fixation of the tester) are compared for all values of $\bar{q}$, using matrix iteration results with $N = 8$. Since the total advance $\Delta\mu = \mu_L - \mu_0$ lies in the range $0 \leqq \Delta\mu \leqq \sigma_d$, for all $\bar{q}$, the quantity shown, $\Delta\mu/\sigma_d$, is the proportion of the possible gain which is realized. Using the diffusion equation approximation we know that, if $p_0 = q_0 = \bar{q}$ and $Mr = Ns$, $\Delta\mu$ is a function of only $Ns = Nia/\sigma_f$ and $\bar{q}$. Therefore $\Delta\mu$ can also be expressed as a function of only $Nia\bar{q}(1-\bar{q})/\sigma_f = Ni\sigma_d/\sigma_f$ and $\bar{q}$, where Ni is under the breeder's control and σ_d/σ_f is a measure of the contribution of the locus in question to the total variability. So in Fig. 5 the limit is expressed as $\Delta\mu/\sigma_d$ for a range of values of $Ni\sigma_d/\sigma_f$. Clearly the most startling aspect of the results is that almost the same advance is made with RST as with RRS for the whole spectrum of parameters. For all but the smallest Ns values an algebraic verification has not been found, since we have no diffusion formula for the limit with RRS. However when second order terms in r and s are ignored, we have from Eq. (19) that with initial equilibrium

$$\mu_L - \mu_0 = a\left[Mr + Ns - \frac{2MN(r+s)}{2M+2N-1}\right]\bar{q}^2(1-\bar{q})^2 . \qquad (27)$$

For $Mr = Ns$, ignoring terms of order $1/N$ relative to 1 and setting $Ns\, a\, \bar{q}^2(1-\bar{q})^2 = Ni\sigma_d^2/\sigma_f$ Eq. (19) reduces to

$$\mu_L - \mu_0 = Ni\sigma_d^2/\sigma_f . \qquad (28)$$

Summing over all overdominant loci at equilibrium, let $\sigma_D^2 = \Sigma\sigma_d^2$ so that the total advance with RRS from equilibrium becomes $Ni\sigma_D^2/\sigma_f$, assuming gene effects are small. Setting $Mr = 0$ in Eq. (20) we obtain the predicted total advance with RST when the tester is fixed instantaneously, which is again $\mu_L - \mu_0 = Ni\sigma_D^2/\sigma_f$. This formula for RST can also be obtained directly from the results of Robertson (1960), who showed that the total advance with selection for additive genes of small effect is $2Ni_m\sigma_A^2/\sigma$, where σ_A^2 is the additive variance and mass selection is practised. With progeny testing, Robertson's formula becomes $Ni\sigma_A^2/\sigma_f$ in the notation of this paper.

Therefore we see that RRS and RST are predicted to give the same total advance if gene effects are small. This advance, $Ni\sigma_D^2/\sigma_f$, is independent of $\bar{q}$, and similarly at a single locus the advance $\Delta\mu$, expressed as a proportion of the possible advance, $\Delta\mu/\sigma_d$, is $Ni\sigma_d/\sigma_f$, again independent of $\bar{q}$. We see in Fig. 5 that, for given $Ni\sigma_d/\sigma_f$, this relation does not hold well if the $\bar{q}$ values are extreme. This is not surprising, since larger values of ia/σ_f (selective values) are required for the same value of $i\sigma_d/\sigma_f$, and the assumptions of small effects are violated. Perhaps a more serious weakness of this theory is that we cannot expect genes of small effect to be segregating at frequencies close to equilibrium if the original populations are of finite size. Hill and Robertson (1968) have studied the distribution of frequency of overdominant genes in finite lines selected on pure performance. The mean frequency in lines still segregating is not the equilibrium frequency, unless $\bar{q} = 0.5$, but is generally intermediate between $\bar{q}$ and 0.5, being closer to 0.5 if gene effects are small.

Fig. 5 shows clearly that when $Ni\sigma_d/\sigma_f$ is greater than one-half or so, a greater proportion of the possible advance from the equilibrium state is made if $\bar{q}$ has intermediate values. If $\bar{q}$ has extreme values, there is a high probability that the favourable allele, initially at higher frequency, will be fixed in both populations by chance.

Although approximately the same total advance is made with RRS and RST from initial equilibrium, the initial rate of advance and thus the overall pattern of change must be greatly different. In Fig. 6 the mean of the cross in succeeding generations is compared for RRS and RST with $\bar{q} = 0.5$ and three values of $Ns(= 4Ni\sigma_d/\sigma_f$ for $\bar{q} = 0.5)$. This figure illustrates the contrast in response rate. If Ns is small, it is shown in the appendix (setting $r = s$, $m = n$, and $p_0 = q_0 = \bar{q}$ in Eq. (A4)) that with initial equilibrium and RRS

$$\mu_t - \mu_0 = Nsa F_t^2 \bar{q}^2 (1 - \bar{q})^2 \text{ approximately}$$
$$= F_t^2 \cdot Ni\sigma_d^2/\sigma_f \tag{29}$$

where $F_t = 1 - (1 - 1/2N)^t$ or $e^{-t/2N}$, approximately, or the inbreeding coefficient estimated from pedigrees. Similarly, with RST, which is effective selection for additive genes, it can be shown with the same assumptions that

$$\mu_t - \mu_0 = F_t \cdot Ni\sigma_d^2/\sigma_f \quad \text{approximately.} \tag{30}$$

Thus the selection advance is proportional to F^2 with RRS and F with RST. Also the half-life of the process, the time taken to get halfway to the limit (Robertson, 1960) will be $t = 2.5N$ cycles or generations approximately, with RRS when $F^2 = 0.5$, and $t = 1.4N$ generations, approximately, with RST when $F = 0.5$. In the example in Fig. 6 with $Ns = 1$

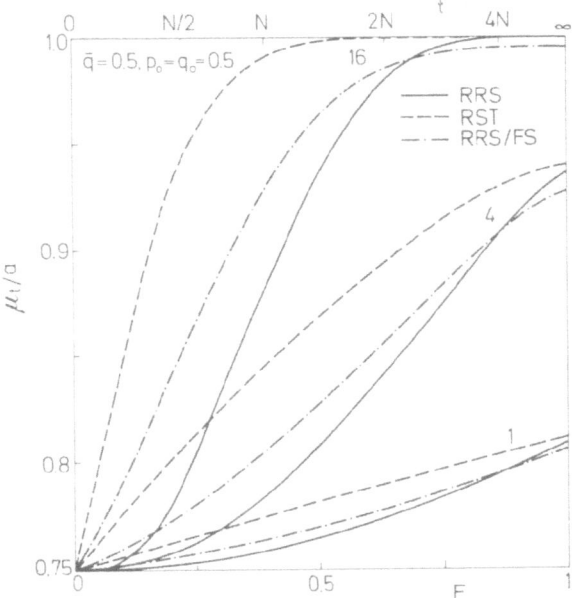

Fig. 6. Progress in the crossbreds from selection with overdominance and $\bar{q} = 0.5$ for three values of Ns and RST, RRS with no prior inbreeding and RRS after one generation of full sibbing. Time is expressed as $F^* = 1 - e^{-t/2N}$ and the crossbred mean as a proportion of the effect a

these results hold fairly well. With RST the half-way point is passed between generations 10 and 11, corresponding to $1.25N$ and $1.375N$ generations respectively; and with RRS the corresponding generations are 19 and 20, or $2.375N$ and $2.5N$. With large Ns values the half-lives are seen to be reduced with both RRS and RST but are always shorter with RST. Again, assuming small gene effects, the response in a single generation is shown in the appendix for RRS and initial equilibrium to be

$$\mu_{t+1} - \mu_t = F_t(1 - F_t)\,i\sigma_d^2/\sigma_f, \quad \text{approximately.} \tag{31}$$

Similarly for RST, or additive selection

$$\mu_{t+1} - \mu_t = (1 - F_t)\tfrac{1}{2}\,i\sigma_d^2/\sigma_f, \quad \text{approximately.}$$

Thus, under these assumptions, the greatest rate of advance is made with RRS when $F_t = 0.5$. Then, incidentally, both RRS and RST have the same predicted rate of advance, but response is fastest with RST at the outset.

As Arthur (1964) has shown, an initial period of inbreeding enables response to be made immediately with RRS from an equilibrium position. It can be shown that if populations X and Y are inbred up to inbreeding coefficients F_0 and G_0, respectively, prior to selection, then the initial rate of advance with equilibrium becomes

$$\mu_1 - \mu_0 = \tfrac{1}{2}a\,[rF_0(1 - G_0) + sG_0(1 - F_0)]\,\bar{q}^2(1 - \bar{q})^2 \tag{32}$$

or if $F_0 = G_0$, and $r = s$,

$$\mu_1 - \mu_0 = F_0(1 - F_0)\,i\sigma_d^2/\sigma_f \tag{33}$$

which is identical to Eq. (31) but since no prior selection is involved, Eq. (33) does *not* require the assumption of small gene effects. A formal proof of Eq. (32) and (33) can be made by similar methods to those used in the appendix; however, the equations can be derived fairly easily in an intuitive manner. We have shown that

$$\delta\mu = \frac{a}{2}\,[rp(1 - p)(q - \bar{q})^2 + sq(1 - q)(p - \bar{q})^2].$$

The average value of $p(1 - p)$ after inbreeding to level G_0 from an initial frequency of $\bar{q}$ is $(1 - G_0)\,\bar{q}(1 - \bar{q})$, for this is the within-line variance of gene frequency. Similarly, the average value of $(q - \bar{q})^2$ is $F_0\bar{q}(1 - \bar{q})$ for, with initial frequency $\bar{q}$, the quantity $(q - \bar{q})^2$ is the between-line variance. The genetic drift occurs independently in the two populations, so that the average value of $p(1 - p)(q - \bar{q})^2$ is $(1 - G_0)\,F_0\,\bar{q}^2(1 - \bar{q})^2$ and Eq. (32) follows immediately.

When gene effects are small, the total advance after inbreeding to level F_0 in each population becomes

$$\mu_L - \mu_0 = (1 - F_0^2)\, N i \sigma_a^2 / \sigma_f \,.$$

For example, after one generation of full sibbing in each population prior to selection, $F_0 = 1/4$. Therefore the early advance is $3/16\, i\sigma_a^2/\sigma_f$, or $3/4$ of the maximum rate at $F = 1/2$, yet the total gain is reduced by only about $1/16$. Results from RRS after a single generation of full sibbing are also included in Fig. 6, so that they can be compared with RST and RRS without prior inbreeding.

So far we have compared RRS and RST with the same values of Ns, yet RST requires only one segregating population and not two. Now it is conceivable, as has been mentioned earlier in the section on complete dominance, that facilities may limit the total number of individuals, $N + M$, which can be maintained. How then should our facilities be utilized? A solution is readily obtained for the model of small gene effects and initial equilibrium. Letting $r = s = ia/\sigma_f$, we have from Eq. (27)

$$\mu_L - \mu_0 = \left(M + N - \frac{2MN}{M + N - 1} \right) i\sigma_a^2/\sigma_f \,. \tag{34}$$

For $M + N$ constant, Eq. (34) has a relative minimum at $M + N$ and is maximized at $M = 0$ or $N = 0$. Thus the RST method is most efficient and the RRS method least efficient under these assumptions, differing by a factor of 2 in predicted advance.

A weakness in our analysis has been the assumption of instantaneous fixation with RST. Again using the model of small effects and initial equilibrium we test this approximation. We assume no selection in the tester and set $r = 0$ and $s = ia/\sigma_f$ in Eq. (27) and obtain

$$\mu_L - \mu_0 = \left(1 - \frac{2M}{2M + 2N - 1} \right) N i\sigma_a^2/\sigma_f \tag{35}$$

where M is the size of the tester. If N is large relative to M, Eq. (35) becomes approximately

$$\mu_L - \mu_0 = (1 - M/N)\, N i\sigma_a^2/\sigma_f \,. \tag{36}$$

For example, with $N = 16$ and $M = 2$ about $1/8$ of the advance possible with instantaneous fixation and RST may not be realized.

Initial Disequilibrium. With overdominance and initial departure from the equilibrium frequency there are so many combinations of parameters that it is difficult to generalize. However in Figs. 7, 8, and 9 some typical results are presented, in which the selection limits predicted from RST and RRS are compared for the intermediate equilibrium frequency, $\bar{q} = 0.5$, with one population at equilibrium (Fig. 7) and neither at equilibrium (Fig. 8), and for a more extreme equilibrium frequency, $\bar{q} = 0.25$ (Fig. 9). In each graph results are given for two values of Ns.

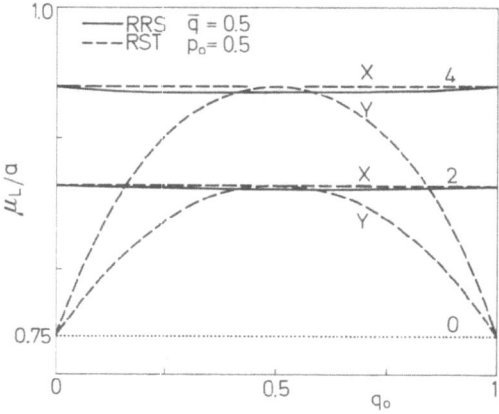

Fig. 7. The selection limit in the crossbreds, expressed as a proportion of the effect, a, with overdominance ($\bar{q} = 0.5$) and two values of Ns. The initial frequency in X is $p_0 = 0.5$ and the initial frequency in Y is q_0. The population in which selection is practised with RST is also shown

From Fig. 7 we find that if $\bar{q} = 0.5$ RST is more efficient if selection is practised in the population nearest the equilibrium frequency. This rule for $\bar{q} = 0.5$ seems to hold quite generally, but has not yielded to algebraic proof. Also, we see in Figs. 7 and 8 that RST is never appreciably superior to RRS when $\bar{q} = 0.5$ and comparisons are made at the same Ns value, and a similar result is observed in Fig. 9 when $\bar{q} \neq 0.5$. However no general rule has been found for identifying which population should be selected in an RST programme. As with partial dominance this depends not only on p_0, q_0, and $\bar{q}$ but also on Ns.

When there is initial disequilibrium, there is less difference in the rate of initial advance with RRS and RST programmes, for additive variance is immediately available with RRS. With $\bar{q} = 0.5$ and $Ns = 4$, some examples are given in Fig. 10. The rate of initial advance is always greater with RST (unless the population in which selection is practised has a very extreme frequency), but some immediate gains are made with

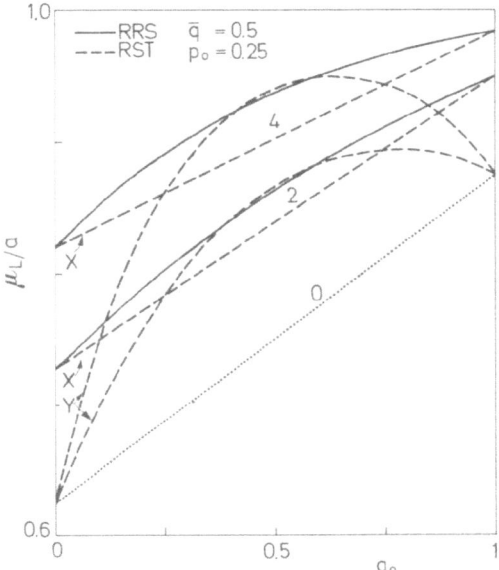

Fig. 8. As Fig. 7 but $\bar{q} = 0.5$, $p_0 = 0.25$

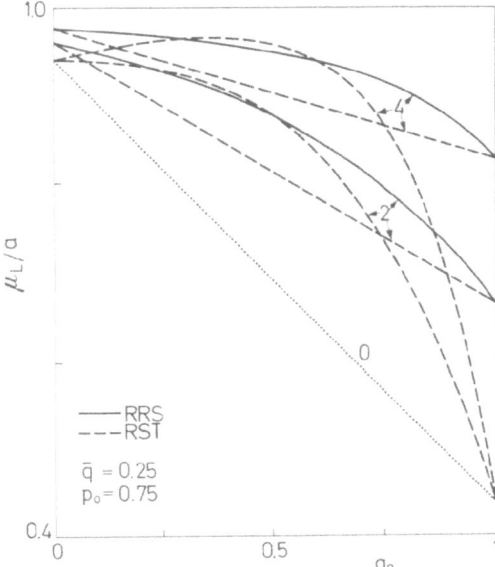

Fig. 9. As Fig. 7 but $\bar{q} = 0.25$, $p_0 = 0.75$

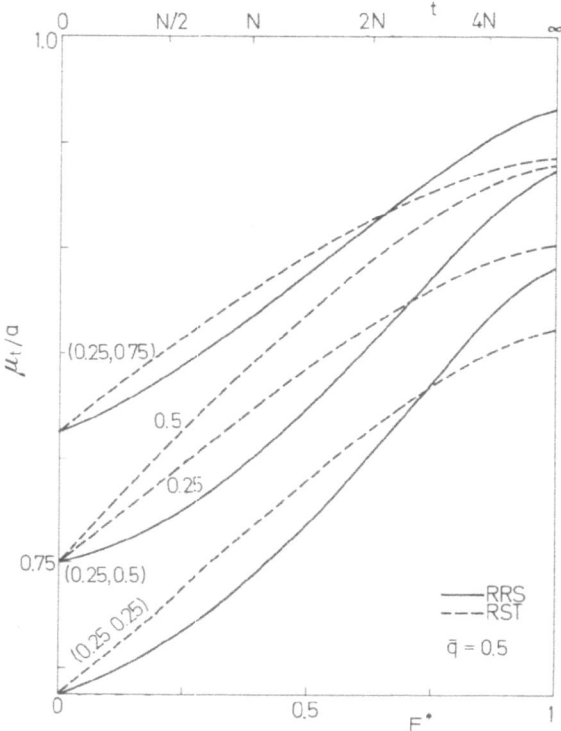

Fig. 10. Progress in the crossbreds from selection with overdominance and $\bar{q} = 0.5$ and $Ns = 4$ for three pairs of initial frequencies. The initial frequency of the selected population with RST is shown. Time is expressed as $F^* = 1 - e^{-t/2N}$ and the crossbred means as a proportion of the effect a

RRS. It is possible to analyze these situations in some detail for the model of small gene effects in the same manner as we have studied complete dominance and overdominance with initial equilibrium, but this will not be undertaken.

Discussion

Although the theory which has been developed is both very approximate and formally restricted to single genes, it is hoped that it gives some information on the problems of selection with subsequent line crossing. The main advance in the theory is, of course, the introduction of finite population size so that the selection limit defined is the expected limit, not the maximum limit possible with fixation of only favourable combinations. Also by introducing finite population size we can draw a contrast between the initial rate of gain and the final limit. In fact the study of initial equilibrium with overdominance requires finite population theory.

A disappointing, but by no means surprising aspect of the results is that it is not possible to reach very general conclusions. We have considered each model of gene action in turn and find, in common with Comstock *et al.* (1949), Dickerson (1952), and Crow (1953), that the optimum breeding system is not the same for each model. For example with partial dominance RST is never very efficient, whereas with overdominance it may be the optimum system both for short term and long term gains. Of course, if epistasis were included in the study, further complications would inevitably be found. In addition some of the conclusions depend upon the size of gene effects on the quantitative trait under selection, and such information is almost completely lacking. Fortunately it appears from the results that the *relative* efficiency of the different methods discussed is not greatly influenced by the magnitude of effects, so in this respect our results may be of as much utility as those from infinite population studies.

There would be little benefit in entering a debate here about the nature of gene action found in practise – such speculations can be left to the breeder when setting up a programme. Unfortunately the experimental data available to him is unlikely to give unequivocal pointers to the genetics of the economic traits in which he is interested. One might argue that more knowledge of gene effects and equilibrium frequencies is necessary before a theory such as in this paper has any merit. However most breeders are used to designing programmes with insufficient information about the parameters, so new theory might still be considered beneficial. Since the relative efficiencies of the breeding systems depend so much on the nature of gene action, it is perhaps not surprising that experiments on different species or strains in the same species have not given clear-cut evidence about the utility of RRS, for example.

No attempt has been made to include in the theory refined variations on mating systems, particulary those possible with plants. The typical scheme adopted has been selection on the basis of progeny test performance, with random mating of individuals both within the strain producing the next generation and between the strains for test crossing. Incorporation of any other system, at least with random mating, should be possible in terms of the effective population size and selection coefficient per cycle, and the method has been outlined earlier in the paper. There are other, quite separate breeding systems, which merit further analysis, perhaps within the framework of this paper. In particular what are the relative efficiencies of programmes based partly or largely on between-line selection? Arthur (1964) studied a model with recurrent cycles of inbreeding and between-line selection, but the gene effects were sufficiently large that the optimum limit was eventually reached.

The analysis has been basically in terms of single genes. However, without tight linkage it should be possible to extend them into a polygenic situation, merely by summation of variances and responses over loci. In order to simplify the analysis, we assumed that selective values remained constant from the early to the late generations, and since $s = ia/\sigma_f$ we are thereby assuming that σ_f also remains constant. As σ_f^2 contains genetic variance, this expectation is unlikely to be realized for both inbreeding and selection will modify the variance so that our results are biased. Qureshi and Kempthorne (1968) and Robertson (1970) use Monte Carlo methods to compute limits in single populations with many loci. They include the case of free recombination and find deviations from single locus theory, depending both on initial frequencies and gene effects. More important in our study is how the relative efficiencies of alternative systems are affected – a similar bias in each system is not important. The approximations may not be too serious as chance fixation of many favourable genes, especially those at extreme frequency, occurs in the early generations of selection, before the variance has changed too much. Inclusion of epistasis or linkage into the theory is in practise quite simple, if laborious, with Monte Carlo methods, but of course the number of possible parameter combinations increases enormously.

Summary

A theoretical comparison is made of alternative breeding systems which utilize only selection within lines to improve a cross between two strains. The schemes considered are selection on pure line performance (PLS) and selection on cross performance either by recurrent selection to an inbred tester (RST) or by reciprocal recurrent selection (RRS). This theory extends earlier comparisons in that the selected lines are assumed to be of finite size and predictions are made of the expected limit rather than the limit possible with fixation of only favourable combinations. A simple model of a single locus with two alleles with specified degree of dominance is used.

The selection limit is defined in terms of the combined parameter Ns, where N is the effective population size for a cycle of progeny testing and selection and $s = i\,a/\sigma_f$ where i is the standardized selection differential,

a the effect of the gene on the quantitative trait, and σ_f the standard deviation of progeny test means. Appropriate values of Ns can be calculated for other schemes such as mass selection and thus the results are quite general. For PLS and RST and for RRS with complete dominance, the limit could be predicted from diffusion equation approximations, and for all methods with small gene effects by simpler approximate methods. Other results were obtained using transition probability matrix iteration.

The optimum selection programme depends on the model of gene action. Comparisons made at the same value of Ns in each selected line gave the following results:

1. With *complete dominance* RRS is more effective than PLS and the efficiency of RST depends on the initial gene frequencies in the two strains. A higher limit with RRS is reached if selection is practised in the population with the higher initial frequency and the other is inbred. Then RST may be almost as efficient as RRS. Although there are some differences in rate of advance between the methods, the ranking does not change appreciably during selection.

2. With *partial dominance* RRS and PLS have similar efficiency, but RST is less efficient, particularly at high Ns values, since the optimum is attained only if both populations are fixed for the favourable allele.

3. With *overdominance and initial equilibrium* in each population PLS is not useful. RRS and RST give essentially the same limit for all values of the equilibrium frequency, but the initial rate of progress is much faster with RST. For genes of small effect the total advance up to cycle t is proportional to F_t with RST and F_t^2 with RRS, and the rates of response are proportional to $1 - F_t$ and $F_t(1 - F_t)$ respectively, where F_t is the inbreeding coefficient from the start of the programme. An initial period of inbreeding up to level F_0 in each line before commencing RRS gives an initial rate of advance proportional to $F_0(1 - F_0)$ and reduces the limit to $1 - F_0^2$, if effects are small.

4. With *overdominance and initial disequilibrium* the relative efficiency of RRS and RST depends on the initial frequencies, and general results are difficult to obtain. There is less difference in the rate of advance between the methods than with initial equilibrium.

Usually it will be possible to attain higher Ns values with PLS than RRS or RST, by using individual selection with reduced generation interval. Also, since only one non-inbred population has to be maintained with RST, it is possible that a larger Ns value can be utilized than with RRS. The rankings of the methods are likely to be affected.

Appendix — Response to Reciprocal Recurrent Selection when Selective Values are Small

From Eq. (8) we have the transition probability matrix D, with elements

$$
d_{(h,i,j,k)} = \binom{m}{j} \left[\frac{h}{m} - \frac{r}{2} \frac{h}{m} \left(1 - \frac{h}{m}\right) \left(\frac{i}{n} - \bar{q}\right) \right]^{j}
$$
$$
\times \left[1 - \frac{h}{m} + \frac{r}{2} \frac{h}{m} \left(1 - \frac{h}{m}\right) \left(\frac{i}{n} - \bar{q}\right) \right]^{m-j}
$$
$$
\times \binom{n}{k} \left[\frac{i}{n} - \frac{s}{2} \frac{i}{n} \left(1 - \frac{i}{n}\right) \left(\frac{h}{m} - \bar{q}\right) \right]^{k} \tag{A1}
$$
$$
\times \left[1 - \frac{i}{n} + \frac{s}{2} \frac{i}{n} \left(1 - \frac{i}{n}\right) \left(\frac{h}{m} - \bar{q}\right) \right]^{n-k}
$$

where $0 \leq h, j \leq m$; $0 \leq i, k \leq n$, and $m = 2M$ and $n = 2N$ in Eq. (8). Thus h and j are the numbers of A_1 alleles in population X and i and k the numbers in population Y at generations t and $t+1$ respectively so that D is square of dimension $(m+1)(n+1)$. A row of the matrix is identified by h and i, a column by j and k. Expanding D into terms up to order r or s, and rearranging, we obtain

$$
d_{(h,i,j,k)} = \binom{m}{j} \left(\frac{h}{m}\right)^{j} \left(1 - \frac{h}{m}\right)^{m-j} \binom{n}{k} \left(\frac{i}{n}\right)^{k} \left(1 - \frac{i}{n}\right)^{n-k}
$$
$$
\times \left[1 + \frac{r}{2}(h-j)\left(\frac{i}{n} - \bar{q}\right) + \frac{s}{2}(i-k)\left(\frac{h}{m} - \bar{q}\right) \right]
$$
$$
+ 0(r^2) + 0(rs) + 0(s^2).
$$

Let us define matrices A, B, C with elements

$$
a_{(h,i,j,k)} = \binom{m}{j} \left(\frac{h}{m}\right)^{j} \left(1 - \frac{h}{m}\right)^{m-j} \binom{a}{k} \left(\frac{i}{n}\right)^{k} \left(1 - \frac{i}{n}\right)^{n-k},
$$
$$
b_{(h,i,j,k)} = a_{(h,i,j,k)} (h-j) \left(\frac{i}{n} - \bar{q}\right),
$$
$$
c_{(h,i,j,k)} = a_{(h,i,j,k)} (i-k) \left(\frac{h}{m} - \bar{q}\right),
$$

so that $D = A + \dfrac{r}{2} B + \dfrac{s}{2} C + 0(r^2) + 0(rs) + 0(s^2)$.

Let us also define vectors of $(m+1)(n+1)$ rows as follows:

$$V_1 : v_{1\,(h,\,i)} = \frac{hi}{mn} = pq\,,$$

$$V_2 : v_{2\,(h,\,i)} = \frac{h(m-h)\,i}{m^2 n}\left(\frac{i}{n} - \bar{q}\right) = p(1-p)\,q(q-\bar{q})\,,$$

$$V_3 : v_{3\,(h,\,i)} = \frac{i(n-i)\,h}{mn^2}\left(\frac{h}{m} - \bar{q}\right) = q(1-q)\,p(p-\bar{q})\,,$$

$$V_4 : v_{4\,(h,\,i)} = \frac{h(m-h)\,i(n-i)}{m^2 n^2} = p(1-p)\,q(1-q)$$

where $p = h/m,\, q = i/n$. It can be shown that

$$A V_1 = V_1\,,$$
$$B V_1 = - V_2\,,$$
$$C V_1 = - V_3\,,$$
$$A V_2 = \frac{m-1}{m}\,V_2 + \frac{m-1}{mn}\,V_4\,,$$
$$A V_3 = \frac{n-1}{n}\,V_3 + \frac{n-1}{mn}\,V_4\,,$$
$$A V_4 = \frac{(m-1)(n-1)}{mn}\,V_4\,.$$

The probability that a cross between X and Y at generation t has progeny $A_1 A_1$ is given by elements of the vector $V_1^{(t)} = D^t V_1$ and the joint probability of fixation of A_1 in both lines by $V_1^{(\infty)} = \lim_{t\to\infty} D^t V_1$. We now use the above relationships to derive these quantities, but ignoring high order terms in r and s. We assume in the following equations that r and s are of similar order of magnitude.

$$V_1^{(1)} = (A + rB + sC)\,V_1 \ = V_1 - \frac{r}{2}\,V_2 - \frac{s}{2}\,V_3 + 0(s^2)\,,$$

$$V_1^{(2)} = (A + rB + sC)\,V^{(1)} = V_1 - \frac{r}{2}\left(1 + \frac{m-1}{m}\right)V_2 - \frac{s}{2}\left(1 + \frac{n-1}{n}\right)V_3$$

$$- \frac{r}{2}\left(\frac{m-1}{mn}\right)V_4 - \frac{s}{2}\left(\frac{n-1}{mn}\right)V_4 + 0(s^2)\,,$$

$$V_1^{(3)} = V_1 - \frac{r}{2}\left[1 + \frac{m-1}{m} + \left(\frac{m-1}{m}\right)^2\right]V_2 - \frac{s}{2}\left[1 + \frac{n-1}{n} + \left(\frac{n-1}{n}\right)^2\right]V_3$$

$$- \frac{r}{2}\left[\frac{m-1}{mn} + \frac{(m-1)^2}{m^2 n} + \frac{(m-1)^2(n-1)}{m^2 n^2}\right]V_4$$

$$- \frac{s}{2}\left[\frac{n-1}{mn} + \frac{(n-1)^2}{mn^2} + \frac{(m-1)(n-1)^2}{m^2 n^2}\right]V_4 + 0(s^2)$$

and so on. In general, if we write

$$V_1^{(t)} = V_1 - \tfrac{1}{2}[\alpha_t r V_2 + \beta_t s V_3 + \gamma_t r V_4 + \delta_t s V_4] + 0(s^2),$$ (A 2)

we obtain, for example, the following recurrence relationships

$$\gamma_{t+1} = \left(\frac{m-1}{mn}\right)\alpha_t + \frac{(m-1)(n-1)}{mn}\gamma_t$$

with initial conditions $\alpha_0 = \beta_0 = \gamma_0 = \delta_0 = 0$.

These have solutions and limiting values

$$\alpha_t = m\left[1 - \left(\frac{m-1}{m}\right)^t\right], \quad \lim_{t\to\infty}\alpha_t = m,$$

$$\beta_t = n\left[1 - \left(\frac{n-1}{n}\right)^t\right], \quad \lim_{t\to\infty}\beta_t = n,$$

$$\gamma_t = \frac{m(m-1)\left[1 - \left(\frac{m-1}{m}\right)^t\right] - mn\left(\frac{m-1}{m}\right)^t\left[1 - \left(\frac{n-1}{n}\right)^t\right]}{m+n-1},$$

$$\lim_{t\to\infty}\gamma_t = \frac{m(m-1)}{m+n-1},$$

$$\delta_t = \frac{n(n-1)\left[1 - \left(\frac{n-1}{n}\right)^t\right] - mn\left(\frac{n-1}{n}\right)^t\left[1 - \left(\frac{m-1}{m}\right)^t\right]}{m+n-1}$$

$$\lim_{t\to\infty}\delta_t = \frac{n(n-1)}{m+n-1}.$$

The vector of probability of fixation of A_1 in both lines is obtained by substitution in Eq. (A2) and is

$$V_1^{(\infty)} = V_1 - \frac{1}{2}\left\{mr V_2 + ns V_3 + \left[mr + ns - \frac{mn(r+s)}{m+n-1}\right]V_4\right\} + 0(s^2).$$

In terms of gene frequencies, the joint chance of fixation $w(p_0, q_0)$ is then

$$w(p_0, q_0) = p_0 q_0 - m\frac{r}{2}p_0(1-p_0)q_0(q_0 - \bar{q}) - n\frac{s}{2}q_0(1-q_0)p_0(p_0 - \bar{q})$$

$$- \frac{1}{2}\left[mr + ns - \frac{mn(r+s)}{m+n-1}\right]p_0(1-p_0)q_0(1-q_0) + 0(s^2).$$ (A 3)

The other joint probabilities of fixation, such as A_2 in each population $x(1-p_0, 1-q_0)$, are obtained using Eq. (A3) and the marginal probabilities of fixation, which can be shown to be

$$u(p_0) = p_0 - m \frac{r}{2} p_0(1 - p_0) (q_0 - \bar{q}) + O(s^2),$$

$$v(q_0) = q_0 - n \frac{s}{2} q_0(1 - q_0) (p_0 - \bar{q}) + O(s^2).$$

The selection limit is given by $\mu_L = a[1 - (1 - \bar{q}) w(p_0, q_0) - \bar{q} x(1 - p_0, 1 - q_0)]$ and, after substituting, the total advance becomes

$$\mu_L - \mu_0 = \frac{1}{2} a \left\{ mr p_0(1 - p_0) (q_0 - \bar{q})^2 + ns q_0(1 - q_0) (p_0 - \bar{q})^2 \right.$$
$$\left. + \left[mr + ns - \frac{mn(r + s)}{m + n - 1} \right] p_0(1 - p_0) q_0(1 - q_0) \right\} + O(s^2).$$

In order to specify the response at intermediate generations it is convenient to let $F_t = 1 - \left(\frac{m - 1}{m} \right)^t$, $G_t = 1 - \left(\frac{n - 1}{n} \right)^t$ so that for example, $\alpha_t = m F_t$, $\gamma_t = \frac{m[(m - 1) F_t - n(1 - F_t) G_t]}{m + n - 1}$ and

$$\mu_t - \mu_0 = \frac{1}{2} a \left\{ mr F_t p_0(1 - p_0) (q_0 - \bar{q})^2 + ns G_t q_0(1 - q_0) (p_0 - \bar{q})^2 \right.$$

$$+ \frac{mr[(m - 1) F_t - n(1 - F_t) G_t] + ns[(n - 1) G_t - m(1 - G_t) F_t]}{m + n - 1} \tag{A4}$$

$$\left. \times p_0(1 - p_0) q_0(1 - q_0) \right\} + O(s^2).$$

If $m = n$ and $r = s$ and n is sufficiently large that $m + n - 1 \sim m + n$,

$$\mu_t - \mu_0 = \frac{1}{2} ans F_t [p_0(1 - p_0) (q_0 - \bar{q})^2 + q_0(1 - q_0) (p_0 - \bar{q})^2$$
$$+ F_t p_0(1 - p_0) q_0(1 - q_0)] + O(s^2) \tag{A5}$$

and

$$\mu_{t+1} - \mu_t = \frac{1}{2} as(1 - F_t) [p_0(1 - p_0) (q_0 - \bar{q})^2 + q_0(1 - q_0) (p_0 - \bar{q})^2$$
$$+ 2 F_t p_0(1 - p_0) q_0(1 - q_0)] + O(s^2). \tag{A6}$$

Proof of Convergence. We have not shown that $w(p_0, p_0) - p_c q_0$ of Eq. (A2) actually converges to the quantity found as r and s tend to zero. We note that D has elements which are polynomials in r and s, hence then are all D^t, including $\lim_{t \to \infty} D^t$. But the elements of $\lim_{t \to \infty} D^t V_1 = V_1^{(\infty)}$ are bounded in the range $0 \leq v_{(h,i)}^{(\infty)} \leq 1$, since $V_1^{(\infty)}$ is a vector of probabilities. Thus we can expand $V_1^{(\infty)}$ in a series of vectors

$$V_1 = E_{00} + r E_{10} + s E_{01} + r^2 E_{20} + rs E_{11} + s^2 E_{02} + \cdots$$

for all $r, s, -\infty < r, s < \infty$, where E_{ij} are vectors of coefficients. Therefore $-\infty < E_{ij} < \infty$ for all i, j, and if r and s are of the same order, say $r = \sigma s$, $0 < \sigma < \infty$

$$\lim_{s \to 0} \left(\frac{V_1^{(\infty)} - E_{00}}{s} \right) = \sigma E_{10} + E_{01} + \lim_{s \to 0} [s(\sigma^2 E_{20} + \sigma E_{11} + E_{02} + \cdots)]$$

$$= \sigma E_{10} + E_{01},$$

and we can ignore all terms of higher order as a first approximation.

References

Arthur, J. A.: Investigation of population structure with recurrent selection. Unpublished Ph. D. Thesis, University of California, Davis (1964).

Bell, A. E., C. H. Moore, and D. C. Warren: The evaluation of new methods for the improvement of quantitative characteristics. Cold Spr. Harb. Symp. Quant. Biol. **20**, 197–211 (1955).

Bowman, J. C.: Selection for heterosis. Anim. Breed. Abstr. **27**, 261–273 (1959).

Comstock, R. E., H. F. Robinson, and P. H. Harvey: A breeding procedure designed to make maximum use of both general and specific combining ability. Agron. J. **41**, 360–367 (1949).

Cress, C. E.: A comparison of recurrent selection schemes. Genetics **54**, 1371–1379 (1966).

— Reciprocal recurrent selection and modifications in simulated populations. Crop Sci. **7**, 561–567 (1967).

Crow, J. F.: Theoretical considerations of reciprocal recurrent selection and recurrent selection compared to other breeding methods. Proc. Amer. Poultry Breeders' Roundtable, 1953.

Dickerson, G. E.: Inbred lines for heterosis tests? *Heterosis*, pp. 330–351, ed J. W. Gowen. Ames: Iowa State College Press 1952.

Ewens, W. J.: Numerical results and diffusion approximations in a genetic process. Biometrika **50**, 241–249 (1963).

Falconer, D. S.: *Introduction to quantitative genetics.* Edinburgh-London: Oliver and Boyd 1960.

Griffing, B.: Theoretical consequences of truncation selection based on the individual phenotype. Aust. J. Biol. Sci. **13**, 307–343 (1960).

Haldane, J. B. S.: A mathematical theory of natural and artificial selection. VII. Selection intensity as a function of mortality rate. Proc. Camb. Phil. Soc. **27**, 131–136 (1931).

Hill, W. G.: On the theory of artifical selection in finite populations. Genet. Res. **13**, 143–163 (1969a).

— The rate of selection advance for non-additive loci. Genet. Res. **13**, 165–173 (1969b).

—, and A. Robertson: The effects of inbreeding at loci with heterozygote advantage. Genetics **60**, 615–628 (1968).

Hull, F. H.: Recurrent selection for specific combining ability in corn. J. Amer. Soc. Agron. **37**, 134–145 (1945).

Kimura, M.: Some problems of stochastic processes in genetics. Ann. Math. Statist. **28**, 882–901 (1957).

— Diffusion models in population genetics. J. Appl. Prob. **1**, 177–232 (1964).

Kojima, K.: Effects of dominance and size of population on response to mass selection. Genet. Res. **2**, 177–188 (1961).

—, and T. M. Kelleher: A comparison of purebred and crossbred selection schemes with two populations of *Drosophila subobscura*. Genetics **48**, 57–72 (1963).

Latter, B. D. H.: The response to artificial selection due to autosomal genes of large effect. I. Changes in gene frequency at an additive locus. Aust. J. Biol. Sci. **18**, 585–598 (1965).

Ohta, T.: Effect of initial linkage disequilibrium and epistasis on fixation probability in a small population, with two segregating loci. Theor. Appl. Genet. **38**, 243–248 (1968).

Qureshi, A. W., and O. Kempthorne: On the fixation of genes of large effects due to continued truncation selection in small populations of polygenic systems with linkage. Theor. Appl. Genet. **38**, 249–255 (1968).

Rasmuson, M.: Reciprocal recurrent selection. Results of three model experiments on *Drosophila* for improvement of quantitative characters. Hereditas **42**, 397–414 (1956).

Robertson, A.: A theory of limits in artificial selection. Proc. Roy. Soc. Lond. B. **153**, 234–249 (1960).

— Selection for heterozygotes in small populations. Genetics **47**, 1291–1300 (1962).

— A theory of limits in artificial selection with many linked loci. (this volume) (1970).

A Theory of Limits in Artificial Selection with Many Linked Loci

A. ROBERTSON

Introduction

Though linkage has been clearly recognized to be an important factor in artificial selection, particularly by Mather and his colleagues, the absence of any predictive treatment of its effect has remained one of the most important gaps in selection theory. In the last decade, there have been several efforts to attack the problem when many loci are involved using computer simulation (Fraser, 1957; Cockerham and Martin, 1958; Gill, 1965; Qureshi and Kempthorne, 1968, and others). These have rather illustrated its importance than increased our fundamental understanding. The problem is certainly very complex. In this paper we deal initially with a rather simplified situation with a set of loci equally spaced along a chromosome, each with two alleles with the same initial allelic frequencies and effects on the character under selection. We further restrict the problem by assuming additive gene action and initial linkage equilibrium. Nevertheless seven parameters are needed to describe the initial state of the population and the selection process. The necessary parameters and the symbols for them are:

(i) The effective population size in Wright's sense of the word (N).

(ii) The number of loci (n).

(iii) The difference between the two homozygotes at a locus in the character selected for (a).

(iv) The initial frequency of the desirable allele at each locus (q).

(v) A measure of the intensity of selection (i).

(vi) The phenotypic variance of the character in the initial population (σ^2).

(vii) The map distance between adjacent loci (c).

We are then interested in the change in the character under selection at different generations with the main emphasis in the present paper on the limit of selection when all loci are fixed, though not necessarily for the desirable allele.

Certainly one of the main reasons for the lack of progress has been a failure to recognize that the population size is one of the most important factors determining the effect of linkage. A theory of selection in finite populations for single loci has only recently been developed. The situation with two linked loci with initial linkage equilibrium is fairly well understood from the work of Latter (1965) and Hill and Robertson (1966). Without the assumption of equivalence between the two loci, the specification of the two-locus problem requires seven parameters, but a satisfactory treatment is possible by a reduction to five parameter combinations. Here the initial frequency of the desirable alleles proves to be of major importance. In almost all other simulation work, with the exception of Qureshi and Kempthorne (1968), initial frequencies of 0.5 have been assumed. Another feature of the work with two loci referred to above is that it was built around a comparison of the extreme situations of no crossing-over ($r = 0$) and free recombination ($r = 1/2$) respectively between the loci. In a surprising number of other studies, no real zero point has been included. Very often simulation runs with a small amount of crossing-over (say $r = 0.005$) have been used for this purpose. In some instances, this may be in effect a long way from the true zero, as will be discussed later in the paper.

This exploration of a seven-dimensional space is not easy. I do not find that the conventional statistical approach to the problem, as used by Qureshi, Kempthorne, and Hazel (1968) in which one takes a set of values of the relevant parameters, does runs with all possible combinations within the set, and analyzes the results in main effects, first order interactions, and so on, contributes much to our understanding of the problem. To say that we "understand" a problem implies that we have found that way of looking at it which gives the simplest possible description of the results – that which has the least number of interactions. We may by a conventional analysis discover that factor A and factor B have significant effects and that there is a significant interaction between them. It is however much more useful to find that an analysis in terms of, say, A^2/B accounts for almost all of the variation due to both factors. In statistical terms, we are seeking the best "re-parameterization". We are only using a computer to find a way of thinking about the problem without needing a computer.

This has been achieved by, first of all, dealing with those special cases which were amenable to algebra on the basis of existing theory to get some clue as to what combinations of parameters were useful and using these as a guide when looking for a pattern in the simulation results. In fact, the parameter combination which now appears most useful in dealing with intermediate amounts of linkage arose from seeing a particular regularity in the results and developing a theoretical explanation for it. The conventional statistical analysis in essence provides a description of a set of results *a posteriori*. Our approach was on the contrary an exploratory and experimental one, never knowing one week what computer runs we would do the next. This raises statistical problems because it means that special attention in terms of extra replications is always given to the exceptions which apparently do not fit into the pattern. Significance tests on such data are of doubtful value. For this reason there is not much statistical analysis in this paper except in cases in which little of such "doctoring" has been done.

Considering first a single chromosome, the theory is developed in terms of the relative advance under selection in the two extreme conditions: on the one hand when the loci are so far apart that they segregate independently and, on the other, when there is no crossing-over at all between them. The two situations are first discussed theoretically. That with independently segregating loci presents few new difficulties. With no crossing-over at all, the many-locus two-allele situation becomes equivalent to a single locus with many alleles with the distribution of "allele" effects approaching normality as the number of constituent loci increases. If L_f is the expected advance at the limit in the character under selection with free segregation and L_0 that with no crossing-over, we are then interested in the variation of L_0/L_f with the many parameters. We then remove the conditions of equality of gene effect and initial gene frequency. Intermediate linkage values are introduced, as well as simultaneous segregation of several chromosomes. This gives a predictive theory with reasonable validity over a wide range of initial conditions. But it must be emphasized that it is not a theory into which the necessary variables can be inserted to make exact predictions but rather an attempt to get an overall simplified view of the situation from which approximations are not absent.

Selection with Independent Segregation of the Loci

For convenience, we shall refer to the set of loci as the "chromosome". Using the symbolism given above, the genetic variance due to all loci, σ_g^{*2}, is $na^2q(1-q)/2$ and will make a proportion of h^{*2} of the phenotypic variance, σ^2. The latter may contain also genetic variance due to loci on other chromosomes.

If the loci segregate independently and we apply a selection differential of i standard deviations, the expected change in gene frequency at each locus in the first generation, δq, is $iaq(1-q)/2\sigma$, when a is small relative to σ. The resultant change from these loci in the character under selection is $na\delta q$ which equals $nia^2q(1-q)/2\sigma$. This, not surprisingly equals $ih^*\sigma_g^*$, the usual formula of quantitative genetics.

Now Kimura (1957) has shown that, if the selective advantage of the gene, s, which equals ia/σ, remains constant to fixation, the chance of fixation of the favourable allele is determined almost completely by Ns and q and is given by $u(q) = \dfrac{1 - e^{-2Nsq}}{1 - e^{-2Ns}}$ approximately. We can use this formula with the reservation that s for any locus will not remain constant as the genetic variance at the other loci will certainly alter, and perhaps also the environmental variance. These effects are not predictable with certainty. The genetic effects will certainly be larger at higher heritabilities. At gene frequencies of the favourable alleles at the other loci above 0.5, both genetic drift and selection will reduce the genetic variance so that the formula will underestimate the probable change. In the first generation, if the frequency, p, at other loci is very small, we know that p (and also $p(1 - p)$) will increase by a proportion $s/2$ due to selection whereas the expected value of $p(1 - p)$ will decrease by a proportion $1/(2N)$ due to drift. There will then be a temporary overall increase in genetic variance at the other loci if $Ns > 1$ and perhaps a consequent underestimation of chance of fixation.

Accepting Kimura's formula, however, we find that if Ns is small, the expected total change in gene frequency tends to $Nsq(1 - q)$, i.e., $2N$ times the change in the first generation. When q is small, the total change may for some values of Ns be greater than $2N$ times the initial change (though never more than $4N$ times it), but the value of the ratio then declines as the expected change reaches the limiting value of $1 - q$. If $u(q) - q$ is plotted against Ns, a series of curves are obtained which have an initial slope of $q(1 - q)$ and a final ordinate of $1 - q$.

How will this be reflected in the character under selection? Suppose the abscissa is altered from $Ns(= Nia/\sigma)$ to Nih^*. This involves a multiplication by σ_g^*/a. We then alter the ordinate from a gene frequency change to one of a change in the character under selection due to all loci, expressed as a proportion of the initial genetic standard deviation. The proportional factor is na/σ_g^*. Under this change of variables, the initial slope of the curves is altered to $q(1 - q) \cdot na^2/\sigma_g^{*2} = 2$, independent of q. The limiting ordinate is now $na(1 - q)/\sigma_g^* = (2n(1 - q)/q)^{\frac{1}{2}}$.

Selection in the Absence of Crossing-Over

Consider now the chromosome segregating as one unit within which there is no crossing over, with the assumption of linkage equilibrium in the initial population of gametes. The genetic variance controlled by the chromosome is, in the early stages of selection, equal to that when the loci are segregating independently. Assume that the gene frequencies at all loci are initially 0.5 and that all loci have an equal effect on the character under selection. Then, even though only a few loci are involved, the distribution of chromosomal effect on the selected character will approach normality. If the initial frequencies are not 0.5, the approach to normality as the number of loci is increased will be slower. We shall deal later with the effect of non-normality of the distribution.

If the distribution of chromosomal effects is normal, we can make certain predictions about the expected limits of selection. When the selection pressures are weak, i.e. when Nih^* is small, we are in a similar situation to that with a single locus with low values of Ns in that the genetic variance will decline by a fraction $1/2N$ each generation. The expected total change will equal $2N$ times the gain in the first generation, or $2Nih^*\sigma_g^*$.

Consider now high values of Nih^* when we select, say, the best 20 individuals each generation. We cannot do better than fix the best chromosome out of the initial sample however intense the selection. The expected superiority of this over the initial mean will depend on the size of the sample. Table 1 shows the expected superiority, in genetic standard deviations, of a line fixed for the gamete with the largest effect on the character from a sample of a given size.

This is constructed from published tables of extreme variates in samples from a normal distribution, multiplied by $\sqrt{2}$ since the line is homozygous for the best gamete.

Table 1. *The expected superiority, in genetic standard deviations, of a line homozygous for the best gamete from an initial sample of a given size*

No. of gametes in initial sample	Expected superiority in genetic standard deviations
10	2.18
20	2.64
40	3.05
80	3.44
160	3.76
1000	4.58

Over the range from 20 to 80, that of typical selection programs, at least in animals, the superiority increases from 2.64 to 3.44. Although there is an increase with population size, the effect is not great. Thus, with the usual population sizes in selection experiments, we can say that the expected advance at high values of Nih^* with no crossing-over will be of the order of $3\sigma_g^*$. In subsequent discussions, the phrase "Nih^* is high" should be taken to imply that L_0 is in this limiting range, i.e. $Nih^* > 5$.

Let us assume for the moment that this factor is really independent of N (which it is almost). If we plot L/σ_g^* against Nih^*, we obtain a curve which has an initial slope of 2 and reaches a final level of about 3. In other words, using the symbol L_0 for the advance with no crossing over and L_f for that with free recombination, we have $L_0/\sigma_g^* = f(Nih^*)$. For the same initial conditions with independent segregation, we know that L_f/σ_g^* is a function of Nih^*, q and n. We can thus make the reasonable conjecture: under these conditions the expected value of L_0/L_f (the effect of completely preventing crossing-over in a situation with free recombination) will be a function of Nih^*, n and q.

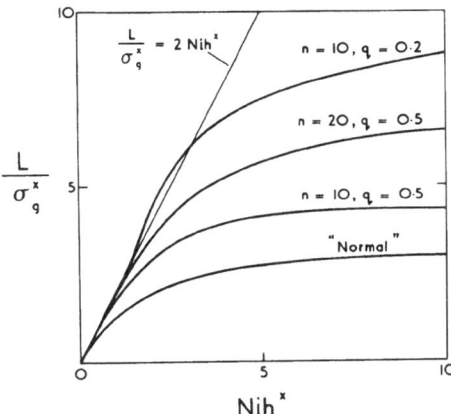

Fig. 1. The expected gain at the limit, in terms of the initial genetic standard deviation, for three situations involving independent segregation of the loci and one ("normal") in which there is no crossing-over and the initial distribution of chromosomal effects on the character selected for is normal

Fig. 1 shows the expected gain in various situations. The curves for free recombination are calculated from the Kimura formula, with $s = ia/\sigma$. That for the "normal" model was obtained by computer simulation by methods described in detail later with no crossing-over with $N = 20$, $n = 100$, and $q = 0.5$.

It is known that, for loci segregating independently, the pattern of change is a function of Ns and q with the time scale expressed in terms of N generations. In the framework of the present paper, this means that the gain for free recombination will be a function of Nih^*, n, q, and t/N. With no crossing-over, a similar generalisation applies to the "normal" model in that the gain in t generations will be a function of Nih^* and t/N.

How will the number of loci and initial gene frequency affect the limits of selection in the absence of crossing-over? Both the skewness and the kurtosis of the distribution will be affected and will be smaller the greater the number of loci involved. The skewness is zero when $q = 0.5$ and at frequencies of the desirable allele below this, the final advance will be greater than expected on the basis of normality and vice versa. The kurtosis (best thought of as the frequency of extreme variates relative to that in a normal distribution with the same variance) will again depend on the gene frequency. Between values of q, roughly 0.2 and 0.8, there is a shortage of extreme deviates and outside this range an excess. The importance of these effects in regard to the validity of the present theory, which rests on the assumption of a normal initial distribution of chromosome effects, is a matter for empirical investigation.

There is one particular effect of skewness which should be discussed here. This arises when the desirable alleles are at such a high initial frequency that, with initial samples of the usual order of size, there is a high probability that the most extreme possible chromosome (one containing the desirable allele at all loci) occurs in the initial sample. If Nih^* is high, this chromosome will probably be fixed even in the absence of crossing-over, so that L_0/L_f will be close to unity. In other words, at high Nih^* values there will be a critical value of q, dependent on the initial sample size and the number of loci, above which L_0/L_f will be unity. The gene frequency which gives such a gamete a 95% or greater chance of occurring in a sample of $2N$ can be calculated from the equation $(1 - q^n)^{2N} = 0.05$.

The critical values are given in Table 2 for various values of N and n.

Table 2. *The initial gene frequency above which a chromosome containing all desirable alleles has a probability of occurring in the initial sample greater than 0.95*

		Number of loci		
		5	10	20
Initial sample	20	0.67	0.82	0.91
of gametes	80	0.51	0.72	0.85

The critical value proves to depend more on the number of loci than on the initial sample size.

Assuming a normal initial distribution of chromosomal effects, we should finally deal with some special cases, when q does not take extreme values:

(i) when Nih^* is small, the expected gain with both free recombination and no recombination will equal $2Nih^*\sigma_g^*$ and L_0/L_f will equal unity. From our simulation results, this is true up to $Nih^* = 0.5$;

(ii) given n and q, there is a value of Nih^* above which, with free recombination, all desirable alleles have a chance of fixation close to unity. L_f then equals $(2n(1-q)/q)^{\frac{1}{2}}\sigma_g^*$ so that L_0/L_f will be of the order of $3(q/2n(1-q))^{\frac{1}{2}}$. As Nih^* increases further the ratio may slowly increase; L_f cannot increase further but L_0 can. This expression then gives a crude minimum to the ratio, dependent only on n and q.

(iii) As n increases, an increasingly longer initial section of the curve of L_f against Nih^* will be linear with slope 2 (see Fig. 1). As the curve of L_0 against Nih^* is also independent of n and q as n increases, it follows that, as n increases, L_0/L_f tends to a limiting form independent of n and q. When Nih^* is small, a treatment using matrix methods shows that L_0/L_f tends to $1 - \frac{2}{3}(Nih^*)^2$. This is in good agreement with computer results at $Nih^* = 0.5$. 1000 replicates with both free and no recombination with $N = 10$, $ih^* = 0.05$, $n = 10$, $q = 0.5$ gave a value of L_0/L_f of 0.84 ± 0.04. At higher values of Nih^*, where L_0 will be in the region of $3\sigma_g^*$, L_0/L_f will tend to $3/(2Nih^*)$ as n increases. At intermediate values of q, L_0/L_f will then tend as n increases to a value dependent only on Nih^*. As the length of the linear part of the L/σ_g^* curve is greater when q is smaller, this situation is more likely to be reached at low values of q. In fact, it has only been met in computer runs with n greater than 20, a value perhaps above the range likely to be encountered in practice.

Extreme Gene Frequencies

We have so far been dealing with intermediate initial frequencies, at which the normal approximation is more likely to hold. It is also possible to deal algebraically with extreme frequencies. At very extreme frequencies ($2Nnp$ or $2Nnq \gg 1$), crossing-over must have no effect. If, for instance, there is never more than one desirable allele at all loci present in the initial sample, the ratio L_0/L_f must be unity. It is possible to give a more precise treatment if at the start few gametes have more than one desirable allele or, at the other extreme, are lacking more than one. If nq is small so that there are very few gametes with more than two desirable alleles, the proportion with two may be written as $(nq)^2/2$. The probability that none of these will occur in a sample of $2N$ is $(1 - (n^2q^2)/2)^{2N} = 1 - Nn^2q^2$, approximately. For this probability to be greater than 0.95, then Nn^2q^2 must be less than 0.05. In the absence of crossing over, the situation is equivalent to that with two alternatives with initial frequencies nq and $1 - nq$, and substitution effect equal to that at a single locus, for which we can use the Kimura formula. The usual equations will apply for free recombination. It must be remembered that, as q approaches zero with Nih^* constant, Ns will increase though Nsq declines, since

$$s = ia/\sigma,$$

$$ih^* = i\frac{a}{\sigma}\left(\frac{nq(1-q)}{2}\right)^{\frac{1}{2}},$$

so that

$$Ns \to Nih^*\left(\frac{2}{nq}\right)^{\frac{1}{2}} \quad \text{and} \quad Nsq \to Nih^*\left(\frac{2q}{n}\right)^{\frac{1}{2}}$$

as q declines.

We thus have

$$L_f = a\left[n\frac{1-e^{-2Nsq}}{1-e^{-2Ns}} - nq\right]$$

which tends to $2Nnsaq$ under these conditions.

$$L_0 = a\left[\frac{1-e^{-2Nnsq}}{1-e^{-2Ns}} - nq\right]$$

Then

$$\frac{L_0}{L_f} = 1 - \frac{e^{-2Nnsq} - ne^{-2Nsq} + n - 1}{2Nnsq}$$

$$= 1 - (n-1)Nsq \quad \text{approximately}$$

$$= 1 - Nih^*\left(\frac{2(n-1)^2q}{n}\right)^{\frac{1}{2}}.$$

Thus, as q approaches zero, L_0/L_f approaches unity as $1 - K Nih^* q^{\frac{1}{2}}$. With $n = 5$, $q = 0.01$, the expression becomes $1 - Nih^*/4$ approximately, so that an appreciable reduction will be found at quite low values of Nih^*.

At high initial values of q we must modify the usual form of Kimura's equation with $p = 1 - q$ to

$$u(q) = \frac{e^{2Ns} - e^{2Nsp}}{e^{2Ns} - 1}.$$

With free recombination between the loci, we have

$$L_f = a\left[n\frac{e^{2Ns} - e^{2Nsp}}{e^{2Ns} - 1} - nq\right]$$

which tends to $na(1 - q)$ as Ns increases.

$$L_0 = a \left[\frac{e^{2Ns} - e^{2Nnsp}}{e^{2Ns} - 1} + (n - 1) - nq \right],$$

$$\frac{L_0}{L_f} = 1 - \frac{e^{2Nnsp} - ne^{2Nsp} + (n - 1)}{(e^{2Ns} - 1) n(1 - q)} \quad \text{approximately}$$

$$= 1 - \frac{2n(n - 1)(Nsp)^2}{e^{2Ns}np} \quad \text{approximately}$$

$$= 1 - \frac{4(n - 1)}{n} (Nih^*)^2 e^{-Nih^*(8/np)^{\frac{1}{2}}}$$

since now $Ns \to Nih^*(2/np)^{\frac{1}{2}}$ and $Nsp \to Nih^*(2p/n)^{\frac{1}{2}}$.

This has quite different properties to the corresponding expression at low frequencies of the desirable allele. Considering the second term as a function of Nih^*, it has the form $x^2 e^{-Kx}$ where $x = Nih^*$. This passes through a maximum when $Kx = 2$, ie $Nih^* = (np/2)^{\frac{1}{2}}$, the value at the maximum being $4e^{-2}/K^2$. L_0/L_f will then pass through a minimum of roughly $1 - 0.28\,np$ when $Nih^* = (np/2)^{\frac{1}{2}}$.

Thus, if we plot L_0/L_f against Nih^*, there is first a decline to a minimum followed by a rise which may, at high gene frequencies, extend up again to unity as we saw earlier from a different point of view. Another view of this can be gotten from Fig. 1. As Nih^* declines from high values with small n, L_f will not at first be affected as it is at the asymptote. L_0 will however decline and so will the ratio. As L_f declines from the asymptotic value, the trend will be reversed and the ratio will rise to unity at low values of Nih^*.

At very low values of q, we saw that L_0/L_f always declines with Nih^*. We now see that, at high values of q, the ratio will, above a certain value of Nih^*, increase as Nih^* increases. This implies that the plots of the ratio against q for different Nih^* values must cross one another at intermediate values of q.

The Simulation Programs

All simulation was done on the Edinburgh University KDF 9 computer. No attempt was made to use the binary nature of a problem involving sets of two alternatives and a gamete consisted of entries of 0 or 1 at n addresses corresponding to loci. In general, the replication was such that the coefficient of variation of final advance was between 5% and 10%. The average degree of replication was of the order of 100–200 though it varied widely, sometimes up to 1000 at small values of Nih^*

and down to 25 for the highest value of N used, 40. In this, computer time was the main limiting factor.

There were several varieties of simulation program varying in the method of selection and linkage simulation. There were three kinds of selection simulation. These were:

(i) "Real" artificial selection. For values of N and i of 20 and 0.8 respectively, for instance, forty diploid individuals were created by random sampling at each locus. The genotypic values were calculated by adding together the effects at all loci and an environmental component added by sampling a normal distribution with the relevant variance. The twenty individuals with the highest phenotypes were taken as the parents of the next generation. Forty progeny were then created by the random choice of 80 gametes from these excluding the possibility of selfing, and so on.

(ii) Selection at the genotypic level. For each of the twenty parental genotypes, the probability of contributing to the next generation can be calculated from its genotypic mean. For an individual with genotype value x in a population with mean $\bar{x}$, this can be written as $erf((x - \bar{x})/\sigma_e)$, where σ_e is the environmental standard deviation. A section of the space between 0 and 1 of length proportional to this probability is then allocated to that individual. A random number between 0 and 1 is generated, and a gamete is obtained from the individual within whose section this falls. The process is continued until $2N$ gametes are obtained. This avoids the production of progeny gametes and saves computing time, especially with high intensities of selection. The probability function corresponds to selection of the highest 50% but this is simply modified.

(iii) Selection at the gametic level. The probability is now calculated from the gametic value and the standard deviation of the inverse error function contains half the genetic variance. The production of a new gamete involves the choice of two parental gametes at random and the formation of a recombinant between them.

The greater part of the simulation was done by the last program. Some genotypic runs were done as a check and a series was also done with the first program, mainly to demonstrate the validity of the $Nih*$ generalization. These latter ones will be presented later for comparison with gametic runs with the same parameter values.

With no crossing-over, a two to three-fold saving of time could be obtained by labelling each gamete with the total value of the genes it contained and ignoring individual loci. Otherwise all loci were checked each generation for fixation. Any locus fixed was removed from the genotype and crossover distances modified. The run ended on the fixation of all loci.

The other main variations were in the simulation of linkage. Two distributions of loci along the chromosome were investigated. At first it was assumed that loci were equally spaced along the chromosome. This was thought to be an unreal model and a further set of runs was done, in which the length of the chromosome was taken as fixed and loci reallocated at random for each replicate assuming an underlying uniform distribution.

A recombinant chromosome was usually formed from two parental chromosomes by a random walk along the two with the following exceptions:

(i) with free recombination, the random walk was done with only a single random number by converting it into a binary sequence and using this as a "mask" (Fraser, 1957).

(ii) with very short chromosome lengths (< 12.5 cM) and equally spaced loci, it was assumed that there was never more than one crossover. A single random number then sufficed to decide whether a crossover would occur and also to locate it. On the equally spaced model, a modification of this might be the best for all chromosome lengths, especially for a large number of loci. A recombinant chromosome is then determined, given the parental chromosomes, by the position and number of the crossovers.

Except for these special methods, the identity of the alternative alleles on the pair of chromosomes was always checked so that a crossover was not considered unless the two parental chromosomes were double heterozygotes at the two loci concerned, with appropriate modification of the recombination fraction i.e. $r(1 + 2) = r(1) + r(2) - 2r(1) r(2)$ where the figures in brackets refer to the regions concerned. It was assumed that there was no interference and that the relationship of recombination fraction (r) to map distance (c) was given by $r = \frac{1}{2}(1 - e^{-2c})$.

Results

Much emphasis has been given in this paper to two simplifications of the problem. The first concerns the description in terms of the parameter combination Nih^* and the second, the simulation of selection at the gametic level using an inverse error function. Table 3 gives evidence on which both these can be judged. It presents results for a series of values of Nih^* for free recombination between the loci and, at the other extreme, no crossing-over at all. The left-hand column of figures of each pair has been calculated from a simulation program of the first type in which progeny genotypes are constructed, turned into pheotypes, ranked, and selected. Within any value of Nih^*, there are thus various combina-

Table 3. *A comparison of computer results using phenotypic and gametic selection. The table gives the expected gain, in terms of the initial genetic standard deviation, for 10 loci with q = 0.2*

Nih^*	N	i	h^*	Free recombination		No crossing-over	
				Phenotypic	Gametic	Phenotypic	Gametic
0.5	5	0.8	0.125	1.13 ± 0.23	1.19 ± 0.12	0.78 ± 0.09	0.95 ± 0.11
	10	0.4	0.125	0.90 ± 0.19	—	0.91 ± 0.11	—
1.0	5	1.6	0.125	2.21 ± 0.19	—	1.60 ± 0.10	—
		0.8	0.25	2.16 ± 0.25	2.26 ± 0.08	1.57 ± 0.09	1.67 ± 0.06
	10	0.8	0.125	2.23 ± 0.18	2.45 ± 0.15	1.63 ± 0.10	1.65 ± 0.14
		0.4	0.25	2.13 ± 0.25	—	1.53 ± 0.10	—
2.0	5	1.6	0.25	4.55 ± 0.14	—	2.39 ± 0.09	—
		0.8	0.5	3.92 ± 0.22	4.12 ± 0.07	2.23 ± 0.09	2.23 ± 0.05
	10	1.6	0.125	4.20 ± 0.40	—	2.35 ± 0.15	—
		0.8	0.25	4.07 ± 0.19	4.46 ± 0.13	2.30 ± 0.08	2.43 ± 0.07
		0.4	0.5	3.90 ± 0.21	—	2.02 ± 0.08	—
4.0	5	1.6	0.5	6.55 ± 0.18	—	3.18 ± 0.07	—
	10	1.6	0.25	6.66 ± 0.27	—	2.80 ± 0.12	—
		0.8	0.5	6.58 ± 0.16	6.47 ± 0.09	2.93 ± 0.07	3.08 ± 0.06
8.0	10	1.6	0.5	8.27 ± 0.11	—	3.88 ± 0.06	—
	20	0.8	0.5	8.56 ± 0.13	8.49 ± 0.13	3.48 ± 0.07	3.55 ± 0.10

Table 4. *The unweighted analysis of the results obtained by simulation using phenotypic selection (Table 3)*

	Free recombination		No crossing-over	
	df	M.S.	df	M.S.
Between Nih^* values	4	89.3179	4	11.3645
Within Nih^* values	11	0.0330	11	0.0211
Expected error		0.0469		0.0089

tions of the constituent parameters. The selection advance is presented in all cases in terms of the genetic standard deviation in the initial population. A description of the results in terms of Nih^* removes almost all the variation. Within values of Nih^*, there would seem to be some real differences in that the advances are less at the lower values of i. A measure of the usefulness of the simplification can be got by an unweighted analysis of variance within and between Nih^* values. This is given in Table 4 in which an approximate error is calculated from the sampling variance of the separate estimates. With free recombination there are no significant differences within Nih^* values but there are suggestions of differences in the absence of crossing-over.

The table also contains, in the right-hand column of each pair, results from the simulation program in which selection is at the gamete level. There are no significant differences between the two methods. This would justify use of the simplified program for the majority of the calculations.

On the basis of the theory presented earlier, the expected advance at low values of Nih^* would be close to $2Nih^*\sigma_g^*$ and would fall off more rapidly when there is no crossing-over. It will be seen, however, that, with free recombination between the loci, the advance when Nih^* is 1 or 2 is somewhat greater than $2Nih^*\sigma_g^*$. This may occur at initial frequencies less than 0.5 as noted earlier.

The main series of computer results for L_0/L_f are presented in Figs. 2–6. Figs. 2, 3, and 4 (calculated with the gametic model with $N = 10$) illustrate the interactions of Nih^* and q at different values of n, for Nih^* up to 4. Because of the computing time necessary, only one set of calculations has been made for $Nih^* < 1$. The points brought out in the theoretical analysis are well illustrated. When q is small, the curves with higher Nih^* values are lower whereas at high values the reverse is true. The curves consequently cross one another. With only five loci, the approach of the ratio to unity above a critical value of q is clearly seen. It was suggested in the theoretical analysis that, as n increased, the ratio would approach a limiting value independent of n and q. The curves for $n = 20$ in Fig. 4 are much flatter at intermediate values of q than are those with five or ten loci. Fig. 5 gives curves calculated at higher

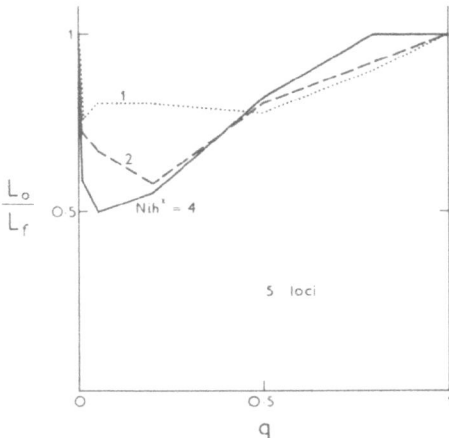

Fig. 2. The ratio of advance at the limit with no crossing-over to that with independent segregation (L_0/L_f) plotted against the initial gene frequency of the desirable allele for five loci. Fig. 2–4 were obtained from computer runs with $N = 10$

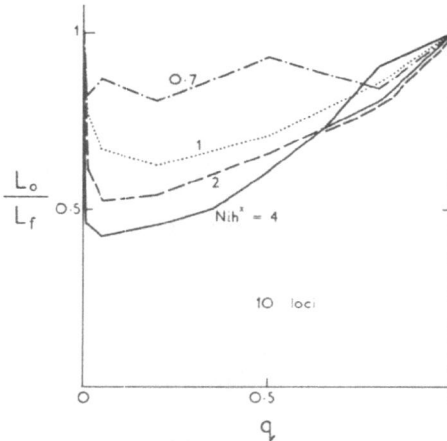

Fig. 3. L_0/L_f for ten loci

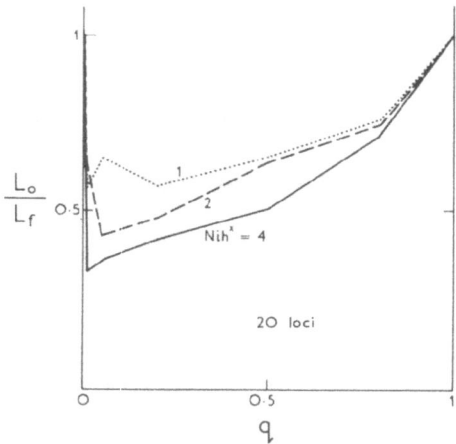

Fig. 4. L_0/L_f for twenty loci

values of Nih^* when $n = 10$. For these, ih^* was kept constant and N increased to 40. The curves are little altered from that with $Nih^* = 4$ except at low initial gene frequencies. The value of q at which the minimum occurs is always below 0.5 and decreases, as does the minimum value of the ratio, as Nih^* increases.

Fig. 6 shows the effect of the number of loci, calculated when $Nih^* = 4$ and $N = 10$. At fixed values of the other parameters, the ratio always declines as n increases.

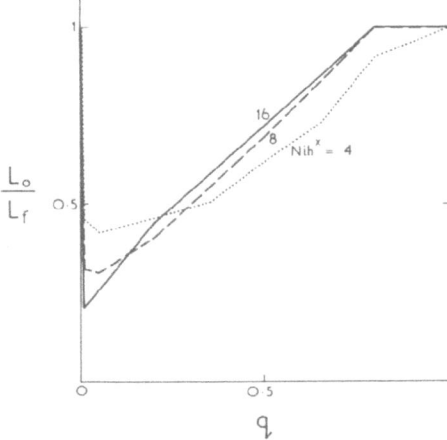

Fig. 5. L_0/L_f at higher values of Nih^*. Calculated for ten loci with $ih^* = 0.4$

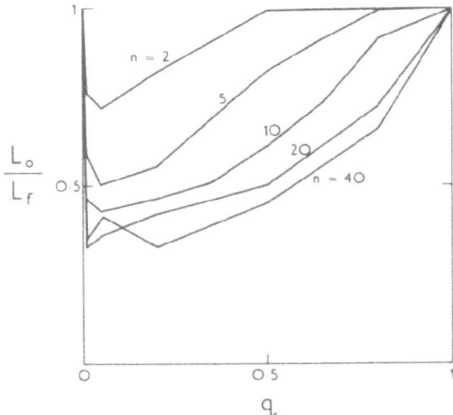

Fig. 6. The dependence of L_0/L_f on the number of loci for $Nih^* = 4$. Calculated with $N = 10$

These curves bear a surprising resemblance to Fig. 1 in Hill and Robertson (1966) which presents the effect of segregation at a locus on the chance of fixation at another closely linked to it. A verbal explanation of the pattern was then given in terms of the effective population size with which the second locus is selected and this has now been given precision by an algebraic treatment (Hill unpublished).

Bearing in mind that n is the number of loci on a single chromosome and not all those affecting the character, the effect of suppressing recombination on the selection limits is surprisingly small. On reflection this is to be expected. In the absence of crossing-over, we can expect a gain of the order of $3\sigma_g^*$, when Nih^* is large. With equal gene effects at all loci, the best that can be done under free recombination is to fix all desirable alleles, with a consequent advance of $(2n(1-q)/q)^{\frac{1}{2}}\sigma_g^*$. When $q = 0.5$, for instance, the absolute limits of advance, in multiples of σ_g^*, are 2, 3.16, 4.47, and 6.32 for 2, 5, 10, and 20 loci respectively. It is not then surprising that L_0/L_f is in our data generally above 0.5 at intermediate gene frequencies.

The Removal of Simplicity

(i) **Inequality of Gene Effects.** When gene effects are not equal and Nih^* is so high that all desirable alleles are fixed with free recombination, the value of n in this connection equals the effective number of loci as defined by Wright (1952) and Mather (1949). Assuming q to be constant, we have $L_f = (1-q)\, \Sigma a$ and $L_0 \sim 3\sigma_g^* \sim 3 \left(\dfrac{q(1-q)}{2} \right)^{\frac{1}{2}} (\Sigma a^2)^{\frac{1}{2}}$ so that L_0/L_f will be proportional to $\dfrac{(\Sigma a^2)^{\frac{1}{2}}}{\Sigma a}$ which will equal $1/\sqrt{n}$ if all a's are equal. We may then write

$$n_e = \frac{(\Sigma a)^2}{\Sigma a^2}$$

which is the same as the Wright and Mather definition. n_e is then the number of loci with equal effects which give the observed ratio of potential limits of selection to the initial genetic standard deviation. Since n_e is always less than the actual number, it follows that the observed value of L_0/L_f will always be higher than that expected from the actual number of loci.

As Nih^* is reduced from these extreme values, a situation may be reached in which, though the gain with no recombination is little affected, the loci at which gene effects are small give only a small proportion of their possible gain with free recombination. Since the curve of chance of fixation against effect is over most of its range concave downwards, it follows that L_f will be less than is expected from the effective number of loci. This can be illustrated by an example. Consider the array of proportional gene effects at 20 loci as 29, 14 ($\times$ 2), 5, 2($\times$ 2), and 1($\times$ 14). These loci give the same initial genetic variance and ultimate limits of selection as five loci with proportional effect 16. With $Nih^* = 4$, the expected change in gene frequency with free recombination from $q = 0.2$ is 0.7 in this latter situation with a corresponding gain of $5 \times 16 \times 0.7 = 56$ units.

With the 20 unequal loci, however, the gain with free recombination was only 46.4 units. The three loci with largest effects had probabilities of fixation of 0.99, 0.89, and 0.84 respectively whereas the 14 small loci had an average value of only 0.243. In other words, the latter, capable of 14/80 or 17.5% of the total gain, contributed only 5% of their potential. The observed value of L_0/L_f was 0.78 ± 0.03 compared to 0.55 ± 0.02 for 5 equal loci and 0.42 ± 0.02 for 20 at this value of Nih^*. At intermediate values of Nih^* the effective number of loci – in this case between two and three – is less than that calculated using the classical formula for the number of loci and much less than the actual number. As a result, L_0/L_f will always be higher than expected from the actual number. The ratios at other initial frequencies are given in Table 5.

Table 5. L_0/L_f $(\times 100)$ for 20 loci with variable effect compared to 5 and 20 equivalent loci respectively at $Nih^* = 4$. The initial genetic variance was the same in all cases. In the variable effect case, one locus had an effect of 29 units, 2 had effect 14, 1 had effect 5, 2 had effect 2, and 14 had effect 1. The 5 constant loci all had effect 16, and the 20 constant loci had effect 8

No. of loci	Initial gene frequency			
	0.05	0.20	0.50	0.80
20 (effect variable)	56 ± 4	78 ± 3	89 ± 3	98 ± 2
5 (effect constant)	52 ± 3	55 ± 2	81 ± 2	100 ± 0
20 (effect constant)	36 ± 3	42 ± 2	50 ± 3	72 ± 3

(ii) **Inequality of Gene Frequencies.** If Nih^* is high and all gene effects are equal, the expected value of L_0/L_f may be written approximately as:

$$3(\tfrac{1}{2} \Sigma q(1 - q))^{\frac{1}{2}}/\Sigma(1 - q)$$

which becomes $3 \left(\dfrac{q}{2n(1 - q)} \right)^{\frac{1}{2}}$ when all initial frequencies are equal. When they vary, it may be rewritten.

$$\frac{3}{\sqrt{2n}} \frac{(\bar{q} - \bar{q}^2 - \sigma_q^2)^{\frac{1}{2}}}{1 - \bar{q}}$$

where $\bar{q}$ and σ_q^2 are the mean and variance of q. If the latter is small, L_0/L_f will be reduced by a fraction $\sigma_q^2/2\bar{q}(1 - \bar{q})$ approximately, since L_0 will not be affected by variation in q.

If q is uniformly distributed between 0 and 1 ($\bar{q} = 0.5$, $\sigma_q^2 = 1/12$), the proportional reduction is 0.17. The greatest effect will be found when the frequency of the desirable allele can take, with equal probability, values either close to 0 or 1. For instance, if it takes values of 0.05 and 0.95, L_0/L_f will be reduced to 0.44 of its expected value when all initial frequencies are 0.5.

Since $u(q)$ plotted against q, given Ns, is always concave downwards, variation in q between loci must lead to a reduction in the average chance of fixation with free recombination below that with constant q, when the latter chance is not unity. This will lead to a decrease in L_f which will have an effect on L_0/L_f opposite in direction to that discussed above, on which it will be superimposed. We took, as an example for simulation, a case with 10 loci with $Nih^* = 4$ and $\bar{q} = 0.5$. When all loci had this frequency, L_0/L_f was 0.62 ± 0.02. But if they were equally spaced in frequency between 0.05 and 0.95, L_0/L_f was 0.70 ± 0.04. Thus, in this instance, when the chance of fixation with free recombination, with all initial frequencies equal to 0.5, is 0.96, the second effect has been greater than the first.

This can be thought of in terms of the average initial frequency when the loci are weighted according to their relative contributions to L_f. If Nih^* is high, this will equal $\Sigma q(1-q)/\Sigma(1-q)$ and will always be less than $\bar{q}$. L_0/L_f will then be reduced. At lower values of Nih^*, when the loci with lower initial frequencies have their chance of fixation reduced while the chance of fixation of those with high frequencies are still close to one, the weighted average and L_0/L_f will increase.

The implications of these calculations will depend on one's views of the probable state of the real world. Variation in gene effect will always increase L_0/L_f and decrease the effect of linkage. I feel that the distribution of gene effect used in the example when the effective number of loci for these proposed was reduced from the actual value of 20 to 5 at very high Nih^* values and to between 2 and 3 when $Nih^* = 4$ may not be unreal. The consequences of variation in initial frequency are difficult to predict. My own view is that this will be less important than variation in gene effect, though on the whole tending to increase L_0/L_f.

Intermediate Linkage Values

Some Theoretical Aspects

Suppose there a large number of loci affecting the character. The response to selection at any time then depends on the total "chromosomal" variation which may be written as

$$\sum_i \tfrac{1}{2} a_i^2 q_i(1 - q_i) + \sum \sum_{i<j} a_i a_j D_{ij}$$

where i, j refer to the loci taken in pairs and D_{ij} measures the linkage disequilibrium between them. The first term may be referred to as the "genic" variance and the second as the "covariance between loci".

It can be shown that, if a sample of $2N$ items are drawn at random from a normal distribution with variance V and are then selected deterministically in an infinite population with a multiplicative effect on selective advantage so that the relative proportion of the item with score x_i at time t is e^{kx_it}, in the early generations the variance of x in the population is given by $V(1 - 1/2N)(1 - 3k^2Vt^2/2N)$. In the absence of recombination, the chromosomal variance declines continuously and must decline more rapidly if subsequent selection is in a finite population. The genic variance is however a function only of the gene frequency and, if the expected change in the latter is small, will decline due to drift. The relative decline appears always to be slower than that in the chromosomal variance, the difference – the term in t^2 in the above equation – being a negative covariance term caused by linkage disequilibrium between the loci. If all loci have the same effect on the character, a state may be reached in which some loci are still segregating but all chromosomes have identical values. The genic variance is then exactly balanced by a negative covariance. With many loci, the linkage disequilibrium between any pair will of course be small.

In any specific instance, the pattern of change of the different variances will depend solely on the values of N and ih^*, if the number of loci is large enough that the chromosomal variation is effectively normal. Let us now introduce a small amount of crossing over but not enough to alter materially the pattern of change in variance and covariance. In any generation, the chromosomal variance will have been increased by crossing-over in the previous generation by an amount $-\sum_{i<j}\sum r_{ij}a_ia_jD_{ij}$,

where r_{ij} is the crossover value between loci i and j, remembering that the expected value of the D_{ij}'s is negative. We can write this as $-\bar{r}\sum_{i<j}\sum a_ia_jD_{ij}$

where $\bar{r}$ is the weighted average crossover distance. If there is no tendency for loci with large effects to cluster together on the chromosome and loci can be assumed to be uniformly distributed along it, then $\bar{r}$ can be shown to equal one-third of the map length of the chromosome, when the latter is small. As, if $\bar{r}$ is small enough, the covariance will be independent of n and q, we may then expect that the increase in variance at any time will be proportional to the length of the chromosome and dependent only on N and ih^* and so will be the increase in the final advance under selection. We may then expect that, when the total amount of crossing over is extremely small, the increase in the final advance under selection will be proportional to the chromosomal length. The constant of proportionality will, of course, depend on N and ih^*. We have earlier argued that these can be taken together as Nih^* and this we shall now do. It also follows that, when l is not small, the change in covariance will be sufficiently described by Nih^*, n, q, and l and that the advance can also be specified in these terms.

It was argued that the pattern of change with selection under the two extreme conditions of no crossing-over and free recombination respectively could be reasonably well predicted if Nih^*, n, and q were known and that the time scale would be proportional to the population size, N. Thus, if we double N but keep Nih^*, n, and q constant, the pattern of change will be repeated but with the time scale doubled. If, with crossing-over, the same total amount of recombination has occurred in this time, we should expect again that the pattern will be repeated. In other words, the amount of crossing-over per generation would have to be halved. It follows that a specification of the problem in terms of Nih^*, n, q, and Nl will probably allow the simplest description of the results.

As n increases, we may expect that the pattern of change in variance and covariance at higher values of Nl will be independent of both n and q and dependent only on Nih^* and Nl. As Nl increases further, L/σ_g^* will then approach the limiting value with free recombination which as discussed earlier is also independent of n and q, when n is high.

Two distributions of the loci along the chromosome were used in simulation. In one the loci were equally spaced, with a distance equal to this spacing spare at each end. The average distance between pairs of loci is then $l/3$. In the other model, a uniform *a priori* distribution of loci along the chromosome was assumed and for each replicate run a new set of locations were generated at random. The expected distance between pairs of loci is then again $l/3$. The two models will clearly be equivalent both when l is large and when it is small. In the latter situation, the expected negative covariance between loci will be independent of the distance between them. However, as l increases, because of crossing over the covariance will be smaller between loci further apart. On the random model a proportion of pairs of loci may be closer together than they could be on the other model. At intermediate values of l, we may then expect to have less response on the random model than on that of equal spacing.

An alternative view of this can be gotten from the fact that, for pairs of loci (cf. Hill and Robertson, 1966) the curve of mean change in gene frequency against distance between loci is concave downwards. Thus variation in distances between adjacent loci would be expected to reduce advance. It was not until a large amount of simulation had been done that it was realized that for a given value of l the equal spacing model was equivalent to the random model in the average distance between loci only when there is space spare at both ends, i.e. if c is the distance between adjacent loci, l must be equal to $(n+1)c$. This accounts for the peculiar chromosome lengths, e.g. 27.5 cM, in many runs.

The regression of advance on Nl when the latter is small was shown above to be independent of n and q. The determination of the initial slope presents some difficulties, however. A simple method would be to estimate the slope in the linear part of the curves close to $Nl = 0$. This proves to be time consuming. For reasons which will appear later, this in essence means working with ratios of advances less than 1.10. With $N = 10$ and $Nih^* = 4$, the coefficient of variation between replicates with little or no crossing-over is of the order of 30%. To estimate a ratio of 1.10 with a standard error of 0.02 requires of the order of 900 runs in total or 15 minutes on our fairly large computer. As it would be necessary to use $N = 40$ to evaluate the slope for $Nih^* = 16$, about one hour's time would be required.

An alternative approach is to fit curves for the variation of L/L_0 with Nl for a given set of conditions where L is the advance with some recombination. Can we then be certain that the same kind of curve will be satisfactory for different values of Nih^*? We have made some exploratory moves in this direction as follows. Any such curve must satisfy the following conditions.

1. As Nl approaches zero, the slope should become constant and the ratio approach unity.

2. As Nl increases, the ratio should increase continually to the asymptotic value of L_f/L_0.

Putting y for L/L_0 and x for Nl, these conditions are fulfilled by equations satisfying

$$\frac{dy}{dx} = k\left(1 - \frac{y-1}{K}\right)^m$$

with k the initial slope, $1 + K$ the final asymptote, and $y = 1$ when $x = 0$. With $m = 1, 2$, and 3 respectively, the solutions may be written

$$y = 1 + K(1 - e^{-kx/K}),$$

$$y = 1 + Kkx/(K + kx),$$

$$y = 1 + K(1 - (1 + 2kx/K)^{-\frac{1}{2}}) \quad \text{respectively.}$$

They differ in the rate at which the slope of the curve falls off as the asymptote is approached. For instance, at the point $x = K/k$, the value at which the prediction from the linear part of the curve equals the asymptotic value, we have progressed 64%, 50%, and 42% of the distance to the latter for $m = 1, 2$, and 3 respectively (see Fig. 7).

The same type of curve should not be expected to fit the results for the equal spacing and random location models. We know that, given

A. Robertson

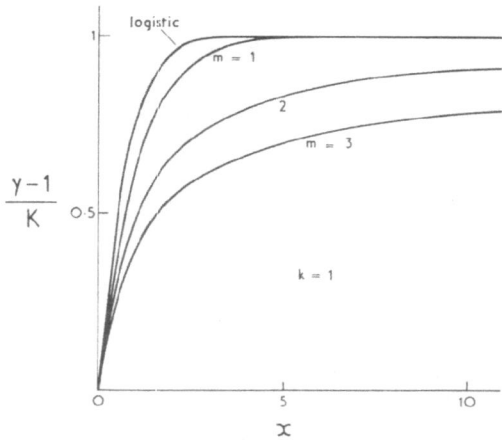

Fig. 7. Four types of relationships considered in the analysis of the effect of increasing crossing-over on the final advance under selection. All curves have the same initial slope and asymptotic value when $x = \infty$

Nih^*, n, and q, the two sets have the same initial slope and asymptotic values but that the intermediate values will be higher for the former than for the latter. Thus, a higher value for m will be required for the random allocation model.

Results

It is perhaps more illuminating to discuss the experimental results in the reverse order. To evaluate the kind of relationship which best described the effect of increasing recombination for given values of the other parameters, curves of the three types were fitted to various sets of runs covering a range of values of Nih^*, n, and q in which recombination alone was altered. Each contained runs covering the entire range of response, with its appropriate standard error calculated from the variation between replicates. To fit each set with a particular value of m then required the calculation of the values of three parameters corresponding to L_f, L_0, and k. This was done by finding the point at which the χ^2 values (sums of squares of deviations divided by sampling variance) was at a minimum. The χ^2 values could then be summed over sets of data, and the different equations compared. Runs with equal spacing of loci and those with a new allocation in each replicate were considered separately. In a sense we are attempting to find the value of m which gives the best fit. The χ^2 values were as follows:

Table 6. *The fitting of different relationships of final advance to chromosome length. The table gives the χ^2 for deviations for the different values of m*

Spacing	Sets	df	Deviations			Total df	
			$m=1$	$m=2$	$m=3$	df	
Equal	11	42	118.66	41.10	56.93	64	6188.04
Random	10	40	118.24	43.40	37.39	60	3741.44

The equal spacing results are best fitted by $m=2$ and the random allocation ones by $m=3$.

We also tried to fit a logistic curve, $y = 1 + K(1 - e^{-2kx/K})/(1 + e^{-2kx/K})$, to these results. As it is more extreme than the equation with $m=1$, the linear prediction equalling the asymptotic value when the ordinate is 76% of the latter, it is not surprising that the fit was poor ($\chi^2 = 191.82$ and 179.95 for the two sets).

Some Properties of the Second Model

As the second model is algebraically simple and fits the equal spacing results well, it is useful to work out some of its consequences. We have

$$L/L_0 = 1 + KkNl/(kNl + K)$$

The predicted extra advance when Nl is small is $kNl \cdot L_0$. If this is a fraction p of the extra advance when there is free combination, i.e., $kNl = pK$, then the complete equation predicts that $p/(p+1)$ of the extra advance with free recombination will be achieved. If $p=1$, $(Nl = K/k)$, for instance, half the possible extra advance will be achieved. If our suppositions that k is dependent only on Nih^* prove correct, it will then follow that, given Nih^*, the value of Nl required to achieve half the extra gain under free recombination will be proportional to K, which is $L_f/L_0 - 1$. With $m=3$, a similar relationship will hold except that now the required value of Nl is $3K/2k$.

How much extra advance can we then get by doubling the amount of crossing-over? The sensitivity (the proportional increase in advance for a given proportional increase in x) will be given by $\dfrac{x}{y}\dfrac{dy}{dx}$. This can be shown to pass through a maximum when $kx = K/(K+1)^{\frac{1}{2}}$, i.e., rather below the half-way value of x given by $x = K/k$. At this point the sensitivity equals

$$\frac{\sqrt{1+K} - 1}{\sqrt{1+K} + 1}.$$

As we have been dealing with values of L_f/L_0 with an upper range of between 2 and 3, this would imply sensitivities of between 0.18 and 0.27. We can estimate the probable effect of doubling the amount of crossing over by multiplying these figures by 0.7 ($= \log_e 2$) to give probable increases of between 14% and 19%.

How long should we expect the linear relationship to hold? This is most easily considered as the extent of deviations from the linear prediction when the latter is expressed as a proportion of the limiting value. These are shown in Table 7.

Table 7. *The proportion of the extra advance achieved on the two models (m = 2 and m = 3)*

| | $(y-1)/K$ | |
kx/K	$m = 2$	$m = 3$
0.1	0.091	0.086
0.2	0.167	0.155
0.5	0.333	0.293
1.0	0.500	0.422
2.0	0.667	0.553
4.0	0.800	0.667
∞	1	1

It will be seen that, with $m = 2$, the increase of y deviates more than 20% from the linear prediction if more than 20% of the possible extra advance has been achieved and for $m = 3$ this deviation occurs rather earlier. This makes direct measurement of the initial slope difficult. The proportional differences between the extra advances on the two models are greatest when $(y-1)/K$ is around 0.5 and are of the order of 16%.

The Interaction of Linkage with the other Parameters

It was pointed out earlier that in the linear part of the curve the increase in response for a given chromosome length should be expected to be independent of n and q. Tables 8 and 9 present some results for $N = 10$, $ih^* = 0.4$, using the random allocation model. It will be seen that the expectation is realized. The average number of replicates used for each point in the two tables was over 100 and the total calculation time on a fairly large computer was over 7 hours.

Table 10 presents an investigation of the adequacy of a description in terms of Nih^* and Nl, when n and q are fixed. The table is designed to show that, given Nih^* and Nl, the advance is the same for pairs of values of N differing by a factor of 2. This approach was used because it is difficult to present such an analysis over a wide range of values of ih^*. We have limited ourselves in the use of the shortened programs to an

Table 8. *The final advance under selection, compared to that with no crossing over as 100, when different number of loci are distributed at random along a given length of chromosome (N = 10, q = 0.2, ih* = 0.4). Note the independence of number of loci when l is small*

Number of loci	Chromosome length in cM				
	6.25	12.5	25	50	∞
2	113 ± 3	118 ± 3	117 ± 3	119 ± 3	124 ± 3
5	116 ± 5	130 ± 5	135 ± 5	148 ± 5	178 ± 5
10	118 ± 6	127 ± 6	141 ± 4	168 ± 7	214 ± 5
20	116 ± 5	130 ± 7	140 ± 7	156 ± 8	235 ± 9
40	118 ± 5	127 ± 7	142 ± 10	179 ± 10	266 ± 12

Table 9. *The final advance under selection, compared to that with no crossing over as 100, when 10 loci with the given initial gene frequency are distributed at random along a given length of chromosome (N = 10, ih* = 0.4). Note the independence of initial gene frequency when l is small*

Initial gene frequency	Chromosome length in cM				
	6.25	12.5	25	50	∞
0.05	123 ± 6	129 ± 7	139 ± 9	167 ± 11	231 ± 12
0.2	118 ± 6	127 ± 6	141 ± 4	168 ± 7	214 ± 5
0.35	123 ± 6	130 ± 6	147 ± 6	160 ± 7	195 ± 7
0.5	115 ± 5	125 ± 6	135 ± 6	141 ± 5	161 ± 6
0.65	111 ± 4	116 ± 4	118 ± 4	126 ± 4	134 ± 4

Table 10. *The final advance, compared to that with no crossing-over as 100, at differen: values of Nih*. The total number of selection runs in each set is also given. Calculated from the equal spacing model with n = 10, q = 0.2*

			Nl					
N	Nih^*	Runs	0.687	1.375	2.75	5.5	∞	k
5	1	4250	111 ± 4	117 ± 4	130 ± 5	121 ± 7	133 ± 7	0.30 ± 0.11
5	2	2700	121 ± 4	123 ± 5	139 ± 3	148 ± 4	172 ± 5	0.30 ± 0.05
10	2	1260	118 ± 5	130 ± 6	144 ± 8	164 ± 9	183 ± 8	0.35 ± 0.07
10	4	1750	118 ± 3	139 ± 4	156 ± 3	166 ± 5	214 ± 5	0.33 ± 0.04
20	4	460	122 ± 6	127 ± 7	146 ± 7	174 ± 8	232 ± 9	$0.2^7 \pm 0.05$
20	8	405	125 ± 4	137 ± 5	161 ± 6	188 ± 9	238 ± 7	0.39 ± 0.05
40	8	175	136 ± 8	141 ± 8	173 ± 13	187 ± 11	252 ± 12	0.48 ± 0.10
40	16	210	131 ± 5	136 ± 6	163 ± 11	171 ± 8	217 ± 5	0.42 ± 0.07

upper value of ih^* of 0.4. It was also in practice very difficult to use low values of ih^* because of the computer time involved. Some runs were made, for instance, with $N = 40$, $ih^* = 0.1$, $Nl = 2$. The majority had not reached fixation at the 125th generation, and the average time per replicate was 2 minutes. The pairs of comparable results in no case differ significantly from one another.

Although the theoretical approach indicated that, given Nih^*, the initial slope of relative advance on Nl should be independent of n and q, it gave no indications as to how the slope should vary with Nih^*. The table presents minimum χ^2 estimates of k, the initial regression of L/L_0 on Nl, calculated on the assumption that $m = 2$ gives an adequate description of the curves for the equal spacing model. The results would suggest that k increases as Nih^* increases but not very strongly. A value of 1/3 for the initial slope would not be greatly out over the range of Nih^* from 1 to 16.

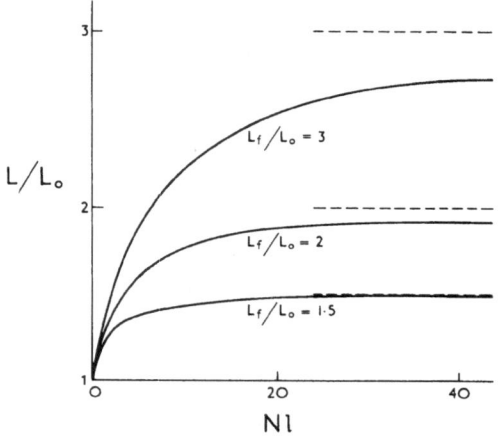

Fig. 8. The relationship between advance at the limit with crossing-over compared to that in its absence, assuming that $m = 2$ gives the best fit for the equal spacing model. All curves have the same initial slope but differ in the limiting value when segregation is independent $(Nl = \infty)$

If we assume that k is not dependent on Nih^*, then we can express L/L_0 as a function solely of Nl and L_f/L_0. The resulting curves, drawn assuming $m = 2$ and $k = 1/3$, are given in Fig. 8. They are all identical when Nl is small but then break away as their particular limiting value is approached. In the introduction, the need to obtain a true zero point, rather than one in which the recombination was very small, was stressed. In the paper by Qureshi, Kempthorne, and Hazel (1968) for instance, the lowest value of Nl used is 1.6 which from Fig. 8 may overestimate the advance with no crossing-over by as much as 50%.

All the runs with intermediate linkage values discussed so far have been done with the model using gametic selection. A few have been done with the first model which simulates selection by constructing a set of diploid progeny, adding an environmental variate and choosing indi-

Table 11. *The final advance with $Nl = 2.75$, compared to that with $Nl = 0$ as 100, calculated from both phenotypic and gametic models ($n = 10$, $q = 0.2$)*

Nih^*	N	i	h^*	Phenotypic	Gametic
	5	0.8	0.25	116 ± 10	130 ± 5
1	5	1.6	0.125	142 ± 18	
	10	0.4	0.25	123 ± 14	
	10	0.8	0.125	132 ± 13	127 ± 10
	5	0.8	0.5	148 ± 7	139 ± 3
2	5	1.6	0.25	157 ± 10	
	10	0.4	0.5	139 ± 11	
	10	0.8	0.25	144 ± 13	144 ± 8
	5	1.6	0.5	168 ± 11	
4	10	0.8	0.5	168 ± 7	156 ± 3

viduals with the highest phenotypes as parents. This needs more time than the two other kinds of program. The results, expressed as a ratio of the advance when $Nl = 2.75$ to that when $Nl = 0$, are summarized in Table 11 in which the corresponding gametic selection results are also given. It will be seen that there is reasonable agreement between the two sets of figures.

The Optimum Intensity of Selection

When individual selection is done from a given total number of measured individuals, it is known that for free recombination the greatest expected final change of gene frequency occurs when half the population

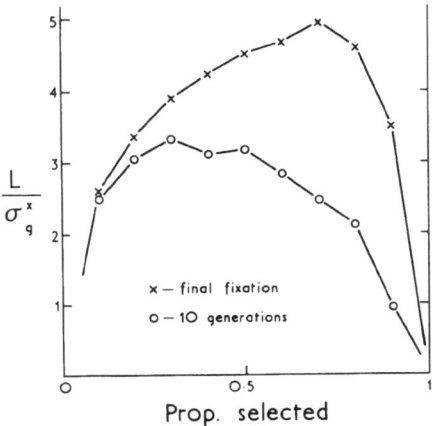

Fig. 9. The expected advance at the limit and the gain after ten generations when different numbers of individuals are selected from twenty measured in each generation ($h^* = 0.5$, $n = 10$, $q = 0.2$, $l = 27.5$ cM)

is selected each generation. Hill and Robertson (1966) showed that for two linked loci in a situation sensitive to changes in recombination value, the optimum proportion selected was shifted slightly to the neighborhood of 0.55. This shift arises from the fact that, though Ni has a maximum when 0.50 are selected and is symmetrical around this point, Nl is higher when a higher proportion of the population is selected. The shift would be expected to be greatest in situations most sensitive to changes in Nl. We saw earlier that this occurs when $L - L_0$ is a little less than $\frac{1}{2}(L_f - L_0)$. We therefore examined such a situation when 20 individuals were selected from in each generation with $n = 10$, $q = 0.2$, $h^* = 0.5$, and $l = 25$ cM. The results are shown in Fig. 9. The expected final advance varies little between proportions selected of 0.5 and 0.8. The response after 10 generations, on the other hand, is largest when about 35% are selected. The proportion selected which gives the greatest final advance is a little higher than for two loci.

The Variance Between Replicates

It is possible to give some general guide lines as to the variance between replicates that might be expected in these results. When Nih^* is low, the selection response is small and variation between replicates comes almost entirely from drift. The variance between replicates at fixation will be twice the initial genetic variance, whatever the linkage situation. This raises a problem in estimating the ratios of advances with any accuracy, as we expect these to be close to unity. Suppose we wish to estimate such a ratio with a coefficient of variation of 5% when $Nih^* = 0.5$. The expected response will then be close to σ_g^*. With R replicates, the expected standard error of this mean will then be $\sigma_g^* \sqrt{\dfrac{2}{R}}$.

If the coefficient of variation of such an estimate is to equal the required value of 3.5%, we require 1600 replicates for each mean.

At high values of Nih^*, the results will be quite different in the two extreme linkage situations.

1. With free recombination, the variance between replicates is almost entirely determined by the mean gene frequency at fixation, bearing in mind that we are then dealing with homozygous populations. If the expected gene frequency of the desirable allele is very close to unity, there will be little variation between replicates compared to that in the initial population, especially if the gene frequency in the latter was in the middle of the range. If, however, we are working with low initial gene frequencies and the chance of final fixation is of the order of 0.5, then the variation between replicates at fixation can be very much greater than the initial genetic variance in the population. To take a case on which

we have worked a great deal, that of $Nih^* = 4$, $n = 10$, the variance between replicates at fixation for different initial gene frequencies is shown in Table 12. As will be seen, with high initial frequencies, the variance between replicates is very low because the chance of fixation is very high. On the other hand, for the lowest frequency, 0.01, the variance between replicates at fixation is more than thirty times the initial genetic variance.

Table 12. *The standard deviation of final advance, expressed as a proportion either of the initial genetic standard deviation or of the mean advance*

	10 loci				20 loci			
	Free recombination		No crossing-over		Free recombination		No crossing-over	
q	S. D/σ_g^*	S. D/L	S. D/σ_g^*	S. D/L	S. D/σ_g^*	S. D/L	S. D/σ_g^*	S. D/L
0.01	5.86	0.57	2.34	0.43	5.43	0.44	1.90	0.48
0.05	2.68	0.31	1.61	0.43	2.98	0.30	1.47	0.42
0.20	1.27	0.20	1.17	0.38	1.50	0.21	1.11	0.36
0.50	0.67	0.16	0.79	0.31	0.74	0.14	0.97	0.37
0.80	0.16	0.07	0.41	0.20	0.30	0.10	0.67	0.31

2. When there is no crossing-over between the loci and the number of loci is large, the situation is almost entirely specified by Nih^*. The variance between replicates at fixation then declines continually as Nih^* increases but only very slowly. For instance, with an initial sample of 10 gametes and a normal distribution of effects, then if the best were certain to be fixed, the expected variance of response would be 68% of the initial genetic variance with a coefficient of variation of 0.38. For a sample of 100 gametes the variance would be reduced to 37% of the initial value with a coefficient of variation of 0.17. With a small number of loci, the initial gene frequency has some effect because of the skewness of the distribution. Table 12 shows the variance of final advance in terms of the initial genetic variance with no crossingover both for 10 loci and for 20 loci. The variance is much less dependent on gene frequency than in the case of free recombination. With 20 loci on the chromosome, the effect of gene frequency is less, as would be expected, especially when the variation is expressed as the coefficient of variation.

When loci varied in their effect on the character, it was found that the variance between replicates was generally rather smaller than when the same variance was produced by an equal number of genes. In the only case studied in detail (that of 20 loci with an effective number equal to 5) it was found that, with $Nih^* = 4$, the variance of response with no crossing-over in the model with variable gene effect was only one-half that in the equivalent one with constant gene effect. The same was found with free recombination though the difference was smaller. This is to be expected since with the variable effect model, the major gene concerned is, irrespective of the degree of linkage, almost certain to be fixed and to contribute little to the variation between replicates. On the other hand, had the selection intensity been smaller and the chance of fixation of the major gene been in the neighbourhood of 0.5 it is probable that the reverse would have been true and that the variation between replicates would have been greater for variable than for constant gene effects. With Nih^* very high, it might also be expected that the constant effect model, with all loci almost certain to be fixed, would show little variance between replicates whereas on the variable effect model the loci with smaller effects would still not have a high chance of fixation and the variance would be higher.

We have not analyzed the effect of intermediate intensities of linkage on the variance between replicates in any detail, though a considerable amount of simulation material is available. In those situations in which variance between replicates was vastly different in the cases of free recombination and no crossing-over, the variance increased continually as recombination increased. But when the variance at the two extremes was comparable, as for instance when $Nih^* = 4, n = 10$, and $q = 0.2$ in the table, we had some evidence that the variance between replicates could be greater for intermediate linkage values. The maximum generally seemed to occur for those linkage values giving an advance about half way between the two extreme cases. In the case mentioned, for instance, when the total length of chromosome was of the order of 50 cM, the relative variance between replicates was 2.12 compared to 1.63 for free recombination and 1.36 for tight linkage. As each point is based on 200 d.f., this effect borders on significance at the 5% level. The same has also been seen in other runs.

With no recombination, the variation between lines at fixation is always less than would be expected from their average gene frequency as has been noted by Qureshi and Kempthorne (1968). This is a consequence of negative linkage disequilibrium between pairs of loci within the hypothetical population formed by crossing such lines. This will be low when Nih^* is small and will increase as Nih^* increases. This disequilibrium can be most easily expressed in terms of the proportion of the genic variation which is immediately available for selection as chromosomal variation on crossing such lines. For a given value of Nih^*, this proportion must for intermediate gene frequencies approach a limiting value as the number of loci increases and the initial distribution of chromosomal variation becomes normal. Table 13 calculated with $n = 10$ and $q = 0.2$ gives for different values of Nih^* the proportion of the genic variation immediately available in the population of lines at fixation

Table 13. *Linkage disequilibrium between lines at fixation, measured as the proportion of the genic variance immediately available for selection as chromosomal variance, in lines themselves selected without crossing-over with $n = 10$, $ih^* = 0.4$, $q = 0.2$*

Nih^*	Proportion of variance available
1	0.69
2	0.48
4	0.43
8	0.25
16	0.17

Table 14. *The effect of position on the change of fixation of the desirable allele for 10 loci equally spaced on a chromosome of length 27.5 cM. $N = 10$. $ih^* = 0.4$, $q = 0.2$. Each point has a standard error of 0.017*

Position	Chance of fixation
1 (end)	0.639
2	0.626
3	0.598
4	0.594
5 (middle)	0.576

with no crossing-over during selection. For intermediate linkage values the proportions were closer to one and paralleled very closely the increase in the advance under selection as recombination is increased. At higher values of Nih^* a high proportion of the potential variation is not immediately available for selection if Nl is small. Selection programs based on crosses between such lines could not therefore be expected to meet the basic assumptions of this present paper. There were some indications of negative disequilibrium between loci even when there was free recombination. This is not altogether surprising as it is known that artificial selection of this type in an infinite population directly causes negative linkage disequilibrium (Nei, 1963). Only one-half of this will be lost by recombination each generation and the persistence of some disequilibrium throughout the selection process will be reflected in the disequilibrium in the population of lines at fixation.

In the absence of crossing-over, the genetic disequilibrium within lines will, for the model in which all genes have the same effects, be the same for all pairs of loci. As, in a chromosome of n loci, there are n squared terms contributing to the chromosomal variance and $n(n-1)$ cross product terms making up the covariance, the genetic disequilibrium between pairs of loci will be small when n is large. At intermediate linkage values, however, all pairs of loci are not equivalent and the degree of disequilibrium, both within lines during selection and between lines at fixation, will be less for loci which are further apart. Correspondingly the chance of fixation of the desirable allele at the different loci, which must be independent of position for the extreme cases of free recombination and no crossing-over, will, in the intermediate situations which are most sensitive to differences in recombination values between the pairs of loci, be least for loci in the middle of the chromosome. Table 14 based on $Nih^* = 4, n = 10, q = 0.2$ and $l = 25$ cM gives the results for the average of 800 replicates which bear out the expectation.

The Gain at Intermediate Generations

We have concentrated on the final chance of fixation. What of the situation at intermediate generations? In the discussions of single loci and of two linked loci (Robertson, 1960; Hill and Robertson 1966) the half life of the selection process (the time for half the final change to occur) has been considered in some detail. When the expected change in gene frequency is small, it was shown that the half life of an additive selection process is close to $1.4N$ generations irrespective of the initial frequency. As the selection pressure increases, the half life declines and does so more at higher initial frequencies.

In discussing the "normal" theory, it was suggested that the selection process is determined by Nih^* if the time scale is measured in multiples of N generations. But we also know that, for high values of Nih^*, the final gain with no crossing-over is of the order of $3\sigma_g^*$, whereas the initial rate of response is $ih^*\sigma_g^*$. The half life at high Nih^* values would therefore be expected to be inversely proportional to ih^*. Examination of computer runs with no crossing-over shows that the half life at higher values of Nih^* is approximately $2/ih^*$ generations (or in terms of the population size, $2N/Nih^*$ generations). As Nih^* declines, the half life tends to $1.4N$ generations.

Discussion in terms of the half life is however a little artificial; the shape of the response curves must be different for different recombination values. If, for instance, the time taken to achieve a small fraction of the total response is taken as the criterion, the results merely reflect differences in final advance because the initial rate of gain is independent of linkage. This point is illustrated in Table 15 which shows the half lives and "90%-lives" as linkage is relaxed with $Nih^* = 4$, $N = 10$, $q = 0.2$. The half lives closely parallel the final advances but the "90%-lives" do not. This confirms the observations of Hill and Robertson and of Latter (1965) that the extra gain on increasing recombination takes place in the later generations.

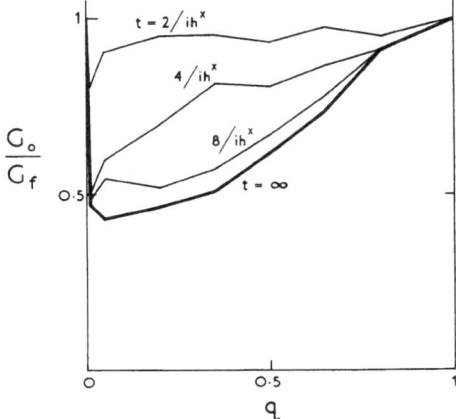

Fig. 10. The ratio of the gain with no crossing-over (G_0) to that with free segregation (G_f) after different numbers of generations of selection ($n = 10$, $Nih^* = 4$)

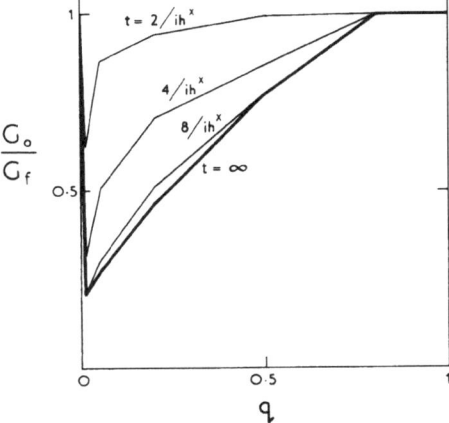

Fig. 11. As for Fig. 10 but with $Nih^* = 16$

Table 15. *The number of generations to achieve 50% and 90% of the final advance for different chromosome lengths ($N = 10$, $n = 10$, $ih^* = 0.4$, $q = 0.2$)*

Length in cM	50% of advance	90% of advance
0	3.6	10.4
6.25	4.7	19.0
25	6.1	22.0
100	7.2	20.7
∞	7.8	21.1

This can be looked at more usefully by asking at what stage in a selection process the effect of linkage will be noticed. Fig. 8 and 9 show the ratio of the gains in the two extreme linkage situations after different numbers of generations for two different combinations of parameters. It appears from these that when Nih^* is high the generalization can be made that, except for very low initial frequencies, linkage has little effect on the response before $2/ih^*$ generations and that, after $8/ih^*$ generations, the ratio approaches its final value. This has been confirmed for five loci and twenty loci in the chromosome. This refers only to the ratio between gains under free recombination and tight linkage respectively. The effects at intermediate linkage values will depend on the amount of recombination. The tighter the linkage, the longer will the gain be identical with that under no crossing-over and the higher the proportion of the extra advance made in the later generations.

Segregation of Several Chromosomes

The extension of these ideas to the whole genotype is surprisingly straightforward. Let us deal with this by gradually introducing more complexity into the situation. If there are several equivalent chromosomes, then the ratio of the overall advances will be the same as for single chromosomes. If H is the haploid number, the expressions for the final advance when Nih^* is high will now be $(2n(1-q)/q)^{\frac{1}{2}} \sqrt{H} \, \sigma_g$ and $3 \sqrt{H} \sigma_g$, approximately, for the extreme cases of free recombination and no crossing-over respectively where σ_g refers to the total genetic variation. This follows since $\sigma_g^2 = H\sigma_g^{*2}$. These will be slight overestimates if, as was discussed earlier, some negative linkage disequilibrium develops between loci on different chromosomes. However the expected values of the ratio of advances would be little different from those for individual chromosomes.

Suppose now, while all loci remain equivalent, there are different numbers on different chromosomes. To what extent can we predict the probable limits from the average variance and number of loci on each chromosome, which will probably be known from the heritability and the haploid number? With intermediate gene frequencies we have dealt earlier with three separate situations:

(i) Linkage so loose that loci are effectively segregating independently. Obviously the chromosomal organization is then irrelevant.

(ii) No crossing over.

If all chromosomes have a normal distribution of effects, then the expected response will be approximately $3 \Sigma \sigma_g^*$. For each chromosome, the response will therefore be proportional to the square root of the number of loci on it whereas for free recombination it will be directly proportional to the number. The more important chromosomes will contribute relatively less with no crossing-over than with free recombination. If there are n_i loci on the i-th chromosome, then it can be shown that

$$\sum_i \sqrt{n_i} = H \sqrt{\bar{n}} \left(1 - \frac{1}{8} V_n/\bar{n}^2 \right)$$

if V_n, the variance of n, is small and $\bar{n}$ is the average number of genes per chromosome. Variation between chromosomes reduces the gain with tight linkage by $1/8$ of the squared coefficient of variation between chromosomes in their contribution to the total genetic variance This effect proves to be surprisingly small. If we assume 10 chromosomes with variance contributions 1, 2, etc., up to 10, the proportional difference between the sum of the square roots and 10 times the square root of the mean is only 4%. The use of $h/\sqrt{H}$ for h^* should not be greatly in error, especially in the context of our ability to detect such small effects in this present analysis. Note that the effect of differences between chromosomes is to decrease L_0/L_f.

(iii) Intermediate linkage values.

With the reasonable assumption that the greater the number of loci the longer the chromosome, the relaxation of linkage will have more effect on the chromosomes with the larger number of loci and the consequence of differences between chromosomes in their contribution to the total genetic variance will be less than when there is no crossing-over.

If the frequency of the desirable alleles is so high that a gamete containing all desirable alleles is fixed even for the largest chromosome, the response on each will be proportional to the number of loci on the chromosome and the chromosomal organization will be irrelevant. At the other extreme of low initial frequencies, the effects may be more important. Some simulation runs have been done with $Nih^* = 4$ and 15 equivalent loci at $q = 0.2$, comparing a situation with 5 chromosomes each with 3 loci with one in which they had 1, 2, 3, 4, and 5 loci respectively. With intermediate linkage, it was assumed that the chromosome lengths were proportional to the number of loci, with the biggest 50 cM long, so that in the variable case, the coefficient of variation of length was 47%. The results given in Table 17, show that variability between chromosomes does reduce the ratio of L_0/L_f a little in this situation.

Introducing the further complication of variable effects and initial frequencies at the different loci, the conclusions should remain true on the average though there would still be room for discrepancies in individual cases. On the assumption that the genetic variance contributed by the different chromosomes is roughly proportional to cytological length, then the use of $h^* = h/\sqrt{H}$ in the formulae may give results very close to

Table 16. *The effect of variation between chromosome on selection limits (compared to that with free segregation as 100). With intermediate recombination, the chromosomes were 30 cM long in the equal chromosome situation and 50, 40, 30, 20, and 10 cms respectively in the other.* $N = 10$, $ih = 0.4$, $q = 0.2$

Recombination	No. of loci on each chromosome	
	3, 3, 3, 3, 3	5, 4, 3, 2, 1
Free	(100)	(100)
Intermediate	95 ± 4	95 ± 4
None	74 ± 3	68 ± 2

the correct ones for mammals. For instance, McConnel, Fechheimer, and Gilmore (1965) give data on chromosome dimensions for swine which indicate a coefficient of variability of chromosome length of 23%. In birds, with a large number of small chromosomes, we may have to be more cautious. Shoffner and Krishan (1965) state that the largest six chromosome pairs of the 39 in the chicken made up 55% of the total length. In general, the higher the haploid number the lower the average values of Nih^* and n over the chromosomes and, as a consequence, the closer to unity the expected value of L_0/L_f.

Discussion

It was stated in the introduction that the basic aim of this analysis was to find the framework which allowed the simplest description of the results. To what extent has this been achieved? As far as concerns the final chance of fixation at the different loci, it has been shown that a description of the results only in terms of Nih^*, Nl, n, and q is adequate. When the chromosome length is small or when the number of loci is high, the first two of these parameter combinations suffice. These two have also the merit that they are immediately observable or guessable with fair accuracy. In the first, Nih^*, the first two factors are imposed by the experimenter and reasonable estimates of h^* can be made from a knowledge of the overall heritability and the number of chromosomes concerned. In most species, evidence will be available as to the probable values of the second, Nl. It should be stressed that the theory will probably not hold above $h^* = 0.5$, not a very limiting condition.

What values of these parameters are we likely to meet? Consider a few of the selection experiments published in the literature. I have myself done many selection experiments for sternopleural bristles in *Drosophila* in which the selection intensity is 10/25 for each sex with an overall heritability of about 0.4. If we assume that the two major autosomes are twice the length of the X chromosome and that the amount of genetic variation on each is proportional to its length, we arrive at values of h^*

of 0.4 for the former and 0.28 for the X chromosome. We can assume N to be 15, somewhat smaller than the actual number of parents, and i will be very close to unity. As a result we arrive at values of Nih^* of 6 for the two major autosomes and 4 for the X chromosome. From the known lengths of the chromosomes and the fact that there is no crossing-over in males in this species, we can estimate Nl to be about 8 for the autosomes and 4 for the X chromosome.

In Falconer's original selection experiments for mice in which selection was for gain from three to six weeks with selection within litters from 12 pairs of parents, we have approximate values of $N = 48$ (because of the equal contributions of all litters to subsequent generations), $i = 0.8$, and $h^* = 0.125$, assuming that the heritability is 0.30 and that the genetic variation is equally distributed over all 20 chromosomes. Nih^* is then 4.8 and Nl of the order of 20–30. In Nordskog's recent experiments (personal communication) on selection for bodyweight in the domestic fowl, Nih^* for the 6 main chromosomes would be about 7 and Nl around 20–30. With such high values of Nl, it is unlikely that recombination was a limiting factor in each of the experiments.

What of the probable values of the other two parameters, n and q? From recent work by Thoday and his group and by workers in this laboratory on Drosophila and from some recent work on wheat, I feel that the effective number of genes on individual chromosomes will probably be less than 10. Remembering that the effective number for present purposes has been shown to be even less than the classical effective number, I feel that we are likely to be dealing with values rather below than above 5 on each chromosome.

We have very little evidence as to the initial frequency of the desirable alleles in our populations. The response to selection for sternopleural bristles in this laboratory would suggest that the mean gene frequency of alleles increasing bristle number is in the neighborhood of 0.4 in the large random breeding population used as source of base material. One also has the common sense feeling that intermediate gene frequencies are much more likely than situations in which the average frequency of all alleles affecting a metric character in a certain direction is, say, 0.01. One important consequence of initial gene frequencies deviating from values of one half is that the effect of modifications of recombination value is then likely to be different for selection in opposite directions.

In dealing with intermediate linkage values, we have obtained expectations on the assumption that the loci are distributed fairly evenly along the chromosome. In any specific practical case, we have to deal with an actual allocation of loci which may be far from random, the more so the fewer the loci. Even though our estimates of the four basic parameter combinations were correct, we might then be astray in predicting the effect of linkage in such a case. Using the random spacing model with 10 loci, the program was modified so that the same allocation of loci was used for 5 successive runs. No significant differences in advances at the limit were found between allocations, but the effort put into this was not great.

What does all this mass of algebra and computer results add up to? I can only present a personal view which is that the effects of linkage (specifically as looked at in terms of suppressing crossing-over on the chromosome) are much less than I had expected and less than might be assumed from discussions of linkage in the literature. This of course depends on what initial situations are likely to be found, but if one accepts the view that the effective number of loci on individual chromosomes is not likely to be large and gene frequencies not to be extreme, it is inescapable that, starting from a situation with initial linkage equilibrium, the effect of modifying linkage intensities will not be particularly large. Of course the necessary condition that linkage equilibrium exists is a very restricting one, in view of the suggestions from some workers that negative linkage disequilibrium is likely to be found between loci affecting quantitative characters. At the very least, however, our presentation is a basis for precise argument on these points.

The results of a series of experiments on the effect of suppressing crossing-over on the response to selection for bristles in *Drosophila melanogaster* are about to be published (McPhee and Robertson). The limits of selection with no recombination were close to those predicted from "normal" theory. With no recombination, the advances at the limit were 0.78 of those with the usual amount of crossing-over for selection upwards and 0.72 clownwards.

Where do we go from here? Although this paper is a long one, some aspects of the results have been dealt with rather briefly. In particular the gain at intermediate generations of selection would perhaps bear more examination as would also the models in which the effects of gene substitutions are not the same at all loci. This is almost certainly closer to the truth than an assumption of equal effects at all loci, and I feel that a more detailed examination of some of the possible models would be valuable.

The theory needs to be extended to situations in which there is linkage disequilibrium between pairs of loci in the initial population. These results have shown that this would certainly be true in crosses between lines which have themselves been selected to fixation from a given initial population. This makes an already complicated problem even more difficult. Preliminary thoughts on this would suggest that at least two more parameters are needed to describe the situation, even in the simple model of a series of equivalent loci. The first of these is the total negative covariance and the second is the distribution of this over the chromosome. For instance, if there were only two kinds of chromosomes available in the initial population, it would make a great deal of difference whether these were 101010 and 010101 or 111000 and 000111 respectively. In the first case, crossing-over in the first generation generates much less gametic variation than it does in the second.

It would be interesting to develop a theory of selection from a situation in which individual loci had themselves a normal distribution of allelic effect. The consequences of this assumption, which appear reasonable at the chromosomal level in the absence of crossing-over, have only been touched upon in this paper. Latter (1968) has considered this model in some of his work.

From the present analysis, it is possible to guess at some of the conclusions. At each such locus, similar arguments will hold as we applied here to chromosomes in the absence of crossing-over. Selection response will then be the same in both directions with and without crossing over and at high values of Nih^*, L_f/L_0 will equal the square root of the number of loci on the chromosome.

In the paper by Hill and Robertson (1966) the concept was used of the reduction in the expected change in gene frequency at fixation at a locus with a small effect, due to tight linkage with a locus with a much larger effect, initial linkage equilibrium being assumed. It was shown that, as the effect at the first locus declined, the proportional reduction in the change of gene frequency tended to a limiting value which was independent of the gene effect and initial frequency at the locus. What is the influence of a chromosome with a normal distribution of "allele" effects on a constituent or tightly linked locus with a small effect? Suppose the conditions are such that the advance for the whole chromosome is $K\sigma_g^*$. Each constituent locus will contribute to this advance in proportion to its contribution to the variance – the usual regression coefficient of a part on the sum of many independent parts. If all constituent loci had such a small effect that the gain was in the linear part of the response curve, the total response from free recombination between all loci will be $2Nih^*\sigma_g^*$ and each will contribute to this in proportion to their variance. It follows that the "normal" locus reduces the chance of fixation of any small part of itself, or of a locus with small effect tightly linked to it, by a factor $K/2Nih^*$ which will decline to zero as Nih^* increases.

Finally a few words on the importance of epistasis in this situation. It has been pointed out by Nei (1963) and Felsenstein (1965) that artificial selection leads to immediate negative disequilibrium between loci. To what extent is such "scalar" epistasis which does not alter the ranking of different genotypes but merely the relative distances between them an important factor in these results. My own view is that it plays a very minor part indeed. In the work on two loci, Hill and Robertson showed that it made little difference whether the genes were assumed to act additively or multiplicatively on the character under selection. In this work with many loci, some runs have been done with no crossing-over in which the selection procedure was modified so that the probability of a gamete being selected was exponentially related to the number of desirable alleles it contained. Selection itself does not then directly cause linkage disequilibrium in progeny if it was absent amongst the parents. Nevertheless in such runs negative disequilibrium built up almost as rapidly as with truncation selection. In the early generations, this apparently increased as the square of the time. The advance at the limit was also little affected. In my view, the essential point is that a population derived from a finite sample has, when recombination is low and selection is practiced, properties quite different from those of the population from which it was derived. In the development of the theory of selection from a normally distributed population of chromosomes (or of alleles at a locus) there would seem to be great scope for better understanding of this problem.

Summary

A general theory of the effect of linkage in artificial selection is presented, assuming additive gene action and initial linkage equilibrium. Concentrating on a set of loci on a single chromosome with identical effects on the character and initial allele frequencies, it is shown that the selection process can be adequately specified by the parameter combinations Nih^*, Nl, n and q where N is the effective population size, i is the selection intensity in standard units, h^{*2} is the heritability of the genetic variation controlled by this chromosome, n is the number of loci, q the initial frequency of the desirable allele, and l the length of the chromosome in cross-over units. The treatment centers round a discussion of selection in the absence of crossing-over on the assumption that, as n increases, the initial distribution of chromosomal effects will be approximately normal. This will require initial allele frequencies not to be too extreme. It is shown that the advance at the limit, when expressed in terms of the initial genetic standard deviation, is a function only of Nih^*. When this is small, the standardized advance is equal to $2Nih^*$ and as Nih^* increases the standardized advance approaches a limit in the neighbourhood of 3. The effect of linkage on the advance is assessed as the ratio of that with no crossing-over to that with free recombination. This ratio, L_0/L_f, can be shown to be a function of Nih^*, n, and q only. When Nih^* is less than one-half, linkage has little effect on the advance. It is shown that, as n increases, the ratio approaches a limiting value dependent only on Nih^* and not on either n or q. When initial gene frequencies are extreme and the initial distribution of chromosomal effects cannot be expected to be normal, it has been possible to deal with some special cases by algebra.

A series of computer runs have been done, the replication being of the order of hundreds. With fixed values of Nih^* and n, it was found that L_0/L_f passes through a minimum when q is less than 0.5. An increase in the number of loci always decreases the ratio. With ten loci on the chromosome, the minima in the ratio were usually above 0.5. At values of q above 0.8, L_0/L_f is close to unity except when n is high. The effects of variability between loci in gene effects and initial frequency have been studied. In general this increases the value of L_0/L_f and consequently reduces the effects of linkage.

Assuming that the loci are evenly scattered along the chromosome, the most useful description of the amount of crossing-over appears to be the chromosome length l, which invariably appeared in the formulae as Nl. It was shown theoretically that the effect of increasing the amount of recombination on the advance at the limit, expressed in terms of the advance with no crossing-over, was proportional to Nl when this is small and was then independent of n and q. The coefficient of proportionality, when estimated from computer runs turned out to be almost independent of Nih^* and of the order of $1/3$. With an equal spacing of loci along the chromosome, it was found that the advance at the limit could be expressed in terms of L_0, L_f, and Nl only. Under conditions likely to be met in practice, the probable increase in gain at the limit from doubling the amount of crossing-over per generation would not exceed $1/6$. The optimum intensity of selection to obtain the maximum advance from a given total number of individuals measured is increased from its value of 0.5 for independently segregating loci. In one situation with ten loci, the optimum proportion selected was 0.65.

The variance between replicate lines at fixation was also discussed. With no crossing-over, the variance between such lines is always less than would be expected on the basis of their mean gene frequencies. This is a consequence of the excess of repulsion over coupling linkages which would occur within a population made by crossing such lines. With intermediate linkage values, the amount of linkage disequilibrium during selection will be greater for loci in the middle of the chromosome. In general, therefore, the chance of fixation of the favorable allele at loci at the middle of the sequence is less than for those at the ends.

The half life of the selection process has also been studied. When Nih^* is high and there is no crossing-over, this equals $2/ih^*$ generations. Up to this time the amount of recombination has little effect on the selection response.

Finally the effect of simultaneous segregation of several chromosomes, contributing different amounts to the genetic variation, was discussed. When all chromosomes are equivalent in the number of loci on them and in their contribution to the total genetic variation, then the ratio of response due to all chromosomes must be the same as that due to single chromosomes. It was found that L_0/L_f was slightly reduced by variation between chromosomes in their contribution to the genetic variance but that the effect was not large at intermediate gene frequencies. Theoretically the consequence of such heterogeneity could be shown to be proportional to the squared coefficient of variation between chromosomes in their contribution to the total genetic variance. Unless this was large, adequate predictions could be made on the assumption that the genetic variation was distributed equally over all chromosomes.

References

Falconer, D. S.: Introduction to quantitative genetics. Edinburgh-London: Oliver and Boyd 1960.

— Selection for large and small size in mice. J. Genet. 51, 470—501 (1953).

Felsenstein, J.: The effect of linkage on directional selection. Genetics 52, 349–363 (1965).

Festing, M. F. W., and A. W. Novdskoj: Response to selection for body weight in chickens. Genetics 55, 219–231 (1967).

Fraser, A. S.: Simulation of genetic systems on automatic digital computers. I. Introduction. Aust. J. Biol. Sci. 10, 484–491 (1957).

Gill, J. L.: Effects of finite size on selection advance in simulated genetic populations. Aust. J. Biol. Sci. 18, 599–617 (1965).

Hill, W. G., and A. Robertson: The effects of linkage on limits to artificial selection. Genet. Res. 8, 269–294 (1966).

Kimura, M.: Some problems of stochastic processes in genetics. Am. Math. Statist. 28, 882–901 (1957).

Latter, B. D. H.: The response to artifical selection due to autosomal genes of large effect. Aust. J. Biol. Sci. 18, 585–598 (1965).

— Simulation studies of artifical selection in animals. Vol. 2, Proc. XII Int. Cong. Genet. Tokyo, 208–209 (1968).

Martin, F. G., and C. C. Cockerham: High speed selection studies. In: Biometrical genetics, pp. 35–45. Ed. by O. Kempthorne. New York: Pergamon Press 1960.

Mather, K.: Biometrical genetics. New York: Dover Publ. Inc. 1949.

McConnell, J., N. S. Fechheimer, and L. O. Gilmore: Somatic chromosomes of the domestic pig. J. Anim. Sci. 22, 374–379 (1963).

Nei, M.: Effect of selection on the components of genetic variance. In: Statistical genetics and plant breeding, pp. 501–515. Ed. by W. D. Hanson and H. F. Robinson. Washington, D. C.: Natl. Acad. Sci. – Natl. Res. Council, Publ. 982, 1963.

Qureshi, A. W., and O. Kempthorne: The role of finite population size and linkage in response to continued truncation selection. Theor. Appl. Genet. 38, 249–255 (1968).

—, —, and L. N. Hazel: Role of finite population size and linkage. Theor. Appl. Genet. 38, 256–263 (1968).

Robertson, A.: A theory of limits in artificial selection. Proc. Roy. Soc. B 153, 234–249 (1960).

Shoffner, R. N., and A. Krishan: The karyotype of Gallus domesticus with evidence for a W chromosome. Genetics 52, 474–475 (1965).

Wright, S.: The genetics of quantitative variability. In: Quantitative inheritance, pp. 5–41. Ed. by E. C. R. Reeve and C. H. Waddington. London: Her Majesty's Stationery Office 1952.

The Evolution of Dominance

J. A. SVED and O. MAYO

Introduction

R. A. Fisher in 1928 introduced his theory of the evolution of dominance. In bare outline the theory states simply that if an unfavourable gene is continuously being produced by mutation, other genes which modify the action of this gene will eventually be selected which will give individuals carrying the gene a phenotype resembling the favoured wild-type phenotype. Since the gene occurs initially to a large extent in the heterozygous rather than homozygous state, it is the heterozygote which is principally subject to this modification. The process is expected to be a very slow one, since the fraction of the population which is exposed to this selection is very small.

It is important to note that it is not dominance itself that is advantageous. Dominance is simply a description of a state, and it evolves because the heterozygote which resembles the normal homozygote is advantageous to one that is not. Also dominance is not necessarily the endpoint of the selective process, but may be only an intermediate stage. When dominance has evolved, the frequency of the mutant homozygote rises to a value which is of the same order of magnitude as that of the original heterozygote. If the mutant homozygote is then also modified to resemble the normal homozygote, the gene no longer has any phenotypic effect, so that, the term dominance becomes inappropriate.

Fisher's theory stimulated considerable discussion and criticism in the years following its introduction, principally, as will be detailed later, by Haldane, Muller, Plunkett, and Wright. It became clear that dominance could evolve in a manner somewhat different from that suggested by Fisher. It is therefore important that a clear distinction be drawn between Fisher's theory, which is often referred to without qualification as *the* theory of the evolution of dominance, and other such theories. Most of the mathematical arguments on the topic are related to Fisher's theory, and will be discussed in Section II. The other, later theories do not lend themselves to mathematical treatment in an equally simple manner, so that most of these arguments, which are summarized in Section III, are purely intuitive. A short account of some of the experimental work relevant to the topic is given in Section IV.

I. The Modification of Dominance

The theory of the evolution of dominance was put forward in 1928 in the absence of any direct evidence that dominance was normally subject to modification. In fact, although the evolutionary aspects constituted the most novel aspects of the theory, this was one of the first statements to the effect that dominance is not necessarily a property of the gene in question but is rather a property of the whole genome. Before this, Wright (1927) had given examples in the guinea pig where dominance at one locus was affected by the genotype at other loci.

The first experiment aimed at studying the possibility of dominance modification was that of Ford (1940) on the current moth *Abraxis grossulariata L.* This has normally a white background colour to its wings, but a variety, called *lutea*, has sulphur yellow colouration instead of white. This difference is mediated by replacement of both alleles at one locus. The heterozygote is intermediate, but close to the normal. The F_2 from a cross between the two types produced widely variable heterozygotes, obscuring segregation, but giving a trimodal distribution. However, heterozygotes could reliably be grouped into eight classes, varying between the homozygote types. Using a stock originating in a single heterozygous female found in the wild, Ford selected heterozygotes for lightness and darkness of colour and bred from them, giving two groups amongst whom bimodality for colour could already be seen. After two further generations of selection, *lutea* had become almost totally dominant to white in one line, and almost totally recessive in the other. At this point the experiment was halted. It is clear that animals could no longer confidently be chosen as heterozygotes, so that any further breeding done would require rejection of many broods because parents were homozygous rather than heterozygous.

The production of dominance depends of course on the modification of the heterozygote occurring in the absence of an equivalent amount of modification of the homozygote. In fact, in neither experiment was the homozygote modified to more than a slight degree. This is presumably a product of two factors: first, that the selection was carried out in heterozygotes and not in homozygotes, and secondly, that the heterozygote, possessing both factors, could be expected on physiological grounds to be more readily modified than the homozygote. While the relative contribution of these two factors is not so important in the present context, it will be seen to be of much more importance in the discussion of the alternative selective theory of dominance modification.

In considering the results of the experiment described above, it should be noted that since "stocks carried in the laboratory are usually [as this was] derived from a few individuals, ... in much larger populations found in the wild, many more suitable modifiers must be available" (Sheppard, 1958). On the other hand, in the wild population there will be a much lower frequency of heterozygotes, so that, as discussed in the following section, a much lower rate of dominance modification is to be expected.

II. Fisher's Theory

Fisher's arguments for the plausibility of the evolution of dominance, summarized in Fisher, 1958, Chapter III, are general ones which are largely independent of the underlying genetic basis for the dominance modification process. Since the principal limiting factor in the process is assumed to be the availability of heterozygotes, Fisher's main concern is to estimate the opportunity for selection of modifiers amongst heterozygotes.

A population is assumed, consisting in large part of individuals which are homozygous for a certain wild-type gene A. [Some justification should be given for the use of the notation $A - a$, which is commonly reserved for genes which are dominant and recessive respectively. Since in the present discussion the heterozygote is assumed to be distinguishable from the wild-type, the A gene is clearly not at the start dominant to a. However, the notation is a convenient one, since the evolution of dominance of A over a is being considered. It is also appropriate to comment here on the use of the terms "dominance" and "recessiveness". The terms were introduced by Mendel to describe the case where the heterozygote was indistinguishable from one homozygote. Following the description of cases where the heterozygote resembled but was distinguishable from the homozygote, the term "incomplete dominance" was introduced. The notion of "degree of dominance" as a quantitative measurement is now commonly used, together with the notion of "dominance" as an all-or none phenomenon. At the same time many mutant genes, for example in *Drosophila melanogaster*, are commonly referred to as "dominant" since the heterozygote is distinguishable from the wild-type homozygote. However, since in most cases the mutant homozygote is distinguishable from the heterozygote, being commonly lethal, such genes should properly be referred to as "intermediate" (see Section IV).] The allele a is continuously being produced by mutation, and eliminated by selection. While because of its low frequency, the gene a is present almost solely in heterozygous form, nevertheless the heterozygote constitutes only a small proportion of the population.

It is assumed that amongst heterozygotes there is a certain amount of heritable variation, or in other words that there are other genes which modify the effect of the heterozygote. Then those genes which increase the fitness of the heterozygote, presumably by modifying its expression towards that of the favoured homozygote, will be at a selective advantage and will increase in frequency. If the population consisted solely of heterozygotes, the increase in the overall frequency of favourable modifiers could be calculated very simply. However, since any modifier genes will occur to a very large extent in AA rather than Aa individuals, the principal concern is to determine to what extent the selective advantage amongst heterozygotes can be effective in increasing the frequency of modifiers in the whole population.

It seems convenient to give a simple example which illustrates the manner in which Fisher takes account of the contribution of the heterozygote to the whole population. In order to avoid all possible complications arising from the inheritance of the modifiers, we consider simply the frequencies of two modifying genotypes before and after selection. M_1 is assumed to be a genotype with a favourable effect on the heterozygote, and M_2 may be thought of as the collection of all alternative genotypes. The frequencies of M_1 and M_2 (Table 1) are p and q before selection $(p + q = 1)$, and the substitution of M_1 for M_2 causes the fitness of the heterozygote to be raised to $1 - s_1$ from a mean of $1 - s_2$. The frequencies of AA and Aa are assumed to be $1 - \alpha$ and α, where α is small.

Table 1. *Frequencies and fitnesses for a simple model of dominance modification. (Fitnesses are shown in the body of the table)*

Genotypes		AA	Aa
	Frequencies	$1 - \alpha$	α
M_1	p	1	$1 - s_1$
M_2	q	1	$1 - s_2$

It is assumed that the modifying genotype is inherited independently from the A genotype. Then the joint genotype frequencies may be written as a product of the marginal frequencies, the frequency of AAM_1 for example being $p(1 - \alpha)$. The frequency of M_1 after selection may now be calculated from Table 1 as

$$f(M_1) = \frac{p(1 - \alpha) + p\alpha(1 - s_1)}{p(1 - \alpha) + q(1 - \alpha) + p\alpha(1 - s_1) + q\alpha(1 - s_2)}$$

$$= \frac{p - p\alpha s_1}{1 - \alpha(ps_1 + qs_2)}.$$

The change in frequency of M_1 is thus equal to

$$\Delta p = \frac{p - p\alpha s_1}{1 - \alpha(ps_1 + qs_2)} - p = \frac{\alpha pq(s_2 - s_1)}{1 - \alpha(ps_1 + qs_2)}.$$

Since α is assumed to be very small, this expression is approximately $\alpha pq(s_2 - s_1)$.

We may now calculate the change in frequency of M_1 over one round of selection in the situation where the population consists solely of heterozygotes. This comes to

$$\frac{p(1 - s_1)}{p(1 - s_1) + q(1 - s_2)} - p = \frac{pq(s_2 - s_1)}{1 - ps_1 - qs_2}.$$

Thus the actual change in frequency of M_1 is equal to the change of frequency in a population consisting solely of heterozygotes multiplied by the frequency of heterozygotes (α), and by the quantity $1 - ps_1 - qs_2$ ($= v$ say), which is the mean fitness of the heterozygote relative to the homozygote. The factor v may be considered necessary because the heterozygote contributes proportionately less than the homozygote to the gene pool of the following generation.

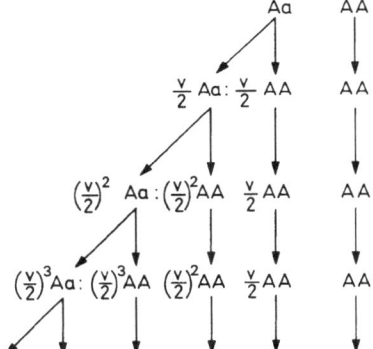

Fig. 1. Relative contributions of heterozygote and homozygote to first, second, third, etc. generations

In the long term, what is important is not so much the relative contributions of the heterozygote and homozygote to the gene pool of the next generation, but rather to the gene pool at a distant time in the future. In order to derive this, we assume for the moment that all heterozygotes have the same fitness v compared to the homozygote. Also we assume that all matings involving heterozygotes are $Aa \times AA$, since the $Aa \times Aa$ mating would be very infrequent in comparison. Then the heterozygote contributes a proportion v in the next generation, of which $v/2$ on the average is to individuals with the genotype Aa, and $v/2$ to AA. In the next generation we may make a similar argument (Fig. 1), so that the $v/2$ heterozygotes of which the original heterozygote is an ancestor will give rise to $\dfrac{v}{2} \cdot \dfrac{v}{2}$ heterozygotes and an equivalent number of homozygotes.

The $v/2$ homozygotes in the following generation on the other hand will give rise to $v/2$ homozygotes again. In the limit, therefore, compared to the homozygote, it is apparent that each heterozygote contributes

$$\frac{v}{2} + \left(\frac{v}{2}\right)^2 + \left(\frac{v}{2}\right)^3 + \cdots = \frac{v}{2-v}$$

to future generations. Fisher derives the same result in a more elegant, although perhaps intuitively less obvious manner, by using a recurrence relation.

It is evident that the argument given above is not exact under all circumstances, particularly in so far as it breaks down to some extent when inherited differences in heterozygote fitness are considered. Then it is not legitimate to give the contribution of heterozygotes in the second generation to heterozygotes or homozygotes in the third as $\left(\frac{v}{2}\right)^2$. Those individuals which contribute more than $v/2$ individuals to the second generation will leave offspring which themselves contribute more than average to the next generation. The factor $\frac{v}{2-v}$ thus underestimates the true value, as will be discussed later.

Using similar arguments to those advanced in connection with the model of Table 1, it is seen that an increase in frequency of modifiers amongst heterozygotes of amount x say, produces in the limit in the overall population a change x reduced by the frequency of the heterozygote (α) and by the factor $\frac{v}{2-v}$. The product $\alpha \frac{v}{2-v}$, which will be denoted as Q, gives the overall contribution of heterozygotes of a particular generation to the gene pool of future generations.

The frequency of the heterozygote (α) is determined by the balance between mutation from A to a, and selection against a. The rate of loss of a genes is approximately $\frac{\alpha}{2}(1 - v)$, which must be equated to the rate of gain of a genes, $viz.$ u, the mutation rate. Thus $\alpha = \frac{2u}{1-v}$. Note that the value of α does not become infinite as v approaches unity, as would be suggested by the formula. When the heterozygote becomes as fit as the normal homozygote ($v = 1$), selection against the mutant homozygote becomes the important factor in keeping α low. By the time this occurs however, dominance must be so far advanced as to be practically complete, so that it seems unnecessary to introduce any correction to the formula to take account of this.

The overall contribution of heterozygotes to future generations now becomes $\dfrac{2u}{1-v} \cdot \dfrac{v}{2-v}$. Then according to Fisher (1958, p. 62): "This quantity [Q] which, when the mutation rate ... is 1 in 1,000,000, rises to about 1 in 5,000 if v is 0.99, represents the rate of progress in the modification of the heterozygote, compared to the rate of progress which would be effected by selection of the same intensity, acting upon a population composed entirely of heterozygotes." This value may therefore be used in conjunction with the results of experiments such as Ford's described above to give a fairly direct estimate of the amount of time required for dominance modification to take place. However, the value of Q appropriate to this case must first be calculated, and in this connection it is not clear that the value of 1 in 5,000 given for the case $v = 0.99$ is really appropriate. For if the value of v is as high as 0.99, the selective intensity tending to modify the heterozygote must be proportionately low, a point made by Crosby (1963). Alternatively, if the quantity Q was as high as 1 in 5,000, the rate of progress achieved by selection would probably be less than one percent of the rate found in the artificial selection experiment. This would indicate that something over one million generations are needed for dominance to evolve in this case.

It might be useful to give an artificial example illustrating the relative rates of modification for cases where the selection against the unmodified heterozygote is low and high. We can consider two populations genetically identical at the beginning in which a pigmentation gene which increases susceptibility to predation is continuously produced by mutation. The two populations differ only in that for one, the predator is assumed to be scarce and for the other, abundant. Thus the mean viabilities of the heterozygote in the two cases might be 0.99 and 0.50 respectively, making the term Q some one-hundred fifty times greater for the former. But owing to the fifty-fold difference in death rates, the selective intensity achieved for modifiers amongst heterozygotes would be about fifty times as high in the latter case, making the initial rate of modification only about three times as high in the former. As dominance modification proceeds, the rates would become more nearly equal in the two populations.

We have not so far considered the relative rates of modification at different stages during the process, although this is a problem closely related to that described above. For as the heterozygote becomes less unfavourable, the quantity Q becomes larger, making possible an increase in the rate of modification. However, it seems very likely that this will be balanced to some extent by a reduction in the selective intensity for modifiers. The actual change in rate of the process, particularly in later generations, might be determined by a rather delicate balance between these two factors.

To estimate the expected selective increase in the frequency of modifiers, it is necessary to consider the amount of "modificatory variance" available. By Fisher's fundamental theorem of natural selection (Fisher, 1958, Chapter II), the rate of change of selective value is equal to the additive genetic variance in selective value, or in this case to the additive modificatory variance (V). This formulation, however, assumes that the selective value is measured in malthusian parameters (see also Wright (1955) for further discussion). In the present case where the selective value is measured on the normal linear scale, the mean change in selective value is equal to the variance divided by the mean, i.e. V/v. (See O'Donald (1967 b) for a derivation of this for the case of dominance modification.)

The overall rate of modification, as measured by the rate of change of v, would then be equal to $Q \cdot V/v$. V will presumably be some fraction less than unity of the phenotypic variance in v amongst individuals within a particular generation. It need not be a constant throughout the process, and its expected value needs to be considered in some detail.

One likely property of V is that it should tend to zero as the mean heterozygote fitness approaches unity. Ignoring for the moment the possibility that the heterozygote can have a higher fitness than the normal homozygote, a mean heterozygote viability of unity implies that all individuals must have viability of unity, so that the variance must be zero. Similarly, the variance must tend to zero as v tends to zero. In between these two values, however, it does not seem possible to derive any general relationship connecting V and v. This is readily seen by making different assumptions about the magnitude of V. For example, if V is very small, then it might be a constant throughout most of the range, tending to zero only at the very end of the process. Under these conditions, there could well be a marked rise in the rate of dominance modification as Q rises (due mainly to the $1 - v$ term in the denominator) unattended by a drop in V except at the very end of the process. However, it should be remembered that the situation of a low value of V is least favourable to the overall theory and seems at variance with the results of experiments such as those by Ford described previously.

If the variance is large, a more reasonable assumption is that V falls steadily after dominance is half-way or so completed, and we can consider the situation where it is proportional to $v(1 - v)$. In this case the $(1 - v)$ term in V cancels with the similar term in the denominator of Q, and $Q \cdot V/v$ is proportional to $\dfrac{2uv}{2-v}$. Thus we can put $\dfrac{dv}{dt} = k \dfrac{v}{2-v}$, from which the time required to change from $v = v_0$ to $v = 1$ is equal to

$\dfrac{1}{k}(v_0 - 1 - 2\log v_0)$: This is plotted in Fig. 2 which is analogous to Fig. 4

of Fisher (1958). The time required to change from $v = v_0$ to $v = v_1$ can also be read from the figure. It is evident that a large time is required in the early stages if the unmodified heterozygote starts with a low selective value. For this reason, Fisher suggests that mutants might in the main be divided into two classes; ones which are severely retarded in fitness in the heterozygote and have not been able to pass through the early stages of the modification process, and more or less fully recessive ones which, having once passed through the early stages of the process, have been able to pass through the latter stages much more quickly.

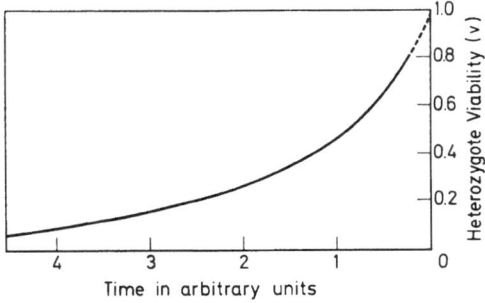

Fig. 2. Graph illustrating the relative amounts of time needed to pass through various stages of dominance modification process. The assumptions for the derivation are given in the text

As mentioned previously the exact shape of the curve in Fig. 2 cannot be interpreted too literally, since particularly in the later stages, the assumptions about the form of the variance become crucial and the formula giving the frequency of the mutant gene tends to break down. Also Fisher considers the possibility that the value of V might be a function of the selective value of the unmodified heterozygote. If the selective value of the unmodified heterozygote is v_0, V might take the form $(v - v_0) \times (1 - v)$ rather than $v(1 - v)$.

One point which has not been considered so far, but which is discussed in some detail by O'Donald (1968), is the basis for the distribution of v. The discussion so far has been solely in terms of the selective value, which must usually only be an indirect measure of some underlying characteristic. The pigmentation example given earlier provides a good example of this, for the primary character is the intensity of pigmentation, while the selective value depends also on the reaction of the predator to the pigmentation. As O'Donald points out, the arguments given by Fisher (1958, p. 15) provide some basis for supposing that there might be a linear relationship between intensity of pigmentation and selective value in this case. If this were not true, however, if there were, for example, a quadratic relationship between v and the intensity of pigmentation, there would no longer be any basis for expecting V to be symmetrical on the scale of v values. It is also necessary to consider the possibility that for physiological reasons it might be easier, or alternatively harder, to make progress in selecting for increased pigmentation earlier rather than later in the process. It is apparent that no simple and general answers can be given to these questions, but that they might all affect the problem of what relative rates of modification are expected at different stages throughout the process.

It should be noted that the arguments given so far have not specified any definite genetic basis for the dominance modifiers. While it would be desirable to have some idea of the underlying genetic basis, it is clear that only in isolated cases (see Section IV) can the actual genes involved be identified. Since little is known about the underlying basis for dominance modification, at the present state of knowledge it seems most realistic to consider the distribution of v as continuous with an envisioned underlying polygenic basis.

The Single Modifier Model

One model where accurate calculations can be made about the selective consequences for genes which cause the modification is the single gene model of dominance modification. Perhaps a majority of the literature on the mathematics of dominance modification has been devoted to this model. In one sense this is unfortunate since this model was apparently never intended to give an accurate description of the genetic basis for the process. The model was introduced by Wright (1929a) in his comments on Fisher's original statement of the theory. Wishing to comment on the extreme slowness of the process, Wright chose an example favourable to the theory, i.e. where there was a gene in the population whose substitution was by itself capable of effecting the entire dominance modification. This was thus intended to give an upper limit to the rate which could be expected for dominance modification (given the other limitations) and was accepted as such by Fisher (1929).

Table 2. *Frequencies and fitnesses for single modifier model. (Fitnesses are shown in the body of the table)*

Genotypes		AA	Aa	aa
	Frequencies	q_A^2	$2q_A(1-q_A)$	$(1-q_A)^2$
MM	q_M^2	1	1	$1-s$
Mm	$2q_M(1-q_M)$	1	1	$1-s$
mm	$(1-q_M)^2$	1	$1-sh$	$1-s$

The calculations for the single locus model will follow essentially those of Mayo (1966). The selective values for the different genotypes are given in Table 2. The selective value of the unmodified heterozygote Aa, i.e. in the presence of the genotype mm, is $1\text{-}sh$, while the gene M causes the heterozygote Aa to resemble the normal homozygote AA and thus to have selective value 1. Assuming that the frequencies of genotypes at the two loci are independent, an assumption which will be discussed below, the genotype frequencies may be written as a product of frequencies at the individual loci:

	AA	Aa	aa
MM	$q_A^2 q_M^2$	$2q_A(1-q_A)q_M^2$	$(1-q_A)^2 q_M^2$
Mm	$2q_A^2 q_M(1-q_M)$	$4q_A(1-q_A)q_M(1-q_M)$	$2(1-q_A)^2 q_M(1-q_M)$
mm	$q_A^2(1-q_M)^2$	$2q_A(1-q_A)(1-q_M)^2$	$(1-q_A)^2(1-q_M)^2$

It follows that the overall mean selective value, v',

$$v' = 1 - 2q_A(1-q_A)(1-q_M)^2\, sh - (1-q_A)^2\, s.$$

Hence, the frequency of M in the next generation is q_M^1, given by

$$q_M^1 = \frac{q_A^2 q_M + 2q_A(1-q_A)q_M + (1-q_A)^2 q_M(1-s)}{v'}.$$

Thus,

$$\Delta q_M = q_M^1 - q_M = \frac{2q_A(1-q_A)q_M(1-q_M)^2\, sh}{v'}.$$

Since $1 - q_A$ is small in comparison to 1,

$$\Delta q_M \doteq 2(1-q_A)q_M(1-q_M)^2\, sh. \tag{1}$$

In substituting for q_A, Wright (1929a) originally assumed that the frequency of the a gene was given by u/sh, the frequency of the unmodified heterozygote given by the balance of mutation and selection. This gives a value of

$$\Delta q_M = 2uq_M(1-q_M)^2. \tag{2}$$

Fisher (1929) pointed out that this does not take into account the rise in frequency of the a gene as selection against the heterozygote becomes less severe. As a first approximation, the frequency of the a gene then becomes

$\dfrac{u}{sh(1-q_M)^2}$. Substituting in Eq. (1) then gives approximately

$$\Delta q_M = 2uq_M. \tag{3}$$

This correction was accepted by Wright (1929b) as valid. Subsequently, Ewens (1965a) claimed that Δq_M should have a form essentially similar to that given in Eq. (2), but later Ewens (1965 b) presented a different solution in the form of a pair of differential equations. Exact computations have been made on this point by Kimura (personal communication and quoted by Ewens, 1967) which are given in Table 3, showing that the

Table 3. *A comparison of predictions from several formulae of the rate of change of dominance with a single modifier at various stages of the process. The computations of this example were made with the initial chromosome frequencies being respectively AM $-$ 0.09998, aM $-$ 0.00002, Am $-$ 0.89982, and am $-$ 0.00018. Thus the initial frequency of the gene M was 10%, while the a gene was started at frequency u/hs. The values of u, h, and s were respectively* $2 \times 10^{-5}, 0.5$ *and* 0.2

Genera-tion	q_M	%Domi-nance	Δq_M $\times 10^5$	Values of $i = \Delta q_M / [q_M(1 - q_M)]$				
				Exact	Form.(2)	Form.(3)	Wright	Ewens
0	0.1000	19.5	0.324	3.599	3.600	4.444	4.441	3.600
100	0.1004	20.7	0.401	4.443	3.598	4.446	4.444	4.444
17400	0.2005	38.1	0.801	4.996	3.198	5.003	4.998	4.996
27500	0.3001	53.0	1.198	5.704	2.800	5.715	5.707	5.706
34800	0.4015	65.8	1.599	6.654	2.394	6.683	6.650	6.650
40400	0.5015	76.3	1.987	7.948	1.994	8.024	7.956	7.952
45000	0.6012	84.8	2.350	9.802	1.595	10.03	9.860	9.854
49000	0.7011	91.4	2.627	12.54	1.196	13.38	12.76	12.62
52800	0.8014	96.7	2.541	15.97	0.794	20.14	16.60	16.14
57800	0.9011	99.0	1.336	14.99	0.396	40.45	15.55	15.20
64100	0.9505	99.7	0.430	9.137	0.198	80.81	9.317	8.652
80000	0.9799	99.96	0.077	3.896	0.080	199.0	3.922	3.912
Col. 1	2	3	4	5	6	7	8	9

latter treatment is essentially correct, while the former is incorrect. Kimura has calculated by iteration the exact values expected at all stages of the process, starting with the initial conditions given in the table. The values of $i = \dfrac{\Delta q_M}{q_M(1 - q_M)}$ found are given in the table in column 5. Values of i expected under formulae (2) and (3) are given in columns 6 and 7. Formula (3) gives an accurate description for the first 50,000 generations until as seen in column 3 more than 90% of heterozygotes are dominant. The other formula is, however, inaccurate at all stages except in the first generation. The agreement in this generation is somewhat artificial since the a gene has been started at a frequency equal to $\dfrac{u}{sh}$ and has not yet had time to increase to the expected frequency of $\dfrac{u}{sh(1 - q_M)^2}$.

Formula (3) starts to become rather inaccurate after 50,000 generations. This is evidently because the frequency of a does not rise indefinitely

high as predicted by the formula $\dfrac{u}{sh(1-q_M)^2}$ as q_M becomes close to
unity, since selection against the homozygote now becomes the dominant
factor in keeping a rare. Wright (quoted by Ewens, 1967) has shown that
this factor may be taken into account giving results (column 8, Table 3)
which are accurate throughout the entire range of Kimura's calculations.
Ewens (1967) has also made the point that the frequency of a will never
quite catch up with the frequency given by the formula $\dfrac{u}{sh(1-q_M)^2}$.
A further small improvement is possible taking this factor into account,
and these results, obtained by numerical solution of Eq. (5) and (7) of
Ewens (1965b), are given in column 9.

It might be questioned whether the formula suggested by Fisher is
not sufficient for all practical purposes. By the time the formula becomes
inaccurate, dominance is more than 90% complete. Most experimental
observations would not be sufficiently accurate to detect such small deviations from complete dominance. In particular, we cannot agree with
Ewens, who has stated (1965b) "it is interesting to note that Fisher
originally developed it [the theory of the evolution of dominance] using
an incorrect formula, and that had he been aware of the correct formula
he would not have advanced his theory at all" and (1967) "from the purely
mathematical point of view, this seems to imply rather less support for
Fisher's theory than has been supposed."

While the above treatment of the single gene model is certainly adequate as shown by the numerical agreement, it may be a little unsatis-
factory conceptually. As pointed out by Lewontin and Kojima (1960),
Kojima and Kelleher (1961), and Bodmer and Parsons (1962), when there
is epistasis the assumption of random combination of genes at the two
loci breaks down, so that working in terms of individual gene frequencies
and multiplying as was done above is not adequate, even if the genes
are on different chromosomes. Ewens' calculations are therefore made
using the gene pairs AM, Am, aM, and am. However, as Mayo (1966)
has shown, the assumptions made to derive an algebraic solution using
this approach make it equivalent to the single gene treatment. Furthermore, the numerical solution given in column 9 of Table 3 apparently
makes the same assumption implicitly.

Kimura (1965) has shown that when epistatic interactions are not too
large and gene frequencies are changing slowly, the deviations from random combination of genes at the two loci are quite predictable. Since
these conditions are apparently fulfilled in the present example, Kimura
(personal communication) has shown that the deviations from random
combination may be taken into account to derive a more comprehensive
theory.

Since Fisher's calculations give the rate of modification for a general model of dominance modification, it is of some interest to try to particularize the result to the single modifier model to compare results from the two approaches. One point which may be noted immediately is that the rate of dominance modification for the single gene model is independent of the selective value of the heterozygote. This is an analogous result to that argued on more general grounds in the previous section. Furthermore, the rate of modification rises as the process proceeds, in a rather similar manner under the two approaches.

It is convenient to compare the changes per generation in the selective value of the heterozygote under the two approaches. For the single gene calculations, the mean selective value of the heterozygote (v) is seen to be $1 - sh(1 - q_M)^2$. Then the change in selective value in one generation is

$$\Delta v = 1 - sh[1 - (q_M - \Delta q_M)]^2 - [1 - sh(1 - q_M)^2],$$

and substituting $\Delta q_M = 2u q_M$, this gives approximately

$$\Delta v = 4 shu q_M (1 - q_M).$$

It is necessary to calculate now the additive genetic variance for the single gene model. The overall variance in heterozygote selective values is seen to be equal to $(1 - q_M)^2 [1 - (1 - q_M)^2] (sh)^2$. This must be partitioned into *additive* and *dominance* components. If this is done using the method of Kempthorne (1957, p. 316), we get for the additive variance $2q_M(1 - q_M)^3 (sh)^2$ and for the dominance variance $q_M^2(1 - q_M)^2 (sh)^2$, summing to the value given above for the overall variance.

We may now substitute in the formula for the change of selective value of the heterozygote given earlier, *viz.*

$$\Delta v = Q \frac{V}{v}$$

$$= \frac{2u}{1 - v} \cdot \frac{v}{2 - v} \cdot \frac{V}{v}$$

$$= \frac{2u V}{(2 - v)(1 - v)} .$$

It is convenient to separate out for the moment the term $2 - v$ in the denominator. Then substituting $1 - sh(1 - q_M)^2$ for v and $2q_M(1 - q_M)^3 \cdot (sh)^2$ for V, we get

$$\Delta v = \frac{1}{2 - v} \cdot \frac{2u \cdot 2q_M(1 - q_M)^3 \cdot (sh)^2}{(1 - q_M)^2 \cdot sh}$$

$$= \frac{1}{2 - v} \cdot 4u \, shq_M(1 - q_M).$$

Thus the results obtained from using the two methods differ by the factor $2 - v$. This is evidently due to the fact that, as mentioned previously, the derivation of Q gives an underestimate since it assumes no heritable differences. The single gene model, of course, departs completely from this assumption, so that it seems that the greatest error which the general approach can give is a factor of 2.

An important assumption in the calculation has not so far been mentioned. The rate of substitution of the modifier has been calculated under the assumption that the only effective selection is that due to its effect on the heterozygote. Wright (1929a, 1934 *et seq.*) has argued that the gene will almost certainly have pleiotropic effects which must lead to other selective pressures greater than the minute value attributable to the modification of the heterozygote. Therefore he questions not so much whether sufficient time is available for the process to take place as postulated by Fisher, but rather whether such small selective pressures can ever be effective in producing evolutionary change. The argument is based on pleiotropic effects which are assumed to lead to fixation of the modifying gene. Fisher (1934) in replying to this argument accepts the probable importance of pleiotropy in this context, and also in the case where the modifier is held at a selective balance through heterozygote advantage. However, he questions whether conditions will be sufficiently stable to allow arguments such as this to be made over evolutionary periods of time and concludes that in general the effect of a selective pressure can be expected to be proportional to its magnitude. It seems difficult to visualize any evidence which could decisively resolve this point.

Evolution of Dominance During Gene Substitution

It is convenient to examine at this point a suggestion made by Haldane (1956). He considered, while on the whole rejecting, the interesting possibility that dominance might sometimes take place during substitution of the gene, when the heterozygote is frequent, rather than at the stage when the heterozygote is rare. The hypothesis was suggested specifically for the case of industrial melanism in the Lepidoptera (Kettlewell, 1956) where it is known that over a period of time dating from the last century the melanic gene replaced its allele. The melanic gene is now dominant, although it is of course not known what was the dominance relationship in the last century before the gene substitution took place. There would obviously have been insufficient time for dominance to have evolved by the process considered by Fisher. Haldane considered the evolutionary hypothesis as a possible alternative to the explanation that dominance might be an inherent property of the pigmentation gene.

It should be stressed that this explanation for dominance does not have universal application. While it is true that every wild-type gene must at some time have been rare, there is little reason to suppose that it replaced any of the alleles now seen at low frequency and to which it is dominant. The explanation is one for dominance of the newly arisen over the ancestral gene, rather than dominance of the favourable over the unfavourable, to which, as Fisher emphasized, his theory applies. In the special case of industrial melanism, the two descriptions are known to coincide.

Haldane's conclusions were based on calculating the total number of heterozygotes which would occur at all stages during the gene substitution. Assuming the heterozygote to be initially at a disadvantage compared to the homozygote, he calculated the likely increase in frequency of a modifier which brought the heterozygote up to the fitness of the homozygote. While it was found that the frequency of an initially rare modifier might be increased, the increase would by no means be sufficient to ensure dominance in the whole population. Confirming this point, Mayo (1966) was able to produce increased dominance in simulation studies with one, two, and ten modifying loci, but the increases were rather small. An important reason for this, as emphasized by Crosby (1963), is similar to that suggested in connection with the previous calculations. If high selective values are involved, so that the intensity of selection for modifiers amongst heterozygotes is high, the gene substitution is completed very quickly. Alternatively, if gene substitution occurs slowly so that there are many heterozygotes, the intensity of selection for modifiers must be low.

The effect of having a modifier linked to the locus was considered by Bodmer and Parsons (1962). They concluded that there would be a larger increase in frequency of the modifier under these conditions, although Ewens (1966) questioned the significance of this increase.

Evolution of Overdominance

Parsons and Bodmer (1961) and Bodmer (1963) have discussed the possibility that overdominance might evolve during gene substitution. During the early stages of any gene substitution, the new gene will be present mainly in the heterozygous state. Therefore, any modifiers will be selected principally for their effect on the heterozygote rather than on the homozygote of the new allele. This argument shows that there is some chance for overdominance to evolve even if the heterozygote is not advantageous a priori. However, the advantage evolved in this manner could only be a very slight one, since Bodmer has shown that when the three genotypes reach the stable equilibrium expected with heterozygote advantage, all three genotypes contribute proportionately equal amounts to the genome of future generations. From this he argues that once an equilibrium is reached, no further improvement of the heterozygote over the homozygotes is to be expected.

O'Donald (1967 b) has considered the evolution of dominance from the point of view of genes which are initially disadvantageous but which evolve dominance and then overdominance. In this case, however, the heterozygote is never as frequent as the advantageous homozygote, so that the frequency argument plays no part in these considerations. The selection of suitable modifiers depends in this case on the heterozygote being inherently modifiable to a higher fitness than the homozygote. No such assumption as this is necessary in explaining the evolution of dominance. In this case there exists an obvious means whereby the heterozygote fitness can be increased, *viz.* by altering the expression of the heterozygote towards that of the homozygote which already enjoys a higher fitness. Further improvement, however, demands that some novelty of the heterozygote exist which can be exploited to produce a higher fitness.

III. Discussion of Other Theories

As previously mentioned, a number of authors including Haldane, Muller, Plunkett, and Wright contributed discussions to the question of how dominance might be explained, many of the arguments given by these authors and also by Fisher being considerably interrelated. Much of the discussion centered around the question of whether dominance has evolved or whether its existence could be predicted simply from a knowledge of how the gene works. Considerable care must be exercised in defining questions such as these, for it must be remembered that all synthetic mechanisms are a product of evolution. While this may at first sight seem a trivial argument, it will be seen to be important under some circumstances.

The argument that dominance is a consequence of the physiology of gene action can be traced back to theories put forward considerably before Fisher's. As pointed out by Wagner and Mitchell (1964, p. 343), "the concept that dominance of a gene is a manifestation of its superior activity has its roots in the presence-absence hypothesis." This hypothesis, introduced by Bateson in 1909, held simply that a recessive character was produced by the absence of a gene, and the corresponding dominant by its presence. While it is now clear that the presence-absence hypothesis is inadequate in its original form, nevertheless studies in bacteria (see e.g. Watson (1965), Chapter 13) have indicated that an analogous situation may exist, since in many cases one gene will code for the production of a functional protein, while its allele will code for a protein lacking that function or for no protein at all. If it is accepted that genes of higher organisms are analogous to those of bacteria, then the essential basis for this argument seems to have been confirmed. It would seem likely, therefore, as argued by Wright (1941) and more recently by Gregg (1967) "that if either the dominant or recessive is inactive it must be the latter."

This does not of course explain by itself why there should in general be dominance rather than intermediacy of the heterozygote. However, this type of argument can be extended, since it seems possible that for purely physiological reasons some sort of law of diminishing returns would be expected, so that the first dose of an active allele would add more than the second. This idea was developed more rigorously by Wright (1934) in terms of rates of chemical reactions catalyzed by particular enzymes.

Perhaps the least proven point in this argument is the supposition that a simple functional versus nonfunctional protein difference is the underlying basis for most simply inherited differences in higher organisms. The effects of a gene substitution are probably observed several levels from the primary gene product, and the biochemical bases for such differences are not known, except for a few cases concerned mostly with colour pigments. It is difficult, for example, to fit the results of Ford's experiment described in Section I into a functional versus nonfunctional scheme, since in this case the heterozygote could be selected to resemble either homozygote. Possibly this could be explained on the basis of selection for some sort of biochemical threshold lying between the normal single and double doses of the active enzyme.

With these reservations, it seems desirable to speculate on the basis of knowledge available at the moment on what is the most likely underlying basis for most mutants. Fisher and Wright seemed in agreement that the genes whose recessiveness needed to be explained were those which were continuously produced by mutation but which nevertheless were probably no longer of any importance in evolution. Presumably many, and perhaps a majority of these, could represent cases with one or more nonfunctional proteins. In such cases, as argued by Wright and Gregg, the direction of dominance is probably predictable, and it is only the magnitude which is subject to modification by evolution.

While these arguments suggest that it is unrealistic to attempt to explain dominance by evolutionary arguments while ignoring all physiological considerations, the reverse is probably equally true. For example, a purely physiological explanation can be given based on the supposition that the homozygote produces twice as much enzyme as the heterozygote, but that the reaction catalyzed by the enzyme does not proceed twice as rapidly in the homozygote as in the heterozygote. But such a scheme ignores one very important factor. The amount of enzyme produced by a given gene or gene pair is probably in most cases strictly regulated by negative feedback mechanisms to coordinate with complex synthetic pathways (see Watson, 1965, Chapter 14), so that it is unlikely that the homozygote does in fact produce twice as much of the enzyme. Such regulatory mechanisms can be of a variety of types, but they must all have an evolutionary origin.

It might be questioned whether, if the homozygote is the normal genotype, the heterozygote is usually capable of reaching an optimal rate of enzyme production. The argument of Haldane (1930) is relevant to this point, since Haldane suggested that the evolution of dominance could be achieved by the strengthening of the wild type allele, so that such optimal levels could be attained even in the heterozygote. The gene in this case would be said to be haplo-sufficient.

It is necessary at this point to emphasize the distinction between dealing with the phenomenon of dominance at the level of the gene, and at the level of the genotype or phenotype. Explanations such as Wright's and Haldane's discussed above are primarily at the level of the gene, but as emphasized by Ford (1964) and Sheppard and Ford (1966), Fisher's theory is directed at the level of the phenotype. (Dominance and recessiveness were defined by Mendel (1865) as properties of genes, but were defined in terms of phenotype, so that they were in fact properties of characters rather than genes.) An essential point of Fisher's theory is that dominance is not an unalterable property of the gene itself, so that the term "dominant gene" for example, must be regarded as merely a convenient manner of stating that a character produced by the gene in question in a particular organism is dominant to the corresponding character produced by its allele.

The distinction between the two levels of argument becomes crucial when we consider the evolutionary hypothesis put forward by Plunkett (1933), and independently but slightly less explicitly by Muller (1932). These authors postulated that dominance might occur through the action of selection to stabilize a particular character. These ideas have been given a firmer foundation by the results of some experiments carried out to investigate the inheritance of certain semistable meristic characters (see e.g. Rendel, 1959a, 1959b, and 1962).

Rendel's experiments were carried out on the scutellar bristles of *Drosophila melanogaster*. The wild type individual of this species almost invariably has four such bristles. Flies with three or five bristles are however, occasionally seen, and using these it is possible with some difficulty to select a line with either less or more than the normal four. Once some progress has been made in selecting away from four, further progress is made much more rapidly. Clearly there is a large amount of genotypic variance affecting this character, but this is hidden to a large extent in the production of the normal phenotype. The scutellar bristle number of *D. melanogaster* is said to be canalized at four.

The sex-linked recessive mutant gene scute (*sc*), when introduced in homozygous or hemizygous condition (i.e. *sc/sc* in females or *sc* in males), causes the average bristle number to be reduced to one or two, with considerably greater variability than possessed by wild-type flies. Stocks with the scute genotype may now be selected for increased bristle number, and a stock with four bristles but still of course having the *sc/sc* genotype is rapidly produced. However, further progress to five or more bristles is again achieved only with considerable difficulty. Clearly during the history of this species there has been selection for modifiers aimed specifically at ensuring the stability of the four bristle phenotype, rather than just the stability of any phenotype produced by the wild type allele at the scute locus. Dominance at this locus cannot necessarily be said to be directly selected, but arises as a consequence of the canalizing selection for four bristles.

The principal point that should be noted about Plunkett's hypothesis is that it avoids the chief obstacle to the acceptance of Fisher's. The system of modifiers needed to produce a canalized phenotype is probably considerably more complex than that needed to ensure dominance at a single locus. But this may be far outweighed by the fact that the deviant genotypes upon which selection may act may constitute a far higher proportion of the population than under Fisher's hypothesis. It is probably not necessary that all of these genotypes be at a low frequency attributable only to mutation, since the character might well be affected by polymorphic loci. Furthermore, the type of system which evolves to ensure the constancy of the phenotype given a constant genotype but a variable environment might also contribute to the suppression of genotypic variability. Note also that a similar argument is applicable to Haldane's hypothesis.

There are several conceptual differences between the two hypotheses. One such difference is that a mutant gene arising for the first time could under Plunkett's hypothesis be completely recessive owing to the canalization developed to suppress variation at other loci affecting the same character. A further, related difference is that Fisher's hypothesis contains an argument which is not present in Plunkett's. Under the latter hypothesis, canalization of development would not *a priori* be expected to reduce the effect of the heterozygote more than the mutant homozygote. Under Fisher's hypothesis on the other hand, modifiers are selected especially for their effect on the heterozygote and might for this reason be expected to affect the heterozygote more than the mutant homozygote. Under Plunkett's hypothesis any preferential response of the heterozygote over the mutant homozygote must be attributed to the inherently greater modifiability of the former genotype. But of course Fisher also placed considerable stress on the greater modifiability of the heterozygote under his hypothesis.

IV. Experimental Work

Fisher (1928, 1930a, and 1958) noted that, in *Drosophila melanogaster* in 1928, the following mutant genes had been observed:

	Recessive	Intermediate	Dominant
Autosomal	130	9	0
Sex-linked	78	4	0

These tabulations may however tend to exaggerate the proportion of completely recessive genes.

To explain these observations, and similar ones in other species, Fisher postulated the evolution of dominance by the very slow accumulation of modifiers. The theory also accounted for the dominance of many mutants in the domestic fowl. When the fowl was being domesticated in primitive times in India, kept hens would often mate with wild cocks, so that, if a man wanted to preserve a new character that had appeared, it would be easier for him to choose characters not completely recessive.

Haldane (1930) pointed out that in the locusts *Paratettix* and *Apotettix* and the fish *Lebistes reticulatus* there were many mutant types "not recessive to the normal type." Fisher (1930b) showed that most homozygotes were at a disadvantage in the locusts, so that, with polymorphisms, it was not correct to write of mutant and normal types. In *Lebistes*, the males were polymorphic for partially sex-linked characters on the Y chromosomes, whereas most females carried the "universal recessive" on their two X chromosomes. In the snail *Helix*, Fisher cited evidence that a pattern polymorphism similar to these others had existed since the Pleistocene epoch. Thus, in all these cases, balanced polymorphism would not lead to recessiveness in heterozygotes.

Fisher (1935 and 1938) carried out a series of experiments on dominant mutant types in the domestic fowl. Using five characters, he bred domestic fowls five times with wild jungle fowls by backcrossing, after which heterozygotes were mated to produce homozygotes to compare with heterozygotes in the wild gene-complex. In every case, the character became less dominant: one, Feathered feet, became almost recessive; and one, Crest, gave rise to a cerebral hernia when homozygous in the wild genotype. This hernia was not found in the crested domestic breeds, so breeding with wild fowls must have removed at least some modifiers of the effects of the gene.

Fisher and Holt (1944) used a dominant character in mice, Danforth's short tail, controlled by a gene (*Sd*) lethal in the homozygous state. It was possible in a brief time by directional selection to make the short tail almost completely recessive, which probably required, Fisher and Holt pointed out, the selection of two sets of modifiers, since the proximal kink in the tail disappeared quickly from the heterozygotes whereas normal development and length of the distal region did not necessarily follow. Further, the modifiers selected tended to suppress the lethal effect of the gene in the homozygote, as Dunn (1942) had found in other experiments on the gene. While this experiment can be regarded as simply a repetition of Ford's work (Section I above), it was carried out on a gene closer in nature to Fisher's idea of a primitive mutant, in that it had substantial deleterious effects, so that natural selection would still be acting on it.

It should be noted that Dunn and Gluecksohn-Schoenheimer (1945) criticized Fisher's and Holt's work severely, on the grounds that they "supposed that the chief factor in viability was tail length so that natural selection for viability aided the selection for longer tails." This was contrary to experimental results achieved by Dunn and Gluecksohn-Schoenheimer, who had found $Sd/+$ to have shorter or longer tails and greater or lesser viability respectively, depending on which of two stocks it was introduced into. However, Fisher's and Holt's hypothesis might well be supposed to have applied in nature, since the "wild-type"mouse does not in fact have a short tail; and Fisher and Holt were selecting for tail length, rather than observing unselected populations.

Harland and Atteck (1941) presented evidence that some characters in cotton have evolved dominance by the accumulation of modifying genes, some by selection of active alleles. Even though these results may have been partly incorrect, as accepted by Fisher (1930a and 1958), the method used, namely putting mutants from one species into the gene complex of another, is a theoretically sound one, as noted in Section I. (It does, of course, have unresolved problems. For example, species which can interbreed are not evolutionarily very distant, so the mutant may have occurred in a common ancestor.)

Clarke and Sheppard (1960) have, in a similar way, considered the evolution of dominance under disruptive selection. Sheppard (1958, p. 64) has pointed out that disruptive selection, that is, selection for two distinct phenotypes, could lead to selection for dominance in a case where the change from one optimum to the other was effected by substitution of one pair of alleles for another. In such a case, there would be three genotypes and only two advantageous phenotypes.

Disruptive selection occurs with Batesian mimicry, and, as Clarke and Sheppard have indicated, there is no reason to suppose that genes improving mimicry will necessarily be dominant originally. (Indeed, examples are known where mimetic forms are recessive; see Ford (1964, Chapters 12 and 13).)

It was found in the polymorphic mimetic butterfly *Papilio dardanus* Brown that, for one case of mimicry, there were almost no intermediates between the nonmimic normal and the mimic, the difference being mediated by a pair of alleles. This could be explained either by complete dominance from the first appearance of the mimetic mutant, with subsequent selection against mutant intermediates or by evolution of dominance for the form by the accumulation of modifiers. The first explanation would predict dominance for the mimic form on moving the allele to another gene-complex; the second would not. Dominance of the mimic form was not observed by Clarke and Sheppard when they put the allele into a geographically distinct race.

The part of this evidence available to him at the time, Fisher (1949) felt, had verified his theory's "only questionable premise, that the reaction of the heterozygote is readily modified by the selection of modifying factors," though it should be noted that Wright (1964) stated that this had never been a major issue in the problem.

Wright (1941) discussed the direct evidence for the hypothesis that dominance is usually predictable from the mode of action of the particular gene. Thus, it had been shown that a deficiency effect usually resembled or enhanced the effect of recessives with loci in the deficient region, and that this was so even in the rare cases where deficiencies could be obtained in a homozygous condition, as well as in the case of "pseudo-dominance" of recessives in heterozygous deficiencies. Wright then noted that "the dominant allele is usually dominant over deficiences," but, as already noted, deficiencies can produce effects in the heterozygous state. Wright felt that better evidence came from cases where three or more doses of an allele with recessive effect could be introduced into an essentially diploid zygote. Then, Wright suggested, if the recessive was a positive effect, three alleles should produce more deviation from type than two, whereas, if the recessive merely was produced by a weaker allele, three alleles should bring the phenotype closer to normal. If the recessive was inactive, three alleles should have the same effect as two. Wright pointed out that most of the evidence supported these latter possibilities.

However, as Wright himself (e.g. 1959) has shown, many multiple allelic series display all possible intermediates between apparently inactive alleles and normal ones with dominant effect. It is clear that a simple hypothesis of activity versus inactivity is inadequate here. Other evidence, from serology, for example, pointed to the fact that alleles in a series could produce qualitatively different products. Muller (1932) discussed, and introduced a notation for, the different kinds of allele; for example, *amorphs* are inactive, and *hypomorphs* less active than normal, and the problem is now much closer to resolution in terms of gene action.

It should be noted that one example where dominance is a primary effect of the gene comes from work on self-incompatibility. Lewis (1947) doubled the chromosome number of *Oenothera organensis* with colchicine, so that the pollen became diploid. It is almost certain that this could never have occurred in the plant's evolution, so that dominance could never have evolved for pollen characters. Incompatibility is gametophytically determined in this species, and Lewis obtained, for example, these results in crosses:

♀		♂ (pollen)	fertile
$S_{2.3.4.4}$	×	$S_{4.6}$	Yes
$S_{2.3.6.6}$	×	$S_{4.6}$	No

Thus S_6 was dominant to S_4 in the pollen. It appears that, in this case, dominance was inherent in the gene product.

In considering the nature of dominance, pleiotropy is very important. Ford (1930 and 1964) and Fisher (1930 b and 1958) considered it essential to the theory of evolution of dominance, and Wright (1964) considered it the greatest objection to the theory.

Wright (1964) held this view largely as a part of his general theory of how evolution acts, stating that there is a "pleiotropic threshold" below which particular contributions to selective values are negligible. The theoretical implications of Wright's views are discussed elsewhere (Section II) but some mention should be made of the known modifiers of dominance which have been observed in connection with other studies, rather than in specific experiments designed to obtain them, which have been discussed above.

Thus, Green (1946) observed the following effects of certain *Minute* alleles "on the dominance of vg^+ over its recessive mutant alleles" (taken from Wagner and Mitchell, 1955):

$vg/+$	% normal flies	$vg^{no2}/+$	% normal flies
Alone	99.2	Alone	95.3
$M(2)1^2$	68.8	$M(2)1^2$	5.4
$M(1)n$	27.6	$M(1)n$	–
$M(3)w$	12.4	$M(3)w$	0.0

(*Minute* flies have less bristles than normal and live longer as they take longer to develop. *Minutes* are thought to be mutations affecting *t*-RNA.) Haldane (1941) has pointed to evidence for the presence of modifiers of the dominance of certain kinds of Friedrich's ataxia and Huntington's chorea, and suggested, as might not be too farfetched, that the latter condition might have been "a disease of infancy in *Sinanthropus*", as modifiers in humans could act to postpone development, making the condition recessive for longer. In Green's work, the other effects of the modifiers were known, whereas, of course, these could not be elucidated by experiment in humans, so that one cannot know whether they would have been selected for other reasons than as modifiers of dominance.

Other cases where dominance modification by single genes has been observed are the examples of Goodwins (1958) [following the work of Wallace (1953)] and Bodmer (1960), working on mice; Clausen (1958) working on ecological races of *Layia platyglossa* (where presence of a central stem was dominant inland where it was required, but not in the maritime region, the difference being mediated by a single gene, absent near the sea); and Hallqvist (1953) who gave an example in barley and another in the lupin *Lupinus angustifolia*.

The other relevant aspect of pleiotropy has been referred to, namely, the fact that some genes have both dominant and recessive effects. For example, Ford (1930) pointed to genes in *Drosophila* which were "almost completely recessive in regard to body colour and eye colour, yet in the shape of the spermatheca the heterozygotes are intermediate." Ford suggested that selection could have acted on the very disadvantageous characters produced by the gene, but not on other, less deleterious ones.

V. Conclusion

The conclusions regarding the evolution of dominance are from most points of view quite straightforward. All the mathematical arguments are in approximate agreement on the basic point that the selective pressure favouring modifiers of the heterozygote expected under Fisher's theory is generally of the order of the mutation rate. The principal argument against the hypothesis is based on the smallness of this figure. This difficulty is apparently removed if, in a manner analogous to that suggested by Plunkett and Muller, a more sophisticated system of modifiers evolves which affects the expression not only of the heterozygote in question but of a much larger class of genotypes.

From the standpoint of dominance itself, it would not seem to be a crucial point whether Fisher's or Plunkett's evolutionary hypothesis, if either, is accepted. Most of the opposition to Fisher's theory (e.g. Wright, 1929a, b, 1934, and 1964) is based on the fact that under Fisher's theory it must be postulated that extremely small selective pressures have been effective in producing evolutionary consequences. If it could be demonstrated that such has in fact happened, this would have an important bearing on evolutionary theory in general. However, it seems unlikely that crucial evidence could ever be obtained on the point of whether modifiers have been selected solely for their effect on an unfavourable and rare heterozygote, or for their effect on a wider class of genotypes.

We are grateful for suggestions from a number of people during the preparation of this review and would particularly like to acknowledge Drs. Walter Bodmer, Motoo Kimura, Peter Workman, and Sewall Wright.

References

Bateson, W.: Mendel's principles of heredity. London: Cambridge Univ. Press 1909.

Bodmer, W. F.: Interaction of modifiers: the effect of pallid and fidget on polydactyly. Heredity **14**, 445 − 448 (1960).

− Natural selection for modifiers of heterozygote fitness. J. Theor. Biol. **4**, 86 − 97 (1963).

−, and P. A. Parsons: Linkage and recombination in evolution. Advan. Genet. **11**, 1 − 100 (1962).

Clarke, C. A., and P. M. Sheppard: The evolution of dominance under disruptive selection. Heredity **14**, 73 − 88 (1960).

Clausen, J.: Gene systems regulating characters of ecological races and subspecies. Proc. 10 th Int. Cong. Genet. **1**, 434 − 443 (1958).

Crosby, J. L.: The evolution and nature of dominance. J. Theor. Biol. **5**, 35 − 51 (1963).

Dunn, L. C.: Changes in the degree of dominance of factors affecting tail length in the house mouse. Amer. Natur. **76**, 552 − 569 (1942).

−, and S. Gluecksohn-Schoenheimer: Dominance modification and physiological effects of genes. Proc. Natl. Acad. Sci. (Wash.) **31**, 82 − 84 (1945).

Ewens, W. J.: A note on Fisher's theory of the evolution of dominance. Ann. Hum. Genet. **29**, 85 − 88 (1965a).

− Further notes on the evolution of dominance. Heredity **20**, 443 − 450 (1965b).

− Linkage and the evolution of dominance. Heredity **21**, 363 − 370 (1966).

− A note on the mathematical theory of the evolution of dominance. Amer. Natur. **101**, 35 − 40 (1967).

Fisher, R. A.: The possible modification of the response of the wild type to recurrent mutations. Amer. Natur. **62**, 115 − 126 (1928a).

− Two further notes on the origin of dominance. Amer. Natur. **62**, 571 − 574 (1928b).

− The evolution of dominance; reply to Professor Sewall Wright. Amer. Natur. **63**, 553 − 556 (1929).

− The genetical theory of natural selection. Rev. 2nd ed. New York: Dover 1930a, 1958.

− The evolution of dominance in certain polymorphic species. Amer. Natur. **64**, 385 − 406 (1930b).

− Professor Wright on the theory of dominance. Amer. Natur. **68**, 370 − 374 (1934).

− Dominance in poultry. Phil. Trans. Roy. Soc. Lond. B **225**, 195 − 226 (1935).

− Dominance in poultry. Feathered feet, rose comb. internal pigment, and pile. Proc. Roy. Soc. Lond. B **125**, 25 − 48 (1938).

Fisher, R. A.: The theory of inbreeding. Edinburgh: Oliver and Boyd 1949.

—, and S. B. Holt: The experimental modification of dominance in Danforth's short-tailed mutant mice. Ann. Eugen. **12**, 102 – 120 (1930).

Ford, E. B.: The theory of dominance. Amer. Natur. **64**, 560 – 566 (1930).

— Genetic research in the *Lepidoptera*. Ann. Eugen. **10**, 227 – 252 (1940).

— Ecological genetics. London: Methuen Press 1964.

Goodwins, I. R.: A further case of dominance interaction in *Mus musculus*. Heredity **12**, 73 – 75 (1958).

Green, M. M.: A study in gene action using different dosages and alleles of vestigial in *Drosophila melanogaster*. Genetics **31**, 1 – 20 (1946).

Gregg, T. C.: Latent neomorphs and the evolution of dominance. Evolution **21**, 850 – 858 (1967).

Haldane, J. B. S.: A note on Fisher's theory of the origin of dominance. Amer. Natur. **64**, 87 – 90 (1930).

— The relative importance of principal and modifying genes in determining some human diseases. J. Genet. **41**, 149 – 158 (1941).

— The theory of selection for melanism in *Lepidoptera*. Proc. Roy. Soc. London B **145**, 303 – 306 (1956).

Hallqvist, C.: Change of dominance in different gene environments. Hereditas **39**, 235 – 240 (1953).

Harland, S. C., and O. M. Atteck: The genetics of cotton. XIX. Normal alleles of the crinkled mutant *Gossypium barbadense L.* differing in dominance potency and an experimental verification of Fisher's theory of dominance. J. Genet. **42**, 21 – 47 (1941).

Kempthorne, O.: An introduction to genetic statistics. New York: John Wiley & Sons 1957.

Kettlewell, H. B. D.: A resume of investigations on the evolution of melanism in the *Lepidoptera*. Proc. Roy. Soc. London B **145**, 297 – 303 (1956).

Kojima, K., and T. M. Kelleher: Changes of mean fitnesses in random mating populations when epistasis and linkage are present. Genetics **46**, 527 – 540 (1961).

Kimura, M.: On the change of population fitness by natural selection. Heredity **12**, 145 – 167 (1958).

— Attainment of quasi-linkage equilibrium when gene frequencies are changing by natural selection. Genetics **52**, 875 – 890 (1965).

Lewis, D.: Competition and dominance of incompatibility alleles in diploid pollen. Heredity **1**, 85 – 108 (1947).

Lewontin, R. C., and K. Kojima: The evolutionary dynamics of complex polymorphism. Evolution **14**, 458 – 472 (1960).

Mayo, O.: On the evolution of dominance. Heredity **21**, 499 – 511 (1966).

Mendel, G.: Versuche über Pflanzenhybriden. (English translation. Ed. by J. H. Bennett, 1965. Edinburgh: Oliver and Boyd) 1865.

Moran, P. A. P.: On the non existence of adaptive topographies. Ann. Hum. Genet. **27**, 383 – 393 (1964).

Muller, H. J.: Further studies on the nature and causes of gene mutations. Proc. 6th Int. Cong. Genet. **1**, 213 – 255 (1932).

O'Donald, P.: On the evolution of dominance, over-dominance and balanced polymorphism. Proc. Roy. Soc. London B **268**, 216 – 228 (1967a).

— The evolution of selective advantage in a deleterious mutation. Genetics **56**, 399 – 404 (1967b).

O'Donald, P.: The evolution of dominance by selection for an optimum. Genetics **58**, 451 – 460 (1968).

Parsons, P. A., and W. F. Bodmer: The evolution of overdominance: natural selection and heterozygote advantage. Nature (Lond.) **190**, 7 – 12 (1961).

Plunkett, C. R.: A contribution to the theory of dominance. Amer. Natur. **67**, 84 – 85 (1933) (abstract).

Rendel, J. M.: Canalization of the scute phenotype of *Drosophila*. Evolution **13**, 425 – 439 (1959a).

— Evolution of dominance. In: The evolution of living organisms. Proc. Symp. Roy. Soc. Victoria, Melbourne. 1959b.

— The relationship between gene and phenotype. J. Theor. Biol. **2**, 296 – 308 (1962).

Sheppard, P. M.: Natural selection and heredity. London: Hutchinson 1958

—, and E. B. Ford: Natural selection and the evolution of dominance. Heredity **21**, 139 – 147 (1966).

Wagner, R. P., and H. K. Mitchell: Genetics and metabolism. New York: John Wiley & Sons 1955, 1964.

Wallace, M. E.: A case of mutual reduction of dominance; observed in *Mus musculus*. Heredity **7**, 435 – 437 (1953).

Watson, J. D.: Molecular biology of the gene. New York: W. A. Benjamin 1965.

Wright, S.: The effects in combination of the major color-factors of the guinea pig. Genetics **12**, 530 – 569 (1927).

— Fisher's theory of dominance. Amer. Natur. **63**, 274 – 279 (1929a).

— The evolution of dominance. Comment on Dr. Fisher's reply. Amer. Natur. **63**, 556 – 561 (1929b).

— Molecular and evolutionary theories of dominance. Amer. Natur. **68**, 24 – 53 (1934).

— The physiology of the gene. Physiol. Rev. **21**, 487 – 527 (1941).

— Classification of the factors of evolution. Cold Spr. Harb. Symp. Quant. Biol. **20**, 16 – 24 (1955).

— A quantitative study of variation in intensity of genotypes of the guinea pig at birth. Genetics **44**, 1001 – 1026 (1959).

— Pleiotropy in the evolution of structural reduction and dominance. Amer. Natur. **98**, 65 – 69 (1964).

Survival of Mutant Genes as a Branching Process*

H. E. SCHAFFER

Introduction

The process of mutation is the ultimate source of the raw materials of evolution. While the majority of mutations may have unfavorable effects on the survival of their bearers, occasionally an advantageous mutation will occur. As this can be considered a rare event at any given locus, it is of interest to study the fate of new alleles of this type, and to see how they contribute to the evolutionary progress of the population.

The descendents of a rare mutation in even a moderately large random mating population will very likely remain in different heterozygous individuals and so have independent fates, at least until the number of descendents increases substantially. Thus, the process of survival of the new mutant gene can be modeled by the occurrence of the mutant in an infinite population of either haploid or diploid individuals. A diagram of the descendents, if any, of this gene would form a figure like a family tree. If the mutant and each of its descendents give rise independently to a family with a random number of offspring, then this process can be studied using the theory of *branching processes* which depends on the *method of generating functions*. [For a basic treatment of probability generating functions and branching processes, see Feller (1957, Chapters 11 and 12). Harris (1963) has a much more comprehensive treatment of the theory of branching processes with a condensed coverage of the basic elements.] Since we will only follow the heterozygotes carrying the mutant gene, for convenience in the following discussions we will refer to the mutant gene as the "individual" being studied.

The two outcomes of the survival process are either that the descendents may eventually become extinct, or that they may increase in number and never become extinct. The probability corresponding to the latter event is called the ultimate survival probability *(u.s.p.)* and is the probability that this mutation will not become lost to the population. The evaluation of the *u.s.p.* will be shown to depend on the probability distribution of the number of offspring of a gene, and especially on the mean

---

    * This paper is dedicated to Professor Th. Dobzhansky in honor of his long leadership in evolutionary genetics.

of this distribution which can be considered to give the selective advantage (or disadvantage) of the mutation.

The work done on the survival of mutant genes has its origin in a problem posed by Galton on the extinction of surnames. Watson (1875) responded to this question and used the method of generating functions in his approach. The formal description of the inheritance of a surname is the same as that of the inheritance of a gene, and the problem of the survival of a mutant gene can be approached in the same manner. This model of a branching process is often referred to as a Galton-Watson process.

The assumption that the generations are discrete is used in almost all of the theory of branching processes. It is often dispensed with in treatments of population dynamics which consider time and gene frequencies to be continuous parameters. When continuous time is introduced to branching processes, and the life spans (time to branching) of individuals are allowed to vary, the branching processes are called age-dependent. This topic is discussed by Bellman and Harris (1952) and Harris (1963, Chapter VI). Age-dependent branching processes become more complex in their structure because, in general, the age structure of the population must be considered as well as the number of individuals. The number of individuals, considered alone, is no longer a Markov process. In the following discussion, discrete generations will always be assumed unless otherwise noted.

Basics of Probability Generating Functions

A probability generating function provides one way of expressing the probability distribution of the number of offspring of an individual. More commonly such a discrete probability distribution is defined by giving its general term. The *probability generating function (p.g.f.)* can yield the probabilities of having each possible family size, but in practice this is inconvenient and the major use of the *p.g.f.* is as a description of the entire distribution and the changes in it which occur with certain manipulations.

The *p.g.f.* for a univariate distribution may be symbolized as follows, where p_i is the probability that an individual has i offspring. There is usually some restriction on the "dummy variable" s such as $|s| \leq 1$:

$$f(s) = p_0 + p_1 s + p_2 s^2 + \cdots + p_i s^i + \cdots. \tag{1}$$

By comparison with a geometric series, Eq. (1) converges absolutely at least for $|s| < 1$. In the cases to be discussed here, a closed expression can be found for $f(s)$.

In a branching process we start with one individual in generation zero. This individual gives rise to a number of offspring to form the first generation. The number of these offspring is a random variable with probability generating function $f(s)$. Each of these offspring then independently gives rise to a number of offspring to form the second generation. If the probability generating function describing the family size of individuals in the first generation is the same for each individual and is $g(s)$, then the probability generating function for the total size of the second generation is $f(g(s))$, i.e., the $p.g.f.$'s have been compounded by substituting $g(s)$ for s in $f(s)$. If the distribution of the number of offspring from a single individual stays the same through the generations then the second generation size has the $p.g.f.$ of $f(f(s))$, the third $f(f(f(s)))$, etc. The $p.g.f.$ for the size of the n-th generation will be $f(s)$ compounded n times which can be symbolized as $f_n(s)$.

The mean of the distribution given by $f(s)$ is the value of the first derivative with respect to s, evaluated at $s = 1$. This is represented as $f'(1)$. Application of the chain rule to the differentiation of $f_n(s)$ yields the mean in the n-th generation $f_n'(1) = f'(1)^n$. In general, the mean of the result of compounding is the product of the individual means. The $p.g.f.$ for the sum of two independent families each with $p.g.f.$ $f(s)$ is obtained by multiplying the two $p.g.f.$'s to give $f(s)^2$. Differentiation yields the mean of $2f'(1)$, corresponding to the sum of the means of the two independent random variables.

The probability that the descendants of a single individual have gone to extinction by the n-th generation is the probability of the zero class in the distribution of the size of the n-th generation. The $p.g.f.$ for this distribution is $f_n(s)$, and as can be seen from Eq. (1), the probability of the zero class can be obtained from any $p.g.f.$ by setting the dummy variable equal to zero. In studying extinction we can then look at $f_n(0)$ for the different generations, and the probability that extinction will eventually occur is the limiting value $\lim_{n \to \infty} f_n(0)$. This is the probability that the descendants of this gene will die out and thus be lost to the population.

The ultimate probability of extinction (the complement of the $u.s.p.$), $\lim_{n \to \infty} f_n(0)$, is equal to the smallest non-negative root of $f(s) = s$. This results from the equality $f(1) = 1$ and from all the derivatives of $f'(s)$ being non negative in the interval $0 \leq s \leq 1$. From this it can also be seen that there is at most one root in the interval $0 \leq s < 1$ (if $f(s) \neq s$) and that the existence of such a root depends on the magnitude of $f'(1)$. There will be no root in the interval $0 \leq s < 1$ unless $f'(1) > 1$, in which case there will be exactly one. The presence or absence of this root can be seen in Figs.1 and 2. Since $f(s) \geq 0$ and $f'(s)$ is increasing, if $f'(1) > 1$ then $f(s)$ must approach the point $(1,1)$ from below the $45°$ line (Fig. 1). Therefore it must cross the line.

This crossing corresponds to a root of $f(s) = s$. If $f'(1) \leq 1$ (Fig. 2) then $f(s)$ must approach from above, so there can be no root.

Certain assumptions about the form of the distribution generated by $f(s)$ are justified biologically and make the theory easier to apply. We always assume that $p_1 < 1$, as this excludes the uninteresting case in which every individual has exactly one offspring and there is no random aspect of the model. For the same reason we usually assume that $p_0 < 1$, although

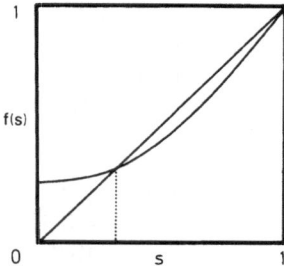

Fig. 1. $f'(1) > 1$. One root exists in the interval $0 \leq s < 1$

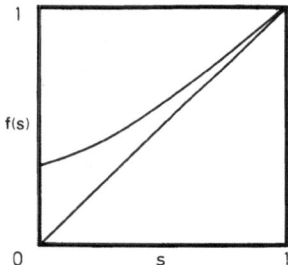

Fig. 2. $f'(1) \leq 1$. No roots exist in the interval $0 \leq s < 1$

in models with more than one locus involved this is not necessary. In commonly used p.g.f.'s, $p_0 > 0$. When $p_0 = 0$ there is a root of $f(s) = s$ at $s = 0$ which corresponds to u.s.p. = 1. Thus there can be no extinction if at least one offspring is always produced. The number of offspring is always limited in nature, so $f(s)$ will always generate a bounded distribution or an approximation to a bounded distribution. Therefore, in cases of biological interest, there will be no conflict with the common theoretical assumption that the mean and variance of the distribution is finite.

Summarizing, the u.s.p. of a gene is zero unless its average number of offspring is greater than one. When the u.s.p. is greater then zero, it is one minus the root of $f(s) = s$ which is less than one. This root can be found by numerical iteration, or approximately by methods mentioned later.

Genetic Applications of Branching Process Theory

Lotka (1931 a, b) investigated the ultimate probability of extinction of a male line of descent (extinction of a surname) using the 1920 population statistics for the United States. From the available statistics he calculated the distribution of sons that a newborn male will eventually have and found that the data were well approximated by a geometric distribution with a modified zero class and with $p.g.f.$

$$f(s) = \frac{0.4828 - 0.0407s}{1 - 0.5586s}. \tag{2}$$

From this it can be found that the average number of sons is 1.175, and the $u.s.p.$ of a male line of descent, starting from a newborn male, can be found (by numerical iteration) to be 0.128.

Such numerical iteration can be tedious, and an approximate formula given by Bartlett (1966) can give the $u.s.p.$ more directly in terms of the mean and variance, m and v, of the distribution. Bartlett's approximation is derived from the case of the mean of the distribution being only slightly greater than unity and gives the

Probability of extinction $= e^{-2(m-1)/v}$.

Since the exponent is small, we can further approximate

$$u.s.p. = 2(m-1)/v. \tag{3}$$

This formula can then be used to estimate the $u.s.p.$ for the $p.g.f.$ given in Eq. (2) for which $m = 1.175$ and $v = 2.768$. Substitution into Eq. (3) yields an $u.s.p. = 0.127$. The probability that n independent particles will all go to extinction is (Probability of extinction)^n. From this the odds are in favor of at least one particle not becoming extinct if more than $\dfrac{0.7}{u.s.p.}$ particles arise independently.

Fisher (1922, 1958) and Haldane (1927) deal with the survival or extinction of the descendants of a new mutation which is either neutral in fitness or has a selective advantage. They consider this mutant gene to be in a heterozygous individual which gives rise to a very large number of progeny (e.g., seeds) of which very few survive to maturity. In such a situation the Poisson distribution is a good approximation to the binomial distribution of the number of "offspring" of the mutant gene. The Poisson distribution with mean m has the $p.g.f.$:

$$f(s) = e^{m(s-1)}. \tag{4}$$

The descendents of a neutral mutation ($m = 1$) will become extinct with probability one, while the mutation with a 1 % advantage will have

$m = 1.01$ and so have an *u.s.p.* greater than zero. Using Eq. (3) with $m = v = 1.01$ gives *u.s.p.* $= 0.0198$ (compared to an exact value of 0.0197). For small $(m - 1)$ and the Poisson offspring distribution Eq. (4), the approximation *u.s.p.* $= 2(m - 1)$ is commonly used.

For either mutation, the probability of extinction in any generation can be computed. There is an approximate formula for the rate of extinction for the case of a neutral gene (i.e., for a branching process with a mean of unity) which is due to Kolmogorov (quoted in Harris, 1963).

$$\text{Probability of survival to the } n\text{-}th \text{ generation } = \frac{2}{n f''(1)}. \qquad (5)$$

Skellam (1948) and Haldane (1948) both investigated the equilibrium distribution of a disadvantageous mutation in a population in which mutation to the disadvantageous allele is taking place. They both reached the same solution. If $\phi(s)$ is the *p.g.f.* of the equilibrium distribution, and if each mutant has a number of offspring with a Poisson distribution with mean $c < 1$, then the *p.g.f.* of the number of descendants in the next generation is $\phi(e^{c(s-1)})$. There are also a number of new mutations with a Poisson distribution with mean λ, and so the *p.g.f.* of the total number of mutations in the next generation is $e^{\lambda(s-1)} \phi(e^{c(s-1)})$, which is equal to the equilibrium *p.g.f.* $\phi(s)$. The solution to this functional equation has not been found explicitly, but Haldane and Skellam give several of its cumulants. It has mean $\lambda/(1 - c)$ and variance $\lambda/(1 - c)(1 - c^2)$.

Most of the applications of branching processes have used the Poisson distribution. As mentioned above, it can be justified for the case of a large number of offspring born with only a small proportion surviving. In the case of the large mammals, the potential number of offspring is not so large, and the proportion surviving will be higher. Kojima and Kelleher (1962) fit the distribution of family sizes taken from the 1950 United States Census data with the negative binomial distribution with mean $= 3$ and variance $= 7$. The *p.g.f.* for the negative binomial distribution with parameters q and r is

$$f(s) = \left(\frac{1 - q}{1 - qs}\right)^r \qquad (0 < q < 1). \qquad (6)$$

The mean is $rq/(1 - q)$ and the variance is $rq/(1 - q)^2$. Note that the mean can be changed by changes in r or q or both, and that for the same change in the mean, changing q has a larger effect on the variance. The variance is always larger than the mean, and in the limit, as the variance approaches the mean, the negative binomial distribution approaches the Poisson, so the negative binomial can be thought of as a generalization of the Poisson distribution with a larger and adjustable variance. The negative binomial also tends to fit unimodal distributions which arise from a mixture of

Poisson distributions with different means. The negative binomial distribution was used as a generalization of the Poisson by Moran (1961) on the basis that deviations from the Poisson are to be expected in the direction of increasing the variance versus the mean of the offspring distribution.

As can be seen from Eq. (3), the *u.s.p.* of a particle is decreased when the variance of its offspring distribution is increased, the mean being kept constant. Therefore, the *u.s.p.* of a particle reproducing with a negative binomial distribution will always be smaller than for the Poisson distribution with the same mean. This relationship is discussed by Kojima and Schaffer (1967) when comparing the *u.s.p.* for genes affecting fertility and viability in a population. In this formulation genes affecting viability vary the value of q in Eq. (6) while genes affecting fertility can change the values of q and r. As a result, for the same increase in the overall average for the generation, an advantageous mutation for fertility will have *u.s.p.* greater or at least equal to that of an advantageous mutation for viability.

Kojima and Kelleher (1962) discuss the average number of descendants of a mutant when the whole population is expanding or contracting. This is of great importance, for in a branching process it is not the relative fitness of the mutant gene that is important but only the distribution of descendants of that gene. Thus, in an expanding population the survival of an advantageous mutation is enhanced. Assuming that the population does not expand forever, the greatest effect on the *u.s.p.* of a gene will come in the initial stages of the survival process when the gene is rare. At that time an expansion or contraction in the population size will have the greatest effect on the *u.s.p.* In later stages, changes in the population size will have less effect because the descendants will probably either have become extinct or have increased considerably in number so as to make their extinction unlikely. It is at this stage that deterministic methods may be appropriate.

As a result of this it would be expected that genetic variability due to new mutations would increase during periods of population expansion and decrease during periods of population contraction.

It was mentioned above that if the offspring distributions for two sequential generations have generating functions $f(s)$ and $g(s)$, then the generating function for the two generations together is $f(g(s))$. This method is called compounding generating functions and can be used to study each generation as a sequence of separate stages, each stage studied with a separate *p.g.f.* Kojima and Schaffer (1964) studied the branching process for two linked mutant genes and characterized each generation into three stages. First, the probability distribution of the number of children born to the family, with generating function $F(s)$. Second, the probability distribution of a child being born in this family with the mutant chromosome. This distribution has generating function $R(s)$ and generates the probabilities of the outcomes which depend on segregation and recombination. Third, the probability distribution of a child carrying the mutant chromosome surviving till sexual maturity, with generating function $W(s)$. The last two stages have only two possible outcomes each, with respect to the presence of the mutant chromosome. Either the mutant chromosome is present, or not. Each such event is called a Bernoulli trial and has a generating function of the form $(1 - p) + ps$, where p is the probability of the outcome being a "success", which in this context is the presence of the mutant chromosome.

The p.g.f. for one generation of the branching process composed of these three sequential stages is obtained by compounding the p.g.f. for each of the three stages and therefore is

$$\theta(s) = F\big(R(W(s))\big). \tag{7}$$

The survival of the descendants with these two linked mutant genes can then be studied as a branching process with p.g.f. $\theta(s)$. In the case where each of the two mutations are by themselves either neutral or disadvantageous, the recombinant types with only one mutant gene will certainly go to extinction. The survival then depends only on $\theta(s)$ for the linked segment, and in particular the mean must be greater than unity for the u.s.p. of the linked genes to be greater than zero. This then is a case of an advantageous epistatic combination of two mutant genes. The mean for $\theta(s)$ is the product of the means of the generating functions compounded to make it and is

$$M\left(\frac{1-\varrho}{2}\right)w$$

where $M = F'(1)$ the average family size, $\dfrac{1-\varrho}{2} = R'(1)$ (ϱ is the recombination fraction) and $w = W'(1)$ the survival fraction.

From this it can be seen that the epistatic advantage and recombination work antagonistically in determining survival. Epistatically favorable combinations will accumulate more rapidly when they have low recombination fractions. This can increase selection for mechanisms reducing recombination, which can also decrease the formation of new gene complexes. Since the survival of epistatic combination depends on the linkage, it follows that they will tend to accumulate closely linked to each other, with tighter linkage being necessary for a smaller epistatic advantage. This phenomenon may be of some general applicability because genes with effects which act multiplicatively (traits such as fertility and viability act multiplicatively in determining overall fitness) act epistatically.

The compounding of different generating functions to represent the stages of the life cycle could as well represent sequential generations in which the conditions have changed so as to give rise to the different distributions. The changed conditions can be changes in the descendants of the original particle, or changes in the environment.

Compounding a repetitive sequence of generating functions could be done to represent the different reproduction of generations in the yearly cycle of seasons. Then, as was done with the life cycle, the repetitive cycle could be represented by one generating function (starting with any season desired) and the branching process could then be studied taking the year as the basic unit, and using all of the theory developed for the case of compounding an unchanging generating function.

Such a cyclical fluctuation of the offspring distribution is discussed by Pollak (1964, 1966) who is also concerned with the differences in the *u.s.p.* depending on which time in the cycle the initial particle is introduced. Wilkinson (1967) deals with random fluctuations in the offspring distributions which he identifies as the result of stochastic environments. We will restrict the discussion of Wilkinson's results to the case of a finite number of possible environments. Wilkinson's results are more general, usually referring to a countable number of environments in which case the conclusions are similar.

There are N possible environments, each with its own offspring distribution for which the generating functions are, $f_i(s)$, $i = 1, \ldots, N$. The *i-th* environment occurs each generation with probability q_i. Starting with one individual in generation zero, the *p.g.f.* for the number of individuals in the *n-th* generation is

$$\theta_n(s) = \sum_{i=1}^{N} q_i \theta_{n-1}(f_i(s)). \tag{7'}$$

The compounding of the $f_i(s)$ into θ represents, as usual, the passage of a generation.

The expected number of offspring from one individual in one generation is

$$m = \sum_{i=1}^{N} q_i f_i'(1) = \sum_{i=1}^{N} q_i m_i \tag{8}$$

which is the arithmetic average of the m_i, the means for each environment. It can then be shown by mathematical induction that the expected number of individuals in the *n-th* generation is

$$\theta_n'(1) = m^n. \tag{9}$$

This result is exactly the same as that expected from a constant environment, as can be seen from substituting the same $f(s)$ for all the $f_i(s)$ in Eq. (7').

In the determination of the *u.s.p.* of a particle in a varying environment, the value of the arithmetic mean, *m*, does not play exactly the same role as with a constant environment. It is still true that $m \leqq 1$ means that *u.s.p.* $= 0$, however it is possible to have an *u.s.p.* $= 0$ even with $m > 1$. That this is so can easily be seen by considering the situation in which one of the environments which occurs with a nonzero probability is lethal. If this environment ever occurs, and it certainly will, the branching process becomes extinct, regardless of the value of *m*.

The quantity which governs the *u.s.p.* is the weighted geometric mean of the distribution means

$$g = \prod_{i=1}^{N} m_i{}^{q_i} . \tag{10}$$

The geometric mean *g* is always smaller than the arithmetic mean *m*, except in the case where all the m_i are equal, and then *g* and *m* are equal.

When $g \leqq 1$, then the *u.s.p.* $= 0$. When $g > 1$, then there is a probability, *u.s.p.* > 0, that the process will not go to extinction. It was mentioned above that when $m \leqq 1$, then *u.s.p.* $= 0$. This is so because when $m \leqq 1$, then also $g \leqq 1$. The dependence on the geometric mean illustrates the multiplicative relationship of the means of the individual distributions when they are compounded. In order for the mean of compounded distributions to be greater than unity, their geometric mean (the *n-th* root of the product of *n* of them) must be greater than unity. For the random selection of a sequence of different possible environments, each of them should occur in the compounding with their relative frequencies, and so they are weighted in the geometric mean Eq. (10) according to these frequencies, the q_i. From this relationship of the geometric mean to the arithmetic mean, it can be seen why the *u.s.p.* is lowered if the offspring distribution is a random variable in successive generations rather than an unvarying distribution with the same (arithmetic) mean. Kimura (1955) discusses the *u.s.p.* when the mean of a Poisson offspring distribution is a random variable. This can be interpreted as the result of stochastic environments. The *u.s.p.* is lowered to approximately

$$u.s.p. = \frac{2(\bar{m} - 1)}{1 + V_m}$$

where $\bar{m}$ and V_m are the mean and variance of the offspring distribution means. If $V_m = 0$ the *u.s.p.* agrees with the approximation discussed in the previous section.

Wilkinson (1967) also discusses the case in which the environments are not chosen entirely at random for each generation, but instead form a Markov chain, with each environment having (possibly) different probabilities of being succeeded by each of the others. Wilkinson proves that the *u.s.p.* is zero whenever the geometric mean of the environmental means is less than unity, regardless of the initial environment. In this case the geometric mean Eq. (10) is computed with the weights, the q_i, being the probabilities of each environment in the steady state distribution for the Markov chain. While this result is similar to that of randomly chosen environments, Wilkinson states that it is still an unanswered question if a geometric mean greater than unity is sufficient to give an *u.s.p.* > 0.

Changes in the offspring distribution can also result from a genetic basis or constitution which changes through time. The survival of inversion chromosomes was investigated in this manner by Ohta (1966) and Ohta and Kojima (1968). A rare inversion in a population effectively ties up a block of genes by preventing recombination. Such inversions have been found and studied in Drosophila populations, and there has been much controversy concerning the mechanisms responsible for the increase in the frequency of the initially occurring inversion and the maintenance of inversions segregating in a population. Ohta and Kojima take the case of a population with a number of loci segregating on one chromosome. There are two alleles per locus, no epistasis and no mutation. Therefore, the population is in a state of change according to the selective values and frequencies of the different genotypes. At this point an inversion occurs which includes the most favorable allele at each locus. The reproduction of the inversion chromosome and its descendants from this point on is a branching process in which the average number of descendants of an inversion heterozygote in each generation depends on the superiority in adaptive value of an inversion heterozygote over that of the rest of the population in that generation. As a result of the selective process (mutation being absent), the noninversion portion of the population is increasing in adaptive value and so the advantage of the inversion is continually decreasing through time. This then is an example of a branching process in which the distribution of offspring changes through time in a regular fashion. In the limit, the advantage of the inversion heterozygote can be shown to vanish as the population reaches a selective optimum. Conditions sufficient for the extinction of an inversion under these conditions are given by Ohta and Kojima, but Pollak (1966) has obtained a stronger result which can be applied to this problem.

Pollak gives a theorem dealing with a case in which the offspring distribution for an individual in each successive generation may be different, and is changing (for the worse) so that as time goes on the limit of the *u.s.p.* of a single individual in later generations is zero. This would be the case if the mean of the offspring distribution is less than or equal to unity in the limit. Then, given that the generating function of the descendants of the initial individual is differentiable in all generations, the *u.s.p.* is zero for the descendants of the initial individual.

The results of Ohta, Kojima, and Pollak then agree that the *u.s.p.* of an initial inversion is zero if this inversion gradually loses all of its advantage. This general statement also applies even if the selective advantage is lost for other reasons, such as the action of recurrent mutation in causing disadvantageous alleles to appear on the inversion chromosomes. With such mutation the initial advantage of the inversion will vanish as the alleles on the inverted segment reach mutation-selection equilibrium, and so the *u.s.p.* is zero for the initial inversion. However, Nei, *et al.* (1967) have shown using a deterministic model appropriate for large but not infinite populations that it is possible for such an inversion to reach a significant frequency in the population before its selective advantage vanishes.

Ohta and Kojima discuss genetic situations in which the selective advantage of the inversion heterozygote would not vanish and which therefore lead to an *u.s.p.* > 0. If the inversion contained a unique advantage, such as an advantageous allele which was not present in the rest of the population, or a favorable position effect due to the inversion, this would give an *u.s.p.* > 0. They also give an example in which such a lasting selective advantage can be secured when an inversion ties together a block of genes with favorable epistatic interaction.

Linked portions of chromosomes and inversions may contain many loci, and still in some cases their survival can be studied using univariate branching processes. The inversion links the loci into a single "supergene" which can be considered a single particle. The inherently bivariate model of two linked genes with epistatic interaction (Kojima and Schaffer, 1964) was placed in a univariate formulation by the assumption that each of the genes individually was neutral or disadvantageous so that recombinant types could be ignored. Without this simplification, there are several types of particles to be counted each generation; the AB, Ab, and aB chromosomes for two linked genes. The reproduction of these and their descendants can be formulated as a multivariate or multitype branching process, with a corresponding multivariate generating function.

Multitype branching processes can be interpreted in many ways, as was the case with the formulation for a single type of particle. Harris (1963, Chapter II) gives a survey of multitype branching processes. Pollak (1964) uses a multitype formulation to study the survival of one mutant gene where there are different ecological niches with migration between them. The different types in this model represent the same mutant gene in the different environments. Pollard (1966) uses the multitype formulation to study a population model suitable for human populations, in which the different types correspond to different age groups. Each type can produce offspring of two types, corresponding to an individual one unit of time older, and a newborn individual. In this way a population which incorporates age-dependent properties can be studied without introducing age dependent branching processes. Pollard (1968) uses a similar multitype formulation to study a model similar to that of Ewens (1966) in which the types correspond to the different genotypes formed by a new mutant gene at one locus together with two alleles segregating at another locus.

We will present as an example of a multitype branching process the formulation of Kojima and Schaffer (1967) of the survival process of two linked mutant genes, where the different types correspond to the different combinations of alleles on a chromosome, i.e., the different genotypes. The multitype branching process involves individuals of different types. each of which can give rise to offspring of one or more different types. The same assumptions of independence hold as in the single type branching process, and so most of the techniques and results carry over. The *p.g.f.* for the distribution of offspring from one individual of a type 1 is (with three possible types)

$$f^1(s_1, s_2, s_3) = f^1(s) = \sum_{i=0}^{\infty} \sum_{j=0}^{\infty} \sum_{k=0}^{\infty} p_{ijk} s_1^i s_2^j s_3^k \qquad (11)$$

where p_{ijk} is the probability of having i, j, k offspring, respectively, of the three types. Compounding to represent generations can be done as before, except that there are more generating functions to substitute: $f^1(s)$ will be substituted for s_1, $f^2(s)$ for s_2, and $f^3(s)$ for s_3, where $f^2(s)$ and $f^3(s)$ generate the offspring distributions of type 2 and type 3 individuals. For the case of an initial individual with two linked mutant genes the formulation is somewhat simplified, because while the AB/ab individual can have mutant offspring of all three genotypes, AB/ab, Ab/ab, and aB/ab, the individuals of types Ab/ab and aB/ab can only pass on their own mutant gene and so can only have one type of mutant offspring. Since every individual has an ab chromosome, this is omitted from the notation and individuals are referred to only by the mutant containing chromosome.

Kojima and Schaffer (1967) use the same division of the life cycle into stages as in Kojima and Schaffer (1964). These are (1) mating and determination of family size, (2) segregation and recombination, (3) infant to adult survival. The generating functions corresponding to Eq. (11) are formed, using the negative binomial distribution for family size, and the *u.s.p.* for each of the types of chromosomes is calculated for several cases.

The Ab and aB types can only produce their own types, so this is a special case of the general multitype branching process (it is not positive regular), but many of the general results can be applied. Let M be the matrix of first moments (means) of the branching process

$$M = [m_{ij}] \qquad (12)$$

where m_{ij} is the expected number of individuals of type j produced by one individual of type i. The largest characteristic root of M then plays the role that the mean of the process did in the univariate case. Every type of individual will have an $u.s.p. = 0$ if, and only if, the largest characteristic root of M is less than or equal to unity. For the branching process of the descendants of an AB individual the matrix M will be of the form

$$
M = \begin{bmatrix}
m(AB, AB) & m(AB, Ab) & m(AB, aB) \\
0 & m(Ab, Ab) & 0 \\
0 & 0 & m(aB, aB)
\end{bmatrix}.
$$

The diagonal elements of this triangular matrix are the characteristic roots, and so if any of them is greater than unity then (since the three elements in the top row are all positive) one or more of the types of mutant chromosomes has an $u.s.p. > 0$ starting with one AB chromosome.

The survival of mutant genes in finite populations has been investigated using branching processes in several approaches. It was mentioned above that the assumption of an infinite population could be justified for a (rare) mutant gene during the initial generations and that by the time the assumption becomes a poor approximation the fate of the gene is probably already settled. For smaller populations such an assumption is not justified even in the initial generations, so changes must be made in the model to take population size into account.

Further Generalizations

A finite diploid population of size N can be approximated by a haploid population of size $M = 2N$. The offspring distribution is usually considered to be binomial or the Poisson approximation. Fisher (1958) uses a diffusion equation to represent such a model and then uses generating functions to obtain a representation of the rate of fixation of segregating alleles. He finds that the $u.s.p.$ (which corresponds to the probability of fixation) for a new mutation with a small selective advantage σ (corresponding to a mean number of offspring $= 1 + \sigma$) is

$$
u.s.p. = \frac{2\sigma}{1 - e^{-2\sigma M}}. \tag{13}
$$

The denominator in Eq. (13) is always less than unity and increases the $u.s.p.$ over the value in an infinite population. For very small σ the $u.s.p.$ approaches a value $1/M$ which is the exact value for the $u.s.p.$ of a neutral mutation in a finite population. Kimura (1962) has extended this type of continuous model to include several modes of gene action. A simulation study for very small diploid populations was carried out by Schaffer (1964), and the results were compared with the theory of Kimura (1962). The correspondence was very close, but one difficulty was noted. Kimura (1962) often interprets the population size parameter M in Eq. (13) to be the variance effective population size rather than the census size of the population. The variance effective population size is the size of a population with binomial sampling of gametes which would give the same (binomial) sampling variance of the gene frequency (Kimura and Crow, 1963). For arbitrarily specified population models, with general offspring distributions, it may be very difficult to analytically determine the effective size, and it may be necessary to resort to simulation methods.

Moran (1960, 1962) has obtained a more exact result corresponding to Eq. (13) for any number of initial mutant genes. If there are k mutant genes initially

$$u.s.p. = \frac{1 - e^{-2\theta k}}{1 - e^{-2\theta k M}} \qquad (14)$$

where θ is

$$\sigma(1 + \sigma)^{-1} \leqq \theta \leqq \sigma .$$

For $\theta = s$, a small positive number and $k = 1$, Eq. (14) is equal to Eq. (13) using the same approximation used in the derivation of Eq. (13), $\exp(-2\sigma) = 1 - 2\sigma$. Pollak (1964) investigates the analogous case for a population subdivided among several niches.

The formula for the $u.s.p.$ given in Eq. (13) and (14) has been further generalized by Moran. Moran (1961) allows the distribution of the number of offspring to depart from the Poisson approximation to the binomial distribution. As a generalization, to allow the variance to be greater than the mean, the negative binomial distribution is chosen. Taking the variance of the offspring distribution into account, an approximation to the $u.s.p.$ is obtained which corresponds to the generalization of Eq. (3). If the offspring distribution has a variance v, the adjustment to be used in Eq. (14) is to substitute s/v for s. With this substitution it can be seen the limit of Eq. (13) or (14), as the population size increases, is Eq. (3).

Moran (1959) also generalizes Eq. (13) and (14) for a model in which there are overlapping generations, and each individual has a life time with a negative exponential distribution. The offspring distribution in this case is again represented by the Poisson approximation to the binomial. The mutant has a small selective advantage which is introduced in this case by increasing the expected lifetime and therefore the expected number of offspring. The $u.s.p.$ derived for this model is the same as that found in Eq. (13) and (14), with the exception that the 2σ in the numerator of each formula is replaced by σ.

The use of branching processes to investigate a finite population model would appear inconvenient because the population size (number of individuals) varies from generation to generation. However, it is possible to define a branching process conditioned on the presence of the same number of individuals in each generation. This was done by Moran and Watterson (1959) and extensively utilized by Karlin and McGregor (1964, 1965, 1968). This results in what Karlin and McGregor call a direct product branching process, which can be formulated as follows.

If there are two types of individuals, A and a, which reproduce according to offspring distributions with generating functions $f(s)$ and $g(t)$, respectively, then the joint generating function for the generation after starting with i type A and j type a is:

$$f^i(s)\, g^j(t)\,. \tag{15}$$

It is from Eq. (15) that the name direct product branching process comes. The probability that there will be k type A and $N-k$ type a (for a total population size of N) is the coefficient of $s^k t^{N-k}$ in the expansion of (15) (compare with Eq. (1) and (11)). The probability that the number of A plus the number of a will sum to a population size of N can be obtained by adding all the coefficients of those terms for which the exponents of s and t add to N. This can be represented as the coefficient of s^N in

$$f^i(s)\, g^j(s)\,. \tag{16}$$

It can be seen that Eq. (16) represents the generating function of the sum of the number of A and a individuals, which is the population size. Using the usual formula for a conditional probability, we obtain

$$\text{Probability} \left\{ \begin{array}{l} k \text{ type } A \text{ offspring} \\ N-k \text{ type } a \text{ offspring} \end{array} \right. \left| \begin{array}{l} i \text{ type } A \text{ parents} \\ j \text{ type } a \text{ parents} \\ \text{with exactly } N \text{ offspring} \end{array} \right\}$$

$$= \frac{\text{coefficient of } s^k t^{N-k} \text{ in } f^i(s)\, g^j(t)}{\text{coefficient of } s^N \text{ in } f^i(s) g^j(s)}\,. \tag{17}$$

The probabilities in Eq. (17) are the elements of the transition matrix for a Markov chain which is referred to as the one induced by the direct product branching process. Once a branching process is reduced to such a Markov chain the very powerful and extensive theory of the behavior of Markov chains can be used. Watterson (1961) and Khazanie and McKean (1966) have investigated such properties as the time to fixation, but this has not been done in the context of branching processes.

Karlin and McGregor (1965) use this approach to formulate a one locus, two allele model of a finite population with the incorporation in the model of selection, mutation, migration, and the drift due to finite population size. They derive the induced Markov chain and show for the case of Poisson offspring distributions in Eq. (17) that the transition probabilities are the binomial probabilities corresponding to those of Wright (1931) and used by him and others in the development of much of the theory of population genetics. Karlin (1968) also gives the transition probabilities, Eq. (17), for the cases where $f(s)$ and $g(s)$ generate either the negative binomial or the binomial distribution.

Karlin and McGregor (1968) define a finite population model with two linked loci and two alleles per locus and with no mutation, migration, or selection. They define how the transition probabilities for the Markov chain can be generated from a branching process with a general offspring distribution, but the results they present for the fixation of these alleles primarily cover the case where the offspring distribution is binomial. They present exact results for the probabilities of fixation of the four types of chromosomes and for the probability of each allele becoming fixed first, as a function of the initial chromosomal composition of the population. They also use the theory of Markov chains to determine the rate of approach to fixation, the rate of loss of each allele, and the change of other characteristics of the population such as the degree of linkage disequilibrium.

Summary

A fundamental assumption in the theory of branching processes is that the particles being studied, genes or chromosomes in our context, act independently in their survival and reproduction. Two genes or chromosomes which are carried by the same individual organism do not possess this independence, and so the application of branching process theory is restricted to those genetic formulations in which an individual carries only one of the particles, or in which this is true to a good approximation.

The early applications of branching process theory satisfied the independence requirements by studying the survival of mutations occuring in very large or infinite populations. The particles considered were mutant genes which, in an infinite population, would (almost certainly) never come together in the same individual. The characteristic of this survival process which has been emphasized has been the ultimate survival probability of the mutation, which is the probability that its descendants will never be lost from the population. In some cases the rate of loss or extinction has also been determined. Generalizations towards more realistic models include treatments of more general offspring distributions, recurrent mutation, and population expansion and contraction. The effects of discrete stages in the life cycle and of cyclical or random environmental fluctuations are incorporated by compounding offspring distributions with different parameters or of different forms to represent successive stages or generations. The reproduction of chromosomes with more than one locus, or of one locus in several different niches are modeled with the use of multitype branching processes. All of these treatments are in the framework of infinite populations, and so satisfy the independence assumption.

Finite populations can be modeled using multitype branching processes in which the process is conditioned on a given finite population size. This conditional process constitutes a Markov chain in which the states correspond to the numbers of the different particle types, with the total number of particles being the population size. This process then can be studied using the theory of Markov chains.

Acknowledgments

The support of Grant GM-11546 of the National Institutes of Health is acknowledged. I would like to thank Drs. J. Bishir and K. Kojima for many years of advice and encouragement.

References

Bartlett, M. S.: An introduction to stochastic processes with special reference to methods and applications. 2nd ed. London: Cambridge Univ. Press 1966.

Bellman, R., and T. E. Harris: On age-dependent binary branching processes. Ann. Math. 55, 280 – 295 (1952).

Ewens, W. J.: The probability of fixation of a mutant: The two-locus case. Michigan State University, Department of Statistics Mimeo RM-157 WJE-1. 1966.

Feller, W.: An introduction to probability theory and its applications. Volume I. 2nd ed. New York: John Wiley & Sons 1958.

Fisher, R. A.: On the dominance ratio. Proc. Roy. Soc. Edinburgh 42, 321 – 341 (1922).

– The distribution of gene ratios for rare mutations. Proc. Roy. Soc. Edinburgh 50, 204 – 219 (1930).

– The genetical theory of natural selection. Rev. 2nd ed. New York: Dover Publ. Inc. 1958.

Haldane, J. B. S.: A mathematical theory of natural and artificial selection, Part V: Selection and mutation. Proc. Cambridge Phil. Soc. 23, 838 – 844 (1927).

– The equilibrium between mutation and random extinction. Ann. Eugen. (Lond.) 9, 400 – 405 (1939).

– Some statistical problems arising in Genetics. J. Roy. Stat. Soc. Ser. B 11, 1 – 14 (1949).

Harris, T. E.: The theory of branching processes. Berlin-Heidelberg-New York: Springer 1963; Englewood Cliffs, N. J.: Prentice-Hall, Inc.

Karlin, S.: Rates of approach to homozygosity for finite stochastic models with variable population size. Amer. Natur. **102**, 443 – 455 (1968).

—, and J. McGregor: Direct product branching processes and related Markov chains. Proc. Natl. Acad. Sci. U.S. **51**, 598 – 602 (1964).

— — Direct product branching processes and related induced Markov chains. I. Calculations of rates of approach to homozygosity. Proceedings of an International Research Seminar Statistical Laboratory. Berkeley: University of California 1963. In: Bernoulli, Bayes, Laplace anniversary volume. Berlin-Heidelberg-New York: Springer 1965.

— — Rates and probabilities of fixation for two locus random mating finite populations without selection. Genetics **58**, 141 – 159 (1968).

Khazanie, R. G., and H. E. McKean: A Mendelian Markov process with binomial transition probabilities. Biometrika **53**, 37 – 48 (1966).

Kimura, M.: Stochastic processes and distribution of gene frequencies under natural selection. Cold Spr. Harb. Symp. Quant. Biol. **20**, 33 – 53 (1955).

— On the probability of the fixation of mutant genes in a population. Genetics **47**, 713 – 719 (1962)

—, and J. F. Crow: The measurement of effective population number. Evolution **17**, 279 – 288 (1963).

Kojima, K., and T. M. Kelleher: Survival of mutant genes. Amer. Natur. **96**, 329 – 346 (1962).

—, and H. E. Schaffer: Accumulation of epistatic gene complexes. Evolution **18**, 127 – 129 (1964).

— — Survival process of linked mutant genes. Evolution **21**, 518 – 531 (1967).

Lotka, A. J.: Population analysis — The extinction of families. — I. J. Wash. Acad. Sci. **21**, 377 – 380 (1931a).

— Population analysis — The extinction of families. II. J. Wash. Acad. Sci. **21**, 453 – 459 (1931b).

Moran, P. A. P.: The survival of a mutant gene under selection. Aust. Math. Soc. J. **1**, 121 – 126 (1959).

— The survival of a mutant gene under selection. Aust. Math. Soc. J. **1**, 485 – 491 (1960).

— The survival of a mutant under general conditions. Proc. Cambridge Phil. Soc. **57**, 304 – 314 (1961).

— The statistical processes of evolutionary theory. Oxford: Clarendon Press 1962.

—, and G. A. Watterson: The genetic effects of family structure in natural populations. Aust. J. Biol. Sci. **12**, 1 – 15 (1959).

Nei, M., K. Kojima, and H. E. Schaffer: Frequency changes of new inversions in populations under mutation-selection equilibria. Genetics **57**, 741 – 750 (1967).

Ohta, T.: A theoretical study of stochastic survival of inversion chromosomes. Ph. D. Thesis, Department of Genetics, North Carolina State University at Raleigh (1966).

—, and K. Kojima: Survival probabilities of new inversions in large populations. Biometrics **24**, 501 – 516 (1968).

Pollak, E.: Stochastic theory of gene frequencies in subdivided population. Ph. D. Thesis, Columbia University (1964).

— Some effects of fluctuating offspring distributions on the survival of a gene. Biometrika **53**, 391 – 396 (1966).

Pollard, J. H.: On the use of the direct matrix product in analyzing certain stochastic population models. Biometrika 53, 397 – 415 (1966).

— The multi-type Galton-Watson process in a genetical context. Biometrics 24, 147 – 158 (1968).

Schaffer, H. E.: The survival of mutant genes in small populations. Ph. D. Thesis, Department of Genetics, North Carolina State University at Raleigh (1964).

Skellam, J. G.: The probability of gene-differences in relation to selection, mutation, and random extinction. Proc. Cambridge Phil. Soc. 45, 364 – 367 (1949).

Watson, H. W.: Probable extinction of families. J. Anth. Inst. 4, 138 – 144 (1875).

Watterson, G. A.: Markov chains with absorbing states: A genetic example. Ann. Math. Statist 32, 716 – 729 (1961).

Wilkinson, W. E.: Branching processes in stochastic environments. Ph. D. Thesis, Department of Statistics, University of North Carolina, Chapel Hill, North Carolina (1967).

Wright, S.: Evolution in Mendelian populations. Genetics 16, 97 – 159 (1931).

The Incomplete Binomial Distribution

C. C. Li

Introduction

Incomplete binomial distributions frequently arise in a "natural" way, largely due to our inadequate means of observation so that certain class or classes of the distribution are either unobservable or cannot be distinguished from some other phenomenon which is not a part of the system under consideration. For instance, if we wish to study the sterility rate in a certain variety of beans, we would count the number of abortive beans as well as the good ones in a random collection of bean pods. The estimate of the rate of sterility derived from these direct counts would, however, be a gross underestimate of the true sterility rate, because the pods with no fertilized beans had failed to develop into full sized pods and had dropped into the field long before our counting was made. That is, the class without any fertilized beans has not been observed, and we have only counted those with at least one good bean in a pod. Thus, our problem is how to estimate the rate of sterility correctly with one class missing from the distribution (which is assumed to be binomial).

An example in diagnostic tests (Mantel, 1951) may also be mentioned to illustrate how an incomplete distribution may arise. Suppose that a diagnostic test yields no false positive result, so that one positive finding is sufficient to indicate definitely that the person under examination is infected by a pathogen, although several other tests on the same person may have yielded negative results. Such negative results (failures to detect the presence of the pathogen) are attributed to the low efficacy of the diagnostic test. By applying this test repeatedly to a number of people we may estimate the efficacy (the probability of yielding a positive result) of the test. Of course, an uninfected person can never yield a positive result, no matter how efficacious a test may be. Hence, a set of tests which are all negative does not tell us whether the person is uninfected in the first place, or infected but the tests failed to detect the pathogen. So, our estimate of the efficacy must be based on the repeated tests on infected persons as indicated by having yielded at least one positive result. Those with no positive results are indistinguishable from the uninfected persons, for whom the test is not intended and cannot be evaluated.

In genetics, and especially in human genetics, a similar situation arises in studying segregation of recessives from heterozygous parents. When the gene A is dominant over gene a, the phenotype of the heterozygote Aa is indistinguishable from that of the homozygote AA. Thus, the three types of mating, $AA \times AA$, $AA \times Aa$, $Aa \times Aa$, are also indistinguishable phenotypically. In studying the segregation of recessives, we are interested only in the mating between two heterozygotes ($Aa \times Aa$) and wish to know what proportion of the offspring will be of the recessive genotype aa and if this proportion differs significantly from the classical Mendelian expectation of 25 %. However, we can identify both parents as heterozygotes only when they have at least one recessive offspring, for the other two types of mating are incapable of producing recessives. Consequently, families of the type $Aa \times Aa$ without any recessive offspring are not directly observable. This class of families will be missing from our data on segregation records.

The bias introduced by the incompleteness of our observation is more serious for small families than for large ones. For instance, if the probability of having a recessive child from two heterozygous parents is $p = 0.25$, then the probability of not having any recessive offspring for families of three children is $(1 - p)^3 = q^3 = (0.75)^3 = 0.42$ approximately. That is, 42 % of the $Aa \times Aa$ families of size 3 has not been included in estimating the proportion of recessives. For families of eleven children, the corresponding probability is $q^{11} = (0.75)^{11} = 0.042$. Now, only about 4 % of the heterozygous families are missing. Therefore the methods of estimating p to be described in the following text are especially important for small families or small number of diagnostic tests on each person. Also, the smaller the value of p, the more serious the bias incurred by the absence of families without any recessive offspring.

Undoubtedly many other similar situations of incomplete observation in various fields may be found; and the problem is a general one: how to arrive at a correct estimate of the parameter p of a binomial distribution when an extreme class is missing. However, instead of using a general and abstract language, the author, being a geneticist, will use genetic terminology and genetic examples. Readers in other fields should have no difficulty in translating the genetic situation into a corresponding one in their own field by simply changing a few technical terms.

As an illustrative example in human genetics, we may consider $Aa \times Aa$ families of four children ($s = 4$). The probability of having a dominant child is $q = 0.75$ and the probability of having a recessive child is $p = 0.25$. Then the families with r recessives per sibship will be distributed according to the expansion of $(q + p)^s = (0.75 + 0.25)^4$; viz.,

No. recessives,	r:	(0)	1	2	3	4	Total, incomplete	Total, complete
per family								
No. families	f:	(81)	108	54	12	1	$175 = n_4$	256
No. recessives	fr:	(0)	108	108	36	4	$256 = r_4$	256

In the above, the number of families, f, has been given in proportional whole numbers for convenience. Throughout this chapter we assume that although the first term of the binomial distribution is missing, *the remaining terms are unaffected*. That is, families of four children with $r = 1, 2, 3, 4$ will be observed in the proportional frequencies $108:54:12:1$. And this is what we mean by an incomplete (or truncated) binomial distribution. This would indeed be the case if all families with any recessive children at all in a community have been ascertained. Hence our assumption is equivalent to "complete ascertainment". If, on the other hand, the ascertainment is incomplete so that families with $r = 4$ recessives are more likely (to an unknown extent) to be recorded than families with $r = 1$ or 2 recessives, then the observed distribution would no longer be proportional to $108:54:12:1$. We shall not deal with such complications in this chapter. The assumption of complete ascertainment is nearly justified for studies of human segregation for certain diseases, especially those in some well defined and small communities. Due to the many practical hazards in sampling human families with a certain disease, the modern trend of study is towards complete ascertainment. In experimental situations such as those concerning the diagnostic tests, the problem of incomplete ascertainment does not arise.

Some of the methods to be described are well known to human geneticists, while others, being comparatively new, may not be so. The organization of this chapter as shown below is essentially chronological.

I. Testing a particular hypothesis.
II. Maximum likelihood estimate.

III. Subdivision of information by first appearance time.
↓
IV. Pooled ratio estimates: successive estimates.

V. Discarding the recessive singletons.
↓
VI. Binomial and Poisson; truncation at any point.

Topics I and II may be found in textbooks of human genetics and are most familiar to the reader. They will be presented in outline form. Topics III and IV on the left and topics V and VI on the right in the diagram above may be read independently of each other, so that the busy reader may proceed to study only the section that interests him.

I. Testing a Particular Hypothesis

In collecting human data on segregation of recessives, Weinberg (1912) noted that the proportion based on direct counts of dominant and recessive offspring is always greater than the expected 25% and realized that this is due to the exclusion of heterozygous families that happen not to have any recessive offspring. Shortly afterwards, Apert (1914) noted the same effect. Their correction method involves simply the restoration of the true total offspring, assuming that the segregation ratio (actually proportion) is in fact 25%. Now, the missing fraction of families of size s is $(0.75)^s$, and

$$1 - (0.75)^s : 1 = t_s : c_s$$

where t_s is the observed (incomplete) total number of children from families of size s, and c_s the theoretical total number of children if our observation were complete (i.e. if the missing families were included). Hence

$$t_s = c_s\{1 - (0.75)^s\} \quad \text{or} \quad c_s = \frac{t_s}{1 - (0.75)^s}.$$

Let r_s (to be distinguished from plain r, the number of recessives per sibship) be the total number of recessives from such families of size s. Then the corrected proportion of recessives is

$$p' = \frac{r_s}{c_s} = \frac{r_s}{\left(\dfrac{t_s}{1 - (0.75)^s}\right)}. \tag{1}$$

Briefly, the method is to replace the observed number of children t_s by the enlarged number c_s. If the particular hypothesis $p = 0.25$ is true, then the value of p' obtained above should not differ from 0.25 significantly. Or, what amounts to the same thing, $(1/4) c_s$ should not differ from r_s significantly. When dealing with families of various sizes, the correction is to be done for families of each size separately and the overall p' value is obtained by pooling the numerators ($\Sigma r_s = R$) and denominators ($\Sigma c_s = C$).

The arithmetic procedure is illustrated in Table 1. For the particular example of red and green colorblindness, perhaps one word of explanation is needed. This form of colorblindness is due to a simple sex-linked recessive gene, and we are studying families of the type $Aa \times AY$; that is, mother is heterozygous but normal, and father (with Y-chromosome) is normal. The probability of having a male child is $1/2$, so that the probability of having a colorblind son is $1/4$. Thus, the size of sibship, s, refers to the number of children: sons *and* daughters in a family. Table 1 shows that the directly observed (crude) proportion of recessives is $\dfrac{330}{1020} = 32\%$ which is much biased upward, but the corrected proportion is

$$p' = \frac{\Sigma r_s}{\Sigma c_s} = \frac{R}{C} = \frac{330}{1319.68} = 0.25$$

almost exactly, indicating the usefulness of such a correction.

Table 1. *The inheritance of red and green colorblindness in man (families of $s = 1$ being omitted from Hogben, 1931)*

Size of sibship s	No. of sibships n_s	No. of color-blind sons r_s	Total no. children $t_s = s n_s$	Restored total no. children $c_s = \dfrac{t_s}{1 - (0.75)^s}$
2	21	26	42	96.00
3	29	38	87	150.49
4	34	50	136	198.95
5	35	59	175	229.60
6	18	24	108	131.40
7	12	28	84	96.96
8	12	29	96	106.68
9	14	32	126	136.22
10	9	20	90	95.31
11	1	4	11	11.49
12	1	3	12	12.40
13	3	13	39	39.93
14	1	4	14	14.25
Total	190	330	1020	1319.68

It may be noted from Table 1 that only families with two or more ($s \geq 2$) children are included, for families with one child (who is necessarily a recessive) provide no information on segregation ratio. This is obvious from the correction procedure. Let $t_1 = r_1$ be the number of such single recessives from the $n_1 = t_1$ families of size one. Then the corrected total number of offspring is always $c_1 = t_1/(1 - 0.75) = 4t_1$, so that the proportion of recessives is always 1/4, or in general, always the same as that originally assumed in the hypothesis. In all subsequent sections, families of size one are also excluded from consideration, because they would cancel out in the procedure of estimation anyway, even if they were included in the beginning.

Authors using this so called *a priori* correction method are Weinberg (1912), Apert (1914), Just (1920), Bernstein (1929), Lenz (1929), Hogben (1931), and possibly others. We may call the twenty years, 1912–1931, the Period I in handling incomplete binomials. Hogben has also derived the variance of $p' = R/C$. It is a special case of a general formula to be given in the following section.

II. Maximum Likelihood Estimate

Period II begins in 1932 when Haldane applied the method of maximum likelihood to estimate the value of p, instead of testing the hypothesis that p is equal to a certain fixed value. Unlike the genetic problem, in evaluating the efficacy of a diagnostic test, there is usually no unique *a priori* value to be assigned to p; its value is to be estimated from the data entirely. Testing a particular hypothesis and the estimation of a parameter are two different, though closely related, problems. Of course, after the estimation has been made, we can always test the significance of the difference between the estimated value and a fixed value of particular interest. Hence, obtaining an estimate is of more general applicability.

Again, let us for the moment concentrate on families of fixed size s. The method of maximum likelihood estimation leads to an equation similar to (1) except that the fixed value 0.75 be replaced by $q = 1 - p$. Thus, the maximum likelihood estimate $\hat{p}$ is the solution of the equation

$$p = \frac{r_s}{c_s} = \frac{r_s}{\left(\dfrac{t_s}{1-q^s}\right)}. \tag{2}$$

The correction (boosting t_s to c_s) depends upon the value of p itself, so that the unknown p appears on both sides of the equation which is of degree $s - 1$. This is the crucial feature that makes the equation difficult to solve.

The equation above may be written in the slightly more convenient form

$$\frac{r_s}{p} = \frac{t_s}{1-q^s}$$

for families of fixed size s. Now, families of different sizes are independent samples. The method of maximum likelihood yields the simple result that the overall estimate from families of various sizes is the solution of the summation of the individual equations: *viz.*,

$$\frac{r_2 + r_3 + r_4 + \cdots}{p} = \frac{t_2}{1-q^2} + \frac{t_3}{1-q^3} + \frac{t_4}{1-q^4} + \cdots$$

or

$$\frac{R}{p} = \sum_s \left(\frac{t_s}{1-q^s}\right) \tag{3}$$

where $R = \Sigma r_s$ is the grand total number of recessives from all the families. This is the well known equation of Haldane (1932, 1938).

Returning to Eq. (2), we see that enlarging the number of children (denominator t_s) is the same as diminishing the number of recessives (numerator r_s). Hence the estimation equation may also be written

$$p' = \frac{r_s(1-q^s)}{t_s}$$

for families of size s. The equation for pooled estimate from families of all sizes becomes

$$p' = \frac{\sum_s r_s(1-q^s)}{\sum_s t_s} = \frac{\sum_s r_s(1-q^s)}{T} \tag{4}$$

where $T = \Sigma t_s$ is the grand total number of children from all families of all sizes. This equation is equivalent to Eq. (3) which is more familiar to geneticists. One procedure is to enlarge each t_s and leave r_s alone; the other is to diminish each r_s and leave t_s alone. The arithmetic labor involved is the same for both procedures. In the following paragraphs we use the form (3), as certain values related to it have already been tabulated by various authors.

With the advent and general availability of high speed computers, a sufficiently good solution of Eq. (3) may be obtained by iteration in real short order. Even without the help of computers, a good solution may also be obtained when certain values are tabulated. Remembering that $n_s s = t_s$, we may rewrite Eq. (3) as follows:

$$R = p \sum_s \left(\frac{n_s s}{1-q^s} \right) = \sum_s \left(n_s \cdot \frac{sp}{1-q^s} \right) = \sum_s n_s \bar{r}_{(s)}$$

where

$$\bar{r}_{(s)} = \frac{sp}{1-q^s} \tag{5}$$

is the average number of recessives per sibship for the incomplete binomial. The value of $\bar{r}_{(s)}$ has been tabulated by Li (1961, p. 66). Then all one has to do is to find the value of p that yields $\Sigma n_s \bar{r}_{(s)} = R$. In practice, we find two adjacent values of p from the $\bar{r}_{(s)}$ table: one yields a total number of recessives larger than the observed R and one yields a total recessive number smaller than R. A simple interpolation yields an intermediate value which may be taken as the approximate maximum likelihood estimate $\hat{p}$.

A numerical example of the interpolation procedure is given in Table 2, data being taken from McKusick et al. (1964) concerning the segregation of a recessive type of dwarfism in the Amish. In the original report, Sibship #23 was given as of size 3, but the first member was a stillbirth of unknown phenotype. In a previous analysis, Li (1965) takes it as a normal; and in a later analysis, Li and Mantel (1968) omit the first birth and take the sibship as of size 2. It makes very little numerical difference either way. In the present example we omit the first birth in order to compare with the results to be obtained by some other methods. The detailed data on the 27 sibships will be given later (Table 8). For the time being we need only the summary as shown in the left half of

Table 2. *Maximum likelihood estimate and its variance obtained by simple interpolation (based on tables of $\bar{r}_{(s)}$ and w_s by Li, 1961, p. 66)*

Size of sibship	No. of sibships	No. of recessives	$n_s\bar{r}_{(s)}$		$n_s w_s$	
s	n_s	r_s	$p = 0.225$	$p = 0.200$	$p = 0.225$	$p = 0.200$
2	2	2	2.25	2.22	7.28	7.72
3	3	6	3.79	3.69	23.32	24.56
4	6	7	8.45	8.13	74.20	77.80
6	4	6	6.89	6.50	91.04	94.96
7	2	6	3.79	3.54	56.92	59.30
8	4	7	8.28	7.69	137.66	143.42
9	2	7	4.50	4.16	81.16	84.62
10	1	2	2.44	2.24	46.90	48.96
11	1	2	2.63	2.41	53.32	55.77
12	1	2	2.83	2.58	59.81	62.71
14	1	1	3.24	2.93	72.87	76.79
Total	$N = 27$	$R = 48$	49.09	46.09	704.48	736.61
Interpolated values			$\hat{p} = 0.216$		$\hat{W} = 716.05$	

Table 2. Then the $\bar{r}_{(s)}$ values at two different values of p are found, and then these values are multiplied by the corresponding observed n_s. Thus, $p = 0.225$ is slightly too high. On the other hand, $p = 0.200$ yields a total of 46.09 recessives which is too low. By linear interpolation we obtain $\hat{p} = 0.216$ approximately.

To find the variance of the estimate, let us again first concentrate on the n_s sibships of the same size s. Straightforward application of the method of maximum likelihood leads to the following "weight" or amount of "information" (reciprocal of variance) from the n_s sibships:

$$W_s = n_s w_s$$

where (lower case)

$$w_s = \frac{s}{pq} \times \frac{1 - q^s - spq^{s-1}}{(1 - q^s)^2} \tag{6}$$

is the information contributed by one sibship of size s and may be tabulated for various values of s and p. Note that the first factor, s/pq, is the weight for a complete binomial. Hence the second factor of Eq. (6) may be regarded as the correction for truncation. The information derived from an incomplete binomial is of course less than that from a complete one. Hence the correction factor must be less than unity; that is, $(1 - q^s)^2 > 1 - q^s - spq^{s-1}$. This inequality, upon simplification, reduces to $sp > q(1 - q^s)$ which further simplifies to

$$s > q(1 + q + q^2 + \cdots + q^{s-1}).$$

The last inequality is obviously true, as q is smaller than unity.

The combined weight from all sibships of all sizes is the sum

$$\hat{W} = \Sigma W_s = \Sigma n_s w_s \qquad (7)$$

where w_s, the weight per sibship, has been tabulated by various authors (e.g. Li, 1961, p. 66). The method of finding the approximate weight by interpolation is shown in the right columns of Table 2. At $p = 0.225$, the total weight would be 704.48; and at $p = 0.200$, the total weight would be 736.61. Then the total weight corresponding to $\hat{p} = 0.216$ may be obtained by interpolation and it turns out to be $\hat{W} = 716.05$. The variance of the estimate is $V(\hat{p}) = \dfrac{1}{\hat{W}} = \dfrac{1}{716}$ and the standard error is s.e. $(\hat{p}) = \dfrac{1}{\sqrt{716}} = 0.0374$. Our final conclusion regarding the estimate is

$$\hat{p} = 0.216 \pm 0.037$$

and it does not deviate significantly from the Mendelian expectation of 0.250. If we set $p = 0.25$ and calculate $\hat{W}_{0.25} = \Sigma n_s w_s$, we obtain the variance of the particular estimate given at the end of Section I,

$$V\left(p' = \frac{R}{C}\right) = \frac{1}{\hat{W}_{0.25}}.$$

The two-stage interpolation method described above (first obtain $\hat{p}$) follows the suggestion of Lejeune (1958) for the ease of understanding. The two tables provided by Lejeune, however, are not the values of $\bar{r}_{(s)}$ in Eq. (5) and w_s in Eq. (6), but their corresponding values without the factor s, so that, in using his table, we have to multiply the entry by number of children $t_s = s n_s$, instead of simply by number of sibships n_s.

Lejeune (1958) was apparently unaware of an earlier work on tabulation and interpolation; Finney (1949) has tabulated the values of w_s in Eq. (6) and another value which he calls bias, $B_s = s^2 p q^{s-2}/(1-q^s)^2$. Starting with an initial trial value of p, we calculate the

$$\text{numerator} = \frac{R}{pq} - \Sigma n_s B_s \qquad (8)$$

and the

$$\text{denominator} = \Sigma n_s w_s = \text{our Eq. (7)}.$$

For the data shown in Table 2 and starting with $p = 0.200$, we obtain $\dfrac{R}{pq} = \dfrac{48}{0.16} = 300$ and, using Finney's table, $\Sigma n_s B_s = 140.74$, so that the numerator is $300 - 140.74 = 159.26$. The denominator is 736.61, as shown at the bottom of the last column of Table 2. Hence the improved estimate is

$$p = \frac{159.26}{736.61} = 0.216$$

in agreement with the one obtained previously (Table 2). The value of $\hat{W}$ for $\hat{p} = 0.216$ is the same as before.

Period II is dominated by the maximum likelihood estimate initiated by Haldane (1932, 1938). In addition to the references already cited, there is a wealth of literature dealing with incomplete ascertainment which was first introduced by Fisher (1934). Elaborations on the subject may be found in Bailey (1951), Smith (1956, 1959), Morton (1959), and very likely some others in fields other than genetical applications. This period (roughly 1932 to early 1960's) is characterized by concentration on the maximum likelihood method of estimating the overall proportion of recessives, without any attempt of looking into some other properties of Mendelian segregation.

III. Subdivision of Information by First Appearance Time

Now we shall describe some recent developments (Period III) in studying the truncated binomial. Unlike peas and flies, human births occur ordinarily in a one-by-one sequence. If a phenotype is entirely due to segregation of a recessive genotype and no other reasons, it should occur in random order within a sibship of any size. The knowledge of the order of human birth enables us to check on this point. To illustrate the method of studying the "appearance time" of the recessives, we may use sibships of size $s = 4$ with $p = 0.25$ as we did in the previous section. The upper portion of Table 3 shows the conventional classification of sibships according to r, the number of recessives in a sibship, but also showing the detailed information on the appearance time of each birth. The sequence of four children is to be read from left to right. Thus we see, for instance, that among the 54 sibships with $r = 2$, the six possible positions for the two recessives are equally likely, there being 9 of each sequence. As far as the estimation of p is concerned, we usually ignore the pattern of the sequences and deal only with the totals 108, 54, 12, 1.

The lower portion of Table 3 shows a reclassification of the same 175 sibships according to the first appearance time (t) of the recessive. If the first child is a recessive, we say $t = 1$, regardless of the phenotypes of the remaining children. Similarly, if the second child is the first recessive in a sibship, we say $t = 2$, regardless of the phenotypes of the subsequent children. When the sibships are so classified, the number of families are 64, 48, 36, 27, for $t = 1, 2, 3, 4$, respectively. As will be shown later (Eq. (14)), there is no loss of information in the new system of classification. It is merely a rearrangement of the same data.

One advantage of classifying the sibships according to first appearance time is that we may now subdivide the data into two parts: one up to first appearance (the "before"-part) and one after first appearance (the

Table 3. *Two systems of classification of 175 sibships of size 4.* ● = *recessive,* ○ = *normal,* f = *number of sibships*

I. Classification by number (r) of recessives in a sibship

f	r = 1	f	r = 2	f	r = 3	f	r = 4
27	●○○○	9	●●○○	3	●●●○	1	●●●●
27	○●○○	9	●○●○	3	●●○●		
27	○○●○	9	●○○●	3	●○●●		
27	○○○●	9	○●●○	3	○●●●		
		9	○●○●				
		9	○○●●				
108		54		12		1	

II. Classification by the first appearance time (t) of a recessive

f	t = 1	f	t = 2	f	t = 3	f	t = 4
27	●○○○	27	○●○○	27	○○●○	27	○○○●
9	●●○○	9	○●●○	9	○○●●		
9	●○●○	9	○●○●				
9	●○○●	3	○●●●				
3	●●●○						
3	●●○●						
3	●○●●						
1	●●●●						
64		48		36		27	

"after"-part). For instance, consider the 48 sibships with $t = 2$. The two parts are:

No.	t = 2	remainders
27	○ ●	○ ○
9	○ ●	● ○
9	○ ●	○ ●
3	○ ●	● ●

Note that the remainders have a complete binomial distribution of degree 2, as the number of families are $27 : (9 + 9) : 3 = 9 : 6 : 1$ for $r = 0, 1, 2$, respectively. It follows that an estimate of p may be obtained from the remainders (the "after"-part) by direct counting of the children and recessives. In this particular case, we obtain $p = \dfrac{8}{32} = 0.25$ correctly.

In general, the complete binomial distribution after the first appearance time t is of degree $s - t$.

We see that the first recessive in the families serves as a boundary line dividing the whole body of data into two parts. From each part an independent estimate of p may be made. This is somewhat analogous to the analysis of variance for a one-way classification. From the entire body of data, one overall estimate of the population variance may be made. If the data are classified into groups (according to some criterion), two independent estimates of the population variance may be obtained: one based on the variation within the groups and another based on the variation between the groups. In our problem, we shall obtain an estimate p_t based on the "variation between t" and another estimate p_0 based on the "variation within t", so to speak. These two estimates of p are uncorrelated.

The estimate (p_0) from the "remainders" (children after the first recessive) for any body of data is simply given by

$$p_0 = \frac{\text{total no. of recessives after 1st recessive}}{\text{total no. of children after 1st recessive}} = \frac{R_0}{T_0}$$

where the subscript 0 denotes "after the first appearance of a recessive". For example, for the data shown in Table 3 (lower portion) we obtain by counting

$$T_0 = 64(3) + 48(2) + 36(1) + 27(0) = 324 ,$$
$$R_0 = 48 + 24 + 9 + 0 = 81 ,$$

so that $p_0 = \dfrac{81}{324} = 0.25$ correctly. The variance of the estimate is simply the binomial variance

$$V_0 = V(p_0) = \frac{pq}{T_0} , \qquad W_0 = \frac{1}{V_0} = \frac{T_0}{pq} . \tag{9}$$

The estimate p_0 and its weight W_0, as shown above, are also the results of maximum likelihood method for the portion of data under consideration.

The portion of data in Table 3 up to the first appearance of a recessive may be summarized as follows:

Appearance time t:	1	2	3	4	total
No. of sibships f:	64	48	36	27	175

from which another estimate p_t is to be obtained. Let $f_1, f_2, ..., f_s$ be the observed number of sibships of size s with $t = 1, 2, ..., s$, respectively. The corresponding probabilities are $p : qp : q^2 p : ... : q^{s-1} p$, constituting a truncated geometric distribution. The probability of observing such a sample is

$$P = \text{Const.} \left(\frac{p}{1-q^s} \right)^{f_1} \cdot \left(\frac{qp}{1-q^s} \right)^{f_2} \cdots \left(\frac{q^{s-1}p}{1-q^s} \right)^{f_s} .$$

Write
$$n_s = n = f_1 + f_2 + f_3 + \cdots + f_s,$$
$$m_s = m = f_2 + 2f_3 + \cdots + (s-1)f_s$$

where n_s, as usual, is the total number of sibships of size s; and m_s is the total number of dominants preceding the first recessive child. The logarithmic likelihood is

$$L = n \log p + m \log q - n \log(1 - q^s),$$
$$\frac{dL}{dp} = \frac{n}{p} - \frac{m}{q} - \frac{nsq^{s-1}}{1 - q^s}.$$

The equation $\dfrac{dL}{dp} = 0$ reduces to (Li, 1966)

$$\frac{m}{n} = \frac{q}{p} - \frac{sq^s}{1 - q^s}$$
$$= \frac{q}{p}\left(1 - \left(\frac{sp}{1-q^s}\right)q^{s-1}\right) \tag{10}$$
$$= \frac{q}{p}\left(1 - \bar{r}_{(s)}q^{s-1}\right)$$

where $\bar{r}_{(s)} = \dfrac{sp}{(1 - q^s)}$ as given in (5). This is the desired estimation equation. Note that m/n is the average number of dominants preceding the first recessive (the average "waiting time"). When s is indefinitely large, $m/n = q/p$ which is the standard result for a geometric distribution without truncation. Hence $\dfrac{sq^s}{(1 - q^s)}$ is the length by which the average waiting time is diminished by truncation at time $t = s$ of the geometric series. In our numerical example, $n = 175$, and $m = 48 + 2(36) + 3(27) = 201$. Substituting in Eq. (10), it may be readily verified that $p_t = 0.25$ is the solution.

The variance of the estimate p_t may also be found by the usual procedure of maximum likelihood method. Thus,

$$W_t = \frac{-d^2 L}{dp^2} = \frac{n}{p^2} + \frac{m}{q^2} - nsq^{s-2}\left[\frac{s - 1 + q^s}{(1 - q^s)^2}\right].$$

Eliminating m by Eq. (10) and simplifying, we obtain

$$W_t = \frac{n}{p^2 q}\left[\frac{(1 - q^s)^2 - s^2 p^2 q^{s-1}}{(1 - q^s)^2}\right]$$
$$= \frac{n}{p^2 q}\left[1 - \left(\frac{sp}{1-q^s}\right)^2 q^{s-1}\right] \tag{11}$$
$$= \frac{n}{p^2 q}\left[1 - \bar{r}_{(s)}^2 q^{s-1}\right]$$

and $V(p_t) = V_t = 1/W_t$. Note that for a geometric distribution without truncation, $V_t = p^2 q/n$. Therefore the factor in brackets represents the percentage loss of information due to truncation.

Combining the information of the two estimates, we obtain the pooled estimate

$$\bar{p} = \frac{W_0 p_0 + W_t p_t}{W_0 + W_t} \tag{12}$$

with variance

$$V(\bar{p}) = \frac{1}{W_0 + W_t}. \tag{13}$$

Now we assert that the pooled estimate Eq. (12) has the same variance as the maximum likelihood estimate Eq. (3) or (4) by demonstrating (Table 4)

$$\hat{W} = W_0 + W_t. \tag{14}$$

Table 4. *Analysis of Information from* $n_s = n$ *sibships of size s*

Source	Estimate	Information (weight)	Reference
After 1st appearance (within given t's)	p_0	$W_0 = \dfrac{T_0}{pq}$	(9)
Before 1st appearance (between the t's)	p_t	$W_t = \dfrac{n}{p^2 q}\{1 - \bar{r}_{(s)}^2 q^{s-1}\}$	(11)
Overall (ignoring the t's)	$\hat{p}$	$\hat{W} = \dfrac{ns}{pq}\left\{\dfrac{1 - q^s - spq^{s-1}}{(1-q^s)^2}\right\}$	(6)

In other words, the two estimates p_0 and p_t have used up all the information provided by the data and there is no loss of efficiency by subdividing the data by t into two non-overlapping portions. The demonstration of Eq. (14) may be facilitated by noting that the total number of children after the first recessive is, for fixed sibship size s,

$$\begin{aligned}
T_0 &= f_1(s-1) + f_2(s-2) + f_3(s-3) + \cdots + f_s(s-s) \\
&= n(s-1) - f_2 - 2f_3 - \cdots (s-1)f_s \\
&= n(s-1) - m \\
&= n(s-1) - n\left(\frac{q}{p} - \frac{sq^s}{1-q^s}\right).
\end{aligned}$$

Substituting this value in the expression for W_0 and simplifying, we see that $W_0 + W_t = \hat{W}$ (Li, 1966).

Practical use of the estimate p_t from families of all sizes would involve the tabulation of the values derived from Eq. (10) and (11) (without the factor n) by s and p. Although these tables may be readily provided by a computer, we shall not use them here as the procedure is no simpler than the overall method of maximum likelihood estimation. Instead, we shall describe a very simple ratio method at little loss of efficiency in the next section. However, in closing, we stress the advantage of paying attention to the appearance time of the recessives. To exaggerate the point, let us suppose that the r recessives of a sibship are always the first r children of that family. Then the distribution of the 175 sibships of size 4 in Table 3 would be as follows:

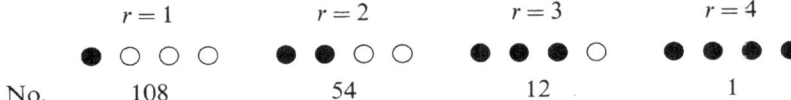

	$r=1$	$r=2$	$r=3$	$r=4$
No.	108	54	12	1

Maximum likelihood estimation Eq. (2), based on the number of recessive yields $\hat{p} = 0.25$ as before; but since $t = 1$ for all families, we have $p_t = 1$, and counting after the first recessive yields $p_0 = \dfrac{81}{525} = 0.154$. The phenomenon certainly cannot be explained by random Mendelian segregation alone. On the other extreme, if the r recessives of a sibship are always the last r children of that family, the distribution would be as follows:

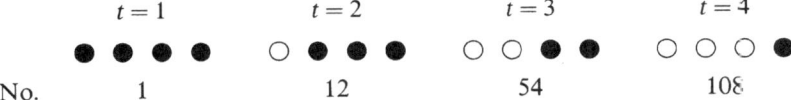

	$t=1$	$t=2$	$t=3$	$t=4$
No.	1	12	54	108

This makes no difference to Eq. (2) which yields $\hat{p} = 0.25$ as before. Then $p_0 = 1$ and there is no meaningful estimate for p_t because $1:12:54:108$ are grossly incompatible with the model $p:qp:q^2p:q^3p$. Unfortunately most published data do not show appearance time. Some unpublished data (personal communication) on human "segregation" have indicated serious discrepancy between p_t and p_0 and deserve further attention.

IV. Pooled Ratio Estimates

The simple and somewhat less efficient estimate to be introduced in this section is to take the place of p_t based on the before-first-appearance portion of data and does not affect p_0 which is based on the after-first appearance portion of data. This is one of the reasons that the overall efficiency of the finally combined estimate remains very high. To illustrate, let us put $p = 0.25$, $s = 4$, $n = 175$ in Eq. (6), (9), (11) for weights (see summary in Table 4). $T_0 = 324$. Thus,

$$
\begin{aligned}
p_0 : W_0 &= 1728.0 \quad 82.64\% \\
p_t : W_t &= \ \ 362.9 \quad 17.36\% \\
\hline
\hat{p} : \hat{W} &= 2090.9 \quad 100.00\%
\end{aligned}
$$

It is seen that more than 82% of the total information has already been used by the estimate p_0. Now suppose that we substitute a simple estimate p_x which is only 75% as efficient as the maximum likelihood estimate p_t. Its weight would be $W_x = 0.75 \times 362.9 = 272.2$. When this inefficient estimate is combined with p_0, the total information would be $272.2 + 1728.0 = 2000.2$ and the overall efficiency would be $\dfrac{2000.2}{2090.9} = 95.66\%$ which is good enough for most practical purposes.

Adopting the notations of the previous section, we see that the number of sibships of given size s with first appearance time $t = 1, 2, \ldots, s$, are

$$
f_1 : f_2 : \ldots : f_s = p : qp : \ldots : q^{s-1}p
$$

where $\Sigma f_t = n_s$. Each of the following ratios of two successive f's yields an estimate of q:

$$
\frac{f_2}{f_1}, \quad \frac{f_3}{f_2}, \quad \ldots, \quad \frac{f_s}{f_{s-1}}.
$$

Hence

$$
q_1 = \frac{f_2 + \cdots + f_s}{f_1 + \cdots + f_{s-1}} = \frac{n_s - f_1}{n_s - f_s}; \quad p_1 = \frac{f_1 - f_s}{n_s - f_s} \tag{15}
$$

provides an estimate of p from the t-distribution of the n_s sibships.

A pooled estimate from families of all sizes $(s = 2, 3, \ldots)$ may be taken as

$$
q_1 = \frac{\Sigma n_s - \Sigma f_1}{\Sigma n_s - \Sigma f_s} = \frac{N_1 - A_1}{N_1 - B_1}; \quad p_1 = \frac{A_1 - B_1}{N_1 - B_1} \tag{16}
$$

where the summation is taken with respect to s. Thus, $N_1 = \Sigma n_s$ is the grand total number of sibships of all sizes, $A_1 = \Sigma f_1$ is the grand total number of sibships of all sizes with $t = 1$, and $B_1 = \Sigma f_s$ is the grand total number of sibships of all sizes with $t = s$, the last child being the first recessive of the family. (The meaning of the subscript 1 attached to N, A, B, and p will be made clear in Table 5.)

The sampling variance of the pooled ratio estimate may be found from standard result. In general, the variance of a ratio $q' = x/y$ is

$$
V(q') = q^2 \left(\frac{V(x)}{E^2(x)} + \frac{V(y)}{E^2(y)} - \frac{2C(x, y)}{E(x)\,E(y)} \right).
$$

In our present case, Eq. (16), $V(x) = \dfrac{A(N-A)}{N}$, $V(y) = \dfrac{B(N-B)}{N}$, and covariance $C(x, y) = \dfrac{-AB}{N}$. Substituting, we obtain (Li, 1964, 1965)

$$V_1 = V(q_1) = V(p_1) = q^2 \left(\frac{A_1 + B_1}{(N_1 - A_1)(N_1 - B_1)} \right). \qquad (17)$$

Its reciprocal, the weight, may be written in a slightly different form:

$$W_1 = \frac{N_1 - B_1}{pq} \left(\frac{A_1 - B_1}{A_1 + B_1} \right)$$

To avoid abstractness, let us examine a numerical example at this stage, using the same 27 sibships shown in Table 2 (full data in Table 8). This time, however, the sibships are classified according to t, instead of according to r, and are shown in Table 5, the last column of which is of course identical with the n_s column of Table 2, with $N = N_1 = 27$. The s and t classification of the sibships yields a triangular array of the number of sibships. It is seen that $A_1 = $ sum of 1st column $= 9$; that is, 9 of the 27 sibships have $t = 1$. Also, $B_1 = $ sum of the outermost diagonal $= 3$; that

Table 5. *Classification of 27 sibships by size s and by first appearance time t*

s \ t	1	2	3	4	5	6	7	8	9	n_s
2	1	1								2
3	3	0	0							3
4	2	1	2	1						6
5	.	.	.	0	0					0
6	1	1	1	0	0	1				4
7	0	1	0	0	1	0	0			2
8	2	0	1	0	1	0	0	0		4
9	0	0	0	2	0	0	0	0	0	2
10	0	0	0	1	0	0	0	0	0	1
11	0	0	1	0	0	0	0	0	0	1
12	0	0	1	0	0	0	0	0	0	1
14	0	0	0	0	0	0	0	0	1	1
Column	9							$B_1 = 3$		$N_1 = 27$
and	A_1									
diagonal		3						$B_2 = 2$		$N_2 = 15$
totals		A_2								
			4					$B_3 = 1$		$N_3 = 10$
			A_3							
				3				$B_4 = 1$		$N_4 = 5$
				A_4						
Total	$A^* = \sum A_i = 19$						$B^* = \sum B_i = 7$			$N^* = 57$

is, there are three sibships, in which the last child is the first (and only) recessive member of that family. These occurred in the three families with $s = t = 2, 4, 6$. Eq. (16) yields the estimate

$$p_1 = \frac{A_1 - B_1}{N_1 - B_1} = \frac{9 - 3}{27 - 3} = \frac{6}{24} = 0.250. \qquad (18)$$

If this is the only substitute for p_t we care to use, its weight may be calculated from Eq. (17):

$$W_1 = \frac{27 - 3}{(0.25)(0.75)} \left(\frac{9 - 3}{9 + 3}\right) = 64.00$$

which may be combined with W_0. But we shall obtain a series of other similar estimates.

Inspection of the triangular array of Table 5 shows that we have used only the first or the outermost "layer" of the body of data. The subscript 1 of A_1, B_1, N_1 refers to the first layer. If we delete the outermost layer we have used, the remaining data form a new and similar triangular array to which we may apply the same argument leading to the pooled ratio estimate. Thus, the sum of the second column (omitting sibships of $s = 3$ and $t = 2$) is $A_2 = 3$, and the sum of the second diagonal is $B_2 = 2$ out of a total of $N_2 = 15$ sibships. A second estimate of p is provided by $p_2 = \dfrac{A_2 - B_2}{(N_2 - B_2)} = \dfrac{1}{13} = 0.077$. Generally,

$$p_i = \frac{A_i - B_i}{N_i - B_i}.$$

Table 5 permits four successive estimates from the four successive layers of the triangular array. It will be shown (page 356) that these estimates are uncorrelated, each being an independent estimate by itself.

A pooled estimate based on the distribution of appearance time is then

$$p_{(t)} = \frac{\Sigma A_i - \Sigma B_i}{\Sigma N_i - \Sigma B_i} = \frac{A^* - B^*}{N^* - B^*} = \frac{19 - 7}{57 - 7} = \frac{12}{50} = 0.240 \qquad (19)$$

where $i = 1, 2, 3, 4$, indicating the layer of the triangular array (Table 5). The individual values of $p_1, p_2, \ldots$ from the various layers are unnecessary, and we shall use the pooled estimate $p_{(t)}$ as a substitute for the maximum likelihood estimate p_t.

The weight associated with $p_{(t)}$ takes the same form as Eq. (17) except A_1 is to be replaced by A^*, B_1 by B^*, and N_1 by N^*. In the present example,

$$W_{(t)} = \frac{57 - 7}{(0.24)(0.76)} \left(\frac{19 - 7}{19 + 7}\right) = 126.52. \qquad (20)$$

The procedure described in the last two paragraphs is considered better than that of calculating each p_i and W_i separately as was done previously (Li, 1965).

Simple counting after the first appearance from the detailed data (Table 8) yields

$$p_0 = \frac{R_0}{T_0} = \frac{21}{96} = 0.21875\,,$$

$$W_0 = \frac{96}{(0.21875)(0.78125)} = 561.74\,.$$

These may then be combined numerically with $p_{(t)} = 0.240$ and $W_t = 126.52$ according to the procedure described previously. However, note that W_0 is based on $p_0 = 0.21875$ and $W_{(t)}$ is based on $p_{(t)} = 0.240$. Theoretically, these weights should be based on a common value of p. The method of doing this is given in the next paragraph. In the present case, however, the difference is very small.

Table 6 *Method of combining p_0 and $p_{(t)}$*

Estimate	Weight	Estimate × weight
$p_0 = \dfrac{R_0}{T_0}$	$W_0 = \dfrac{1}{pq} \cdot T_0$	$\dfrac{1}{pq} \cdot R_0$
$p_{(t)} = \dfrac{A^* - B^*}{N^* - B^*}$	$W_{(t)} = \dfrac{1}{pq} \cdot \dfrac{(N^* - B^*)(A^* - B^*)}{(A^* + B^*)}$	$\dfrac{1}{pq} \cdot \dfrac{(A^* - B^*)^2}{(A^* + B^*)}$
Total	$\sum W$	$\sum pW$

The method of combining p_0 and $p_{(t)}$ according to their respective weights is given in Table 6, from which we obtain, after cancelling the common factor $1/pq$,

$$\bar{p} = \frac{\sum pW}{\sum W} = \frac{R_0 + \dfrac{(A^* - B^*)^2}{(A^* + B^*)}}{T_0 + \dfrac{(N^* - B^*)(A^* - B^*)}{(A^* + B^*)}} = \frac{21 + 5.5385}{96 + 23.0769} = 0.22287. \quad (21)$$

Note that $\dfrac{5.5385}{23.0769} = 0.240 = p_{(t)} = \dfrac{12}{50}$, being given a smaller weight in comparison with $\dfrac{21}{96}$. Now, $\bar{p}$ is our final estimate of p based on both "before" and "after" the first appearance of a recessive.

Substituting the common value $\bar{p} = 0.22287$ in the expressions for weights,

$$W_0 = 554.27$$
$$W_{(t)} = 113.24$$
$$\overline{W_0 + W_{(t)} = 687.51} .$$

Comparing this with the maximum likelihood weight $\hat{W} = 716.05$ (Table 2), we see that the relative efficiency of the proposed method is

$$\frac{W_0 + W_{(t)}}{\hat{W}} = \frac{687.51}{716.05} = 96\%$$

which is sufficient for practical purposes. In summary, the proposed method merely involves the classification of sibships in the manner of Table 5; the remaining labor is simple arithmetic. We may note that $p_0 = 0.21875$ is very close to the maximum likelihood estimate $\hat{p} = 0.216$, but $p_{(t)} = 0.240$ is higher than $\hat{p}$. This is partially due to the fact that in Table 5 we have not utilized the sibship of $s = 14$ with $t = 9$. The inclusion of this late appearance would have lowered the estimate somewhat.

The only remaining technical point that needs clarification is the independence of the ratio estimates from the successive layers of the triangular array of sibships as assumed in the preceding part of this section. Although it may seem intuitively reasonable, an analytical demonstration is desirable. It is more convenient to work directly with q rather than with p. It is sufficient to show the independence of the ratios for sibships of any fixed size s. To illustrate, let $s = 4$ and f_t = number of sibships with first appearance time t. The ratios taken from the first and second layers involve respectively

$$q_1 = \frac{f_2 + f_3 + f_4}{f_1 + f_2 + f_3} = \frac{n - f_1}{n - f_4}, \qquad q_2 = \frac{f_3}{f_2} .$$

All we need to show is that q_1 and q_2 are uncorrelated. Note that the sum $f_2 + f_3$ is a common component of both the numerator and the denominator of q_1, and the information on their ratio has not been utilized. Using the delta method for large samples (Li, 1966),

$$dq_1 = \frac{(n - f_4)\, d(n - f_1) - (n - f_1)\, d(n - f_4)}{(n - f_4)^2}$$

$$= \frac{-(n - f_4)\, df_1 + (n - f_1)\, df_4}{(n - f_4)^2}$$

since $dn = 0$. Similarly,

$$dq_2 = \frac{f_2 df_3 - f_3 df_2}{f_2^2}.$$

The numerator of the product $dq_1 dq_2$ is, after multiplying out,

$$-nf_2 df_1 df_3 + f_2 f_4 df_1 df_3 + nf_2 df_3 df_4 - f_1 f_2 df_3 df_4 ,$$
$$+ nf_3 df_1 df_2 - f_3 f_4 df_1 df_2 - nf_3 df_2 df_4 - f_1 f_3 df_2 df_4 .$$

Taking the expectations, $E(df_i df_j) = \text{Covariance} (f_i, f_j) = \dfrac{-f_i f_j}{n}$. Substituting in the expression above, we see that the terms cancel out pairwise, so that

$$E(dq_1 dq_2) = \text{Covariance}(q_1, q_2) = 0 \tag{22}$$

implying that the two estimates are uncorrelated. This finding may be generalized to any size s. For instance, when $s = 6$, it may be shown in exactly the same way that the following three estimates are uncorrelated:

$$q_1 = \frac{f_2 + \cdots + f_6}{f_1 + \cdots + f_5}, \qquad q_2 = \frac{f_3 + f_4 + f_5}{f_2 + f_3 + f_4}, \qquad q_2 = \frac{f_4}{f_3}.$$

This completes the justification of the method of "peeling the onion" layer by layer as was done in Table 5. Further discussions of this method may be found in Smith (1968).

V. Discarding the Recessive Singletons

If we ignore the appearance time pattern and are only interested in obtaining an overall estimate of p, there exists an extraordinarily simple method with nearly full efficiency, rendering the solution of Eq. (3), (4), or (5) unnecessary. Several simple estimates are known, but they are all of low efficiency. The one to be introduced in this section, however, has almost the same variance as that of maximum likelihood estimate.

As a preliminary observation, let us turn to Table 3 once more. There are $n = 175$ sibships of size $s = 4$. The total number of children is $t = 4 \times 175 = 700$, of whom 256 are recessives. Note that there are $4 \times 27 = 108$ sibships, each with a single recessive. We shall call these the "recessive singletons". If these 108 singleton recessives are deleted from data, these 108 sibships will be of size 3 without any recessives. The other sibships (54, 12, 1, with $r = 2, 3, 4$, respectively) remain the same as before. Now, a recounting of the recessives and children among the resulting data will give directly the value of p: thus,

$$p = \frac{\text{No. recessives}}{\text{No. children}} = \frac{256 - 108}{700 - 108} = \frac{148}{592} = \frac{1}{4}.$$

This is a general and most remarkable property of the binomial distribution. The method of estimating p then merely involves the simple counting of recessives and children after discarding the recessive singletons from data.

Table 7. *Binomial distribution* $(q + p)^s$ *with and without the first term* q^s

No. recessives in a sibship r	Frequency of sibships f	No. of recessives fr	Correspondence to observation
0	q^s	0	(unobserved)
1	spq^{s-1}	spq^{s-1}	j_s singletons
2	$\binom{s}{2}p^2q^{s-2}$	$s(s-1)p^2q^{s-2}$	
3	—	—	
.			
s	p^s	sp^s	
Complete total	1.00	sp	
Incomplete total	$1 - q^s$	sp	r_s recessives
Incomplete total children		$s(1 - q^s)$	t_s children

A general demonstration of the property observed above is given in Table 7. From the bottom two lines it is seen that $\dfrac{r_s}{t_s} = \dfrac{p}{(1 - q^s)}$ is the maximum likelihood Eq. (2) for sibships of size s. Now, the estimate by the method of discarding the recessive singletons is

$$p' = \frac{r_s - j_s}{t_s - j_s} \tag{23}$$

where j_s is the number of recessive singletons in sibships of size s. This estimate in the long run tends to equal (see correspondence in Table 7)

$$\frac{sp - spq^{s-1}}{s(1 - q^s) - spq^{s-1}} = \frac{p(1 - q^{s-1})}{1 - q^{s-1}} = p.$$

This gives the theoretical basis for proposing the estimate Eq. (23).

Since the properties observed above are true for all values of s, a pooled estimate of p from sibships of all sizes is (Mantel, 1951; Li and

Mantel, 1968)

$$p'' = \frac{\Sigma r_s - \Sigma j_s}{\Sigma t_s - \Sigma j_s} = \frac{R - J}{T - J} \qquad (24)$$

where $J = \Sigma j_s$ is the grand total number of recessive singletons, and $R = \Sigma r_s$ and $T = \Sigma t_s$ have the same meaning as defined before. Gart (1968), independently, has also arrived at the estimate Eq. (24).

The full data of the 27 sibships we have been using as examples are given in Table 8. From the grand totals of that table,

$$p'' = \frac{R - J}{T - J} = \frac{48 - 14}{172 - 14} = \frac{34}{158} = 0.215 \qquad (25)$$

Table 8. *The 27 sibships with recessive dwarfism (Ellis-van Creveld syndrome), rearranged from McKusick et al., 1964*

Family No.	Ist time t	● = dwarf, ○ = normal	Size s	Recessive r	Singleton j
14	1	●○	2	1	1
23	2	○●	2	1	1
10	1	●○○	3	1	1
1	1	●○●	3	2	
16	1	●●●	3	3	
20	1	●○○○	4	1	1
15	1	●●○○	4	2	
5	2	○●○○	4	1	1
9	3	○○●○	4	1	1
21	3	○○●○	4	1	1
2	4	○○○●	4	1	1
27	1	●○○○○○	6	1	1
3	2	○●●●○○	6	3	
8	3	○○●○○○	6	1	1
6	6	○○○○○●	6	1	1
7	2	○●○○○●●	7	3	
17	5	○○○○●●●	7	3	
19	1	●○○○○○○○	8	1	1
29	1	●○●○○○○○	8	2	
30	3	○○●○○○○○	8	1	1
13	5	○○○○●●●○	8	3	
4	4	○○○●●○○●○	9	3	
24	4	○○○●○●●●○	9	4	
28	4	○○○●○○○○○●	10	2	
25	3	○○●○○○○○●○○	11	2	
12	3	○○●○○●○○○○○	12	2	
11	9	○○○○○○○○●○○○○○	14	1	1
Total			$T = 172$	$R = 48$	$J = 14$

in close agreement with the maximum likelihood estimate $\hat{p} = 0.216$ obtained by interpolation (Table 2). Application of Eq. (24) to some other data where the value of J is shown also yields a value very close to the maximum likelihood estimate (Li and Mantel, 1968; Gart, 1968).

The variance of p'' in Eq. (24) takes the usual form for a ratio estimate (see p. 352) and it involves $V(R)$, $V(J)$, and $C(R, J)$. But $V(R) = \Sigma \, V(r_s)$, etc. and $V(p'')$ is cumbersome to calculate. An alternative method is to calculate the variance of p' in Eq. (23) for sibships of a fixed size s and then combine weights of sibships of various sizes, as was done with the maximum likelihood estimation. Li and Mantel (1968) have shown that the difference in these two procedures is only a fraction of one percent under ordinary circumstances. So, in the following paragraph we proceed to find $V(p')$ for sibships of the same size s.

For convenience we shall calculate the various quantities on a per sibship basis, and the subscript s of Eq. (23) may be dropped. Also, since we consider only one sibship, $t = s$, so that Eq. (23) becomes

$$p' = \frac{r - j}{s - j} = \frac{a}{b}$$

where $a = r - j$ and $b = s - j$ for further brevity. The distribution of a and b and the various quantities needed for calculating $V(p')$ are listed in Table 9. Substituting in the standard formula for a ratio estimate we

Table 9. *Distribution, variance, and covariance of $a = r - j$ and $b = s - j$ for a sibship of size s*

Recessive, r	1	2	$\cdots$ s	Total or mean
Singleton, j	1	0	$\cdots$ 0	
Probability, P	$\dfrac{spq^{s-1}}{1-q^s}$	$\dfrac{\binom{s}{2}p^2 q^{s-2}}{1-q^s}$	$\cdots$ $\dfrac{p^s}{1-q^s}$	1.00
$a = r - j$	0	2	$\cdots$ s	$\sum aP = E(a)$
$b = s - j$	$s - 1$	s	$\cdots$ s	$\sum bP = E(b)$
ab	0	$2s$	$\cdots$ s^2	$\sum abP = E(ab)$

For brevity we write $P_1 = spq^{s-1}/(1 - q^s)$.

$$E(a) = \frac{sp}{1 - q^s} - P_1, \qquad V(a) = \frac{spq + s^2 p^2}{1 - q^s} - P_1 - E^2(a),$$

$$E(b) = s - P_1 = E(a)/p, \qquad V(b) = V(j) = P_1(1 - P_1),$$

$$C(a, b) = sE(a) - E(a)\,E(b) = E(a)\,[s - E(b)] = E(a) \cdot P_1.$$

obtain

$$
\begin{aligned}
V(p') &= p^2 \left\{ \frac{V(a)}{E^2(a)} + \frac{V(b)}{E^2(b)} - \frac{2C(a, b)}{E(a)\,E(b)} \right\} \\
&= \frac{pq}{s} \times \frac{(1-q^s)\,[1-q^s+(s-2)\,pq^{s-1}]}{(1-q^{s-1})^2}.
\end{aligned}
\tag{26}
$$

Note that the first factor, $\frac{pq}{s}$, is the variance for a complete binomial distribution. Hence, the second factor, which can readily be shown to exceed unity, may be regarded as the correction (enlargement) for truncation and discarding of the singletons. The weight w' to be attached to one sibship of size s is the reciprocal of Eq. (26) and has been extensively tabulated by Li and Mantel (1968). The total weight from all sibships of all sizes is then $W' = \Sigma\, n_s w'_s$.

The calculation of the variance of $p'' = 0.215$ as applied to the 27 sibships (Table 8) is shown in Table 10. The total weights at $p = 0.21$ and $p = 0.22$ are first obtained and then by interpolation we obtain the total weight $W' = 696.44$ at $p = 0.215$. The relative efficiency is approximately (because W' at $p = 0.215$ and $\hat{W}$ at $p = 0.216$)

$$
\frac{W'}{\hat{W}} = \frac{696.44}{716.05} = 97.3\%.
$$

Standard errors are s.e. $(\hat{p}) = 0.0374$ and s.e. $(p') = 0.0379$. The conclusion is that the method of discarding the recessive singletons is both simple

Table 10. *Calculation of total weight for 27 sibships of various sizes. Values of entry w', reciprocal of (26), are taken from tables of Li and Mantel (1968)*

s	n_s	at $p = 0.21$		at $p = 0.22$	
		w'_s	$n_s w'_s$	w'_s	$n_s w'_s$
2	2	3.76	7.52	3.68	7.36
3	3	7.90	23.70	7.74	23.22
4	6	12.41	74.46	12.18	73.08
6	4	22.55	90.20	22.19	88.76
7	2	28.14	56.28	27.71	55.42
8	4	34.04	136.16	33.53	134.12
9	2	40.21	80.42	39.61	79.22
10	1	46.61	46.61	45.90	45.90
11	1	53.20	53.20	52.37	52.37
12	1	59.93	59.93	58.95	58.95
14	1	73.68	73.68	72.33	72.33
Total	27	$W' = 702.16$		$W' = 690.73$	
Interpolation		at $p = 0.215$,		$W' = 696.44$	

and of high efficiency. Gart (1968) has shown that it is superior to all other simple estimates which have been proposed.

The estimates obtained by the method of maximum likelihood and Mantel's method of discarding recessive singletons are both biased for small samples. However, Thomas and Gart (1970) point out that the bias of Mantel's estimate is less, sometimes much less, than that of maximum likelihood's.

Table 11. *The estimates of p by Mantel's and maximum likelihood method from four sibships, each of size three, and their average value $\bar{p}$, and bias $\bar{p} - 0.2500$*

Type and frequency		No. of recessives	Mantel's estimate	Maximum likelihood estimate
X^4	531,441	4	0	0
Y^4	6,561	8	0.6667	0.6340
Z^4	1	12	1.0000	1.0000
$4\,X^3Y$	708,588	5	0.2222	0.2155
$4\,X^3Z$	78,732	6	0.3333	0.3820
$4\,X\,Y^3$	78,732	7	0.5455	0.5180
$4\,XZ^3$	108	10	0.8182	0.8292
$4\,Y^3Z$	2,916	9	0.7500	0.7362
$4\,YZ^3$	36	11	0.9167	0.9161
$6\,X^2Y^2$	354,294	6	0.4000	0.3820
$6\,X^2Z^2$	4,374	8	0.6000	0.6340
$6\,Y^2Z^2$	486	10	0.8333	0.8292
$12\,X^2YZ$	78,732	7	0.5000	0.5180
$12\,X\,Y^2Z$	26,244	8	0.6364	0.6340
$12\,X\,YZ^2$	2,916	9	0.7273	0.7362
$(37)^4 =$	1,874,161	Average, $\bar{p}$	0.2328	0.2284
		Bias	-0.0172	-0.0216

To illustrate how the exact bias may be determined, suppose that we have four sibships, each of size three. Let X, Y, Z denote the sibship containing 1, 2, 3 recessives, respectively. For convenience, let us also attach a dual meaning to each of the symbols, so that $X = 27$, $Y = 9$, $Z = 1$, the relative frequencies of the three types of sibships of size three. The four sibships we actually observe may consist of any combination of the three types of sibships. The combination and its relative frequency is given by the expansion of $(X + Y + Z)^4$. Then for each combination, the estimate of p is calculated, as shown in Table 11. The average value of such estimates minus the true value 0.2500 is the bias of the estimator. We see that both methods underestimate, but Mantel's method does so to a lesser extent than maximum likelihood.

For larger values of s and n_s, the bias may be calculated in a similar manner by a high speed computer (Thomas and Gart, 1970). Of particular interest to geneticists is the question: which case incurs a smaller b as —

four sibships of size three, or three sibships of size four? The answer is provided by the following results for true $p = 0.25$.

size of sibship	no. of sibships	no. of children	bias of estimate Mantel's	max. lik.
3	4	12	-0.0172	-0.0216
4	3	12	-0.0134	-0.0203

The following excerpted from Thomas and Gart (1970):

4	10	40	-0.00404	-0.00628
8	5	40	-0.00177	-0.00482

It is seen that with a fixed number of children, the fewer-larger sibships give less bias than the more-smaller ones.

VI. Binomial and Poisson: Truncation at Any Point

In all previous sections we have been dealing with an incomplete binomial distribution with only the first term $(r = 0)$ missing. The principle of discarding the recessive singletons, however, may be extended to incomplete binomial distributions with truncation at any point. For instance, if the first two terms $(r = 0, 1)$ are missing, the estimate would be simply

$$p' = \frac{\text{total recessives} - \text{doubletons}}{\text{total children} - \text{doubletons}}.$$

To illustrate, suppose that the sibships with singletons are removed (or missing) from the data of Table 8. Then, each of the remaining 13 sibships has at least two recessives. Direct count shows that the total number of children is $T = 97$, total number of recessives is $R = 34$, of whom $D = 12$ are doubletons in six families. Hence

$$p' = \frac{R - D}{T - D} = \frac{34 - 12}{97 - 12} = \frac{22}{85} = 0.2588$$

provides an estimate of p. This value, incidentally, is very close to the Mendelian expectation of 0.250. Its variance has not been investigated.

Table 12 gives a full numerical illustration of the general property for the binomial distribution $(0.75 + 0.25)^5$ with successive truncations. The correct estimate for each situation is given in the last column.

Mantel's (1951) original argument, based on the concept of "effective trials", is not limited to binomial distributions. In fact, the method of discarding singletons (or doubletons, etc.) is equally applicable to Poisson distributions. We shall not develop this subject any further except to point out that as s becomes indefinitely large, the denominator of the estimate no longer needs correction (Li and Mantel, 1968, p. 79). The

Table 12. *Successive truncations of a binomial distribution with* $q = 3/4$, $p = 1/4$, *and* $s = 5$, *and the corresponding estimate of* p

r	0	1	2	3	4	5	Sibships n	children t	Estimate
f	243	405	270	90	15	1	1024	5120	$p' = \dfrac{1280 - 0}{5120 - 0} = \dfrac{1}{4}$
fr	0	405	540	270	60	5		1280	
f		405	270	90	15	1	781	3905	$p' = \dfrac{1280 - 405}{3905 - 405} = \dfrac{1}{4}$
fr		405	540	270	60	5		1280	
f			270	90	15	1	376	1880	$p' = \dfrac{875 - 540}{1880 - 540} = \dfrac{1}{4}$
fr			540	270	60	5		875	
f				90	15	1	106	530	$p' = \dfrac{335 - 270}{530 - 270} = \dfrac{1}{4}$
fr				270	60	5		335	
f					15	1	16	80	$p' = \dfrac{65 - 60}{80 - 60} = \dfrac{1}{4}$
fr					60	5		65	

following numerical example of a Poisson distribution with the first term ($r = 0$) missing illustrates the procedure.

r:	1	2	3	4	5	6	Total
f:	6.00	3.00	1.00	0.25	0.05	0.01	$10.31 = N$
fr:	6.00	6.00	3.00	1.00	0.25	0.06	$16.31 = R$

The distribution is constructed on the basis $\lambda = sp = 1$. Now, our estimate yields

$$\lambda' = \frac{R - J}{N} = \frac{16.31 - 6.00}{10.31} = 1.00$$

correctly. Furthermore, if we also delete the term $r = 1$ from the table above, we would have $N = 4.31$, $R = 10.31$, and the number of doubletons $D = 6$. The estimate would be

$$\lambda' = \frac{R - D}{N} = \frac{10.31 - 6.00}{4.31} = 1.00 .$$

The situation is quite analogous to that of a binomial.

Summary

The methods of estimating the parameter p of a binomial distribution whose first term is missing have been reviewed essentially in chronological order and in terms of applications in human genetics. In Period I (1912–1931) the technique employed may be called an *a priori* correction method; that is, a particular value of p is first assumed, say $p = 0.25$ and $q = 0.75$, and the correction is then made by the factor $1 - (0.75)^s$. This is equivalent to testing the agreement of observations with the particular hypothesis $p = 0.25$. Period II (1932–1960's) begins with the application of the maximum likelihood method of estimation by Haldane in 1932, without assuming any particular value of p beforehand. The correction factor is then $1 - q^s$, leading to a complicated estimation equation. Most of the work in this period concentrates on the practical methods of obtaining an approximate solution of the estimation equation and the variance of the estimate. The methods are mostly iteration and interpolation by the aid of certain tabulated quantities.

In Period III (contemporary, since mid 1960's) there are two new developments. One is to subdivide the body of data into two parts by the first recessive member that appeared in each sibship. The portion prior to the first recessive member may be called the "before" part (the b-part) and the portion after the first recessive may be designated as the "after" part (the a-part). From each part an independent estimate of p and its variance may be obtained. It has been shown that if from the a-part the estimate is p_0 with the weight W_0 and from the b-part the estimate is p_t with W_t, then the combined weight $W_0 + W_t = \hat{W}$, the full weight for the maximum likelihood estimate. Since p_t is comparatively difficult to obtain, a simpler substitute $p_{(t)}$ based on pooled ratios has been proposed. A detailed numerical example has been worked out to illustrate the procedure.

A second recent development is the revival of a pleasingly simple estimate (Mantel, 1951) that remained unknown to geneticists for many years. The method merely involves the direct counting of children and recessives after discarding the recessive singletons from the families, there being no equation to solve. This method is superior to all other known substitute estimates in both simplicity and efficiency. Tables for calculating the variance have been provided by Li and Mantel (1968). Extensions of the method of discarding the singletons to discarding the doubletons, etc. and to Poisson distributions have also been mentioned.

References

The following listing is intended to be an aid for further reading, especially by geneticists, and is far from being exhaustive. However, further references may readily be found from the few cited below.

Section I

Apert, E.: The laws of Naudin-Mendel. J. Hered. **5**, 492–497 (1914).

Bernstein, F.: Variations und Erblichkeitsstatistics. Handbuch Vererbgswiss., p. 96. Berlin: Gebr. Bornträger 1929.

Hogben, L.: The genetic analysis of familial traits. I. Single gene substitutions. J. Genet. **25**, 97–112 (1931).

Just, G.: Der Nachweis von Mendel-Zahlen usw. Arch. mikr. Anat. **94**, 604–652 (1920).

Lenz, F.: Methoden der menschlichen Erblichkeitsforschung. Handbuch der hygienischen Untersuchungsmethoden, Bd. 3, S. 700. Jena 1929.

Weinberg, W.: Methode und Fehlerquellen der Untersuchung auf Mendelsche Zahlen beim Menschen. Arch. Rass. u. Ges. Biol. **9**, 165–174 (1912).

Section II

Bailey, N. T. J.: The estimation of the frequencies of recessives with incomplete multiple selection. Ann. Eugen. (Lond.) **16**, 215–222 (1951).

Finney, D. J.: The truncated binomial distribution. Ann. Eugen. (Lond.) **14**, 319–328 (1949).

Fisher, R. A.: The effect of methods of ascertainment upon the estimation of frequencies. Ann. Eugen. (Lond.) **6**, 13–25 (1934).

Haldane, J. B. S.: A method for investigating recessive characters in man. J. Genet. **25**, 251–255 (1932).

— The estimation of the frequencies of recessive conditions in man. Ann. Eugen. (Lond.) **8**, 255–262 (1938).

Lejeune, J.: Sur une solution »a priori« de la methode »a posteriori« de Haldane. Biometrics **14**, 513–520 (1958).

Li, C. C.: Human genetics, principles and methods, Chap. 5. New York: McGraw Hill Co., 1961.

McKusick, V. A., J. A. Egeland, R. Elderidge, and D. E. Krusen: Dwarfism in the Amish. Part I. The Ellis-van Creveld syndrome. Bull. Johns Hopkins Hosp. **115**, 306–336 (1964).

Morton, N. E.: Genetic tests under incomplete ascertainment. Amer. J. Hum. Genet. **11**, 1–16 (1959).

Smith, C. A. B.: A test for segregation ratios in family data. Ann. Hum. Genet. **20**, 257–265 (1956).

— A note on the effects of method of ascertainment on segregation ratios. Ann. Eugen. (Lond.) **23**, 311–323 (1959).

Sections III and IV

Li, C. C.: Estimate of recessive proportion by first appearance time. Ann. Hum. Genet. **28**, 177–180 (1964).

— Segregation of the Ellis-van Creveld syndrome as analyzed by the first appearance method. Amer. J. Hum. Genet. **17**, 343–351 (1965).

— A new method of studying mendelian segregation in man. Proc. Symp. on Mutation in population, pp. 155–166. Prague: Publ. House, Czech. Acad. Sci. 1966.

Smith, C. A. B.: Testing segregation ratios. In: Haldane and modern biology, pp. 99–130. Ed. by K. R. Dronamraju. Baltimore, Md.: Johns Hopkins Press 1968.

Sections V and VI

Gart, J. J.: A simple nearly efficient alternative to the simple sib method in the complete ascertainment case. Ann. Hum. Genet. **31**, 283–291 (1968).

Li, C. C., and N. Mantel: A simple method of estimating the segregation ratio under complete ascertainment. Amer. J. Hum. Genet. **20**, 61–81 (1968).

Mantel, N.: Evaluation of a class of diagnostic tests. Biometrics **7**, 240–246 (1951).

Thomas, D. G., and J. J. Gart: The small sample performance of some estimators of the truncated binomial distribution. Jour. Am. Stat. Assoc. **65** (in press), (1970).

Evolutionary Significance of Linkage and Epistasis★,★★

K. Kojima and R. C. Lewontin

Introduction

Linkage is the linear sequence of a group of non-allelic genes which forms a physically disjunct group from other linked groups of genes. The tie between closely located genes is stronger than that between genes far apart on the linear sequence. For this reason, closely linked genes are more likely to take similar evolutionary paths than those linked loosely. Epistasis is the functional interdependence or interaction of non alleles. When an effect of interaction between particular non-alleles is favorable under selection, the chance of survival of these alleles is increased in comparison to the chance for other non-alleles at those loci which do not interact favorably. Since linkage and epistasis represent physical and functional aspects of non-alleles, the joint consideration of these two factors is a necessary step for the examination of evolutionary problems of more than one locus.

The objective of this chapter is to discuss some problems which must be taken into account when linkage and epistasis in Darwinian fitness are considered at the level of populations. In order to avoid lengthy and involved arguments, all the populations to be discussed are assumed to have infinite size and to go through only random mating. Further assumed are that natural selection is effective only at the diploid level and there is no difference between individuals of different sexes with respect to selection as long as the individuals are of the same genotypes.

Linkage Disequilibrium

The association of non-alleles in a given population may be measured by the degree of deviations of gametic frequencies of these non-alleles from the expected gametic frequencies when the association is independent. For a pair of gene loci, Geiringer (1944) was probably the first to introduce an explicit formulation of such a measure.

★ Kojima acknowledges the support provided by Public Health Service Grant GM-15769 and AT-(40-1)-3681, and Lewontin that of AEC Contract AT (11-1) 1437.

★★ This paper is dedicated to Professor Th. Dobzhansky for his seventieth birthday celebration, and his long lasting leadership in experimental population genetics.

Consider non-alleles A and B at two different loci A and B, with the frequencies of p_A and p_B. When the association is independent, the gametic frequency of AB is expressed as

$$p_{AB} = p_A p_B . \tag{1}$$

A measure of lack of independence is

$$D_{ABt} = g_{ABt} - p_{ABt} \tag{2}$$

where g_{ABt} is the actual frequency of the AB gamete in a given population at time t and p_{ABt} is the product of p_{At} and p_{Bt} as in Eq. (1). The measure, D_{ABt}, is called "linkage disequilibrium". Jain and Allard (1966) thought that linkage disequilibrium was a misnomer and the term *gametic phase unbalance* might be more appropriate for D.

With random mating, i.e., random union of gametes, D in one generation can be related to D in any other generation. When there is no selection,
$$D_{ABt} = (1 - R_{AB}) D_{AB(t-1)}$$
so that
$$D_{ABt} = (1 - R_{AB})^t D_{AB(t-i)} \tag{3}$$

where R_{AB} is the recombination fraction between the A and B loci. Eq. (3) says that linkage disequilibrium after one generation is reduced by the recombination fraction times the previous D or is equal to the previous D multiplied by the portion of gametes which are non recombinants.

The parameter D may be defined in a few different ways. It is the degree of excess or deficiency of coupling phase double heterozygote (AB/ab) over repulsion double heterozygote (Ab/aB) at the time of zygote formation. According to this definition,

$$D_{AB} = g_{AB} g_{ab} - g_{Ab} g_{aB} \tag{4}$$

where subscripts denote gametic types and time subscript t is dropped. Another definition is that it is the covariance of allele distribution in gametes. This can be demonstrated as follows. Consider a 2 x 2 table for the allele association.

	A	a	
B	$g_{AB} = p_A p_B$ $+ D$	$g_{aB} = (1 - p_A) p_B$ $- D$	p_B
b	$g_{Ab} = p_A (1 - p_B)$ $- D$	$g_{ab} = (1 - p_A)(1 - p_B)$ $+ D$	p_b
	p_A	p_a	

The cell values are g_{AB}, g_{Ab}, g_{aB} and g_{ab}, and the marginal sums should be respective allele frequencies. Now assign the value of 1 to alleles A and B and that of 0 to alleles a and b. The marginal average of assigned values for locus A is p_A and that for locus B is p_B. The covariance is computed by

$$g_{AB}(1 \times 1) + g_{Ab}(1 \times 0) + g_{aB}(0 \times 1) + g_{ab}(0 \times 0) \\ - p_A p_B \tag{4'}$$

which reduces to D as defined in Eq. (2).

Pairwise linkage disequilibria are sufficient for most analyses of a multilocus system. However, a higher order linkage disequilibrium may exist and can be defined in a similar manner as in Eq. (2). For three loci A, B, and C,

$$D_{ABC} = g_{ABC} - p_A p_B p_C \tag{5}$$

where the generation subscript was dropped, and g_{ABC} is the frequency of gamete ABC and p_C is the frequency of allele C at the third locus C. D_{ABC} includes all pairwise D's and linkage disequilibrium specific to the three loci. Thus, a formal relation is

$$D_{ABCt} = p_A D_{BCt} + p_B D_{ACt} + p_C D_{ABt} + \Delta_{ABCt} \tag{6}$$

where Δ_{ABCt} is the deviation after accounting for all pairwise D's in the same generation.

The relation of Δ_{ABCt} to $\Delta_{ABC(t-1)}$ is much the same as that of D_t to $D_{(t-1)}$ for a pair of loci. The kinds of recombination involved in this relation are:

No recombination: $(1 - R_{AB})(1 - R_{BC}) = r_{00}$

Recombination in $A - B$
but no recombination in $B - C$; $R_{AB}(1 - R_{BC}) = r_{10}$

No recombination in $A - B$
but recombination in $B - C$; $(1 - R_{AB}) R_{BC} = r_{01}$

Recombination in $A - B$ and in $B - C$
or no recombination in $A - C$; $R_{AB} R_{BC} = r_{11}$.

Let g's be the actual frequencies of gametes specified by subscripts. Then,

$$g_{ABCt} = r_{00} g_{ABC(t-1)} + r_{10} p_A g_{BC(t-1)} + r_{01} p_C g_{AB(t-1)} + r_{11} p_B g_{AC(t-1)} \tag{7}$$

where t and $(t-1)$ are for the generation specifications for two successive generations. Eq. (7) holds true, since the gametic frequencies are the same as in the previous generation if there is no recombination, and with recombination the recombinants are from different uniting gametes in the previous generation and their numbers are proportional to their frequencies of occurrence. Now, the left-hand side of Eq. (7) is from Eq. (5) and (6),

$$g_{ABCt} = p_A p_B p_C + p_A D_{BCt} + p_B D_{ACt} + p_C D_{ABt} + \Delta_{ABCt}. \tag{8}$$

The right-hand side of Eq. (7) is equal to

$$r_{00}\{p_A p_B p_C + p_A D_{BC(t-1)} + p_B D_{AC(t-1)} + p_C D_{AB(t-1)}$$
$$+ \Delta_{ABC(t-1)}\} + r_{10} p_A \{p_B p_C + D_{BC(t-1)}\}$$
$$+ r_{01} p_C \{p_A p_B + D_{AB(t-1)}\} + r_{11} p_B \{p_A p_C + D_{AC(t-1)}\}$$

which reduces to

$$p_A p_B p_C + p_A(1 - R_{BC}) D_{BC(t-1)} + p_B(1 - R_{AC}) D_{AC(t-1)}$$
$$+ p_C(1 - R_{AB}) D_{AB(t-1)} + r_{00} \Delta_{ABC(t-1)} \tag{9}$$

where $R_{AC} = R_{AB} + R_{BC} - 2 R_{AB} R_{BC}$ is the recombination fraction between the A and C loci with the coincidence value of unity. Comparing Eq. (8) and (9),

$$\Delta_{ABCt} = r_{00} \Delta_{ABC(t-1)} = (1 - R_{AB})(1 - R_{BC}) \Delta_{ABC(t-1)}. \tag{10}$$

This relation can be extended for any number of random mating cycles; thus

$$\Delta_{ABCt} = r_{00}^i \Delta_{ABC(t-i)} \tag{11}$$

which corresponds to Eq. (3).

Frequency Changes of Selectively Neutral Alleles through Linkage

The gametic frequency changes in a general model of two loci, each with two alleles, were analytically formulated for the first time by Kimura (1956) using Malthusian fitness values in the continuous time scale. Various modifications and improvements have been made in this formulation (e.g. Lewontin and Kojima, 1960; Felsenstein, 1965; Kimura, 1965). For the present purpose, Lewontin and Kojima's formulation with Wrightian fitness model in the discrete time scale will be used. Consider a 4 x 4 table representing the union of four types of gametes AB, Ab, aB and ab in a random mating population (Table 1). Individual W's have two subscripts; the first 2, 1 or 0 is for AA, Aa or aa and the second 2, 1 or 0 for BB, Bb or bb, respectively. The marginal means are obtained by the corresponding W's and g's. For example,

$$\bar{W}_{AB} = g_{AB} W_{22} + g_{Ab} W_{21} + g_{aB} W_{12} + g_{ab} W_{11}.$$

Table 1

	g_{AB}	g_{Ab}	g_{aB}	g_{ab}	marginal mean
g_{AB}	W_{22}	W_{21}	W_{12}	W_{11}	$\bar{W}_{AB}$
g_{Ab}	W_{21}	W_{20}	W_{11}	W_{10}	$\bar{W}_{Ab}$
g_{aB}	W_{12}	W_{11}	W_{02}	W_{01}	$\bar{W}_{aB}$
g_{ab}	W_{11}	W_{10}	W_{01}	W_{00}	$\bar{W}_{ab}$
marginal mean	$\bar{W}_{AB}$	$\bar{W}_{Ab}$	$\bar{W}_{aB}$	$\bar{W}_{ab}$	$\bar{W}$

The population mean is

$$\bar{W} = g_{AB}\bar{W}_{AB} + g_{Ab}\bar{W}_{Ab} + g_{aB}\bar{W}_{aB} + g_{ab}\bar{W}_{ab} \,.$$

The gametic frequency changes are given by

$$
\begin{aligned}
\Delta g_{AB} &= (1/\bar{W}) \{ g_{AB}(\bar{W}_{AB} - \bar{W}) - R W_{11} D \} \,, \\
\Delta g_{Ab} &= (1/\bar{W}) \{ g_{Ab}(\bar{W}_{Ab} - \bar{W}) + R W_{11} D \} \,, \\
\Delta g_{aB} &= (1/\bar{W}) \{ g_{aB}(\bar{W}_{aB} - \bar{W}) + R W_{11} D \} \,, \\
\Delta g_{ab} &= (1/\bar{W}) \{ g_{ab}(\bar{W}_{ab} - \bar{W}) - R W_{11} D \}
\end{aligned}
\tag{12}
$$

where R and D refer to locus A and locus B. The detailed derivation of equations in Eq. (12) is found in Lewontin and Kojima (1960). The gametic frequency changes in random mating populations of more than two loci were derived by Lewontin (1964a).

Now consider the case of locus B being selectively neutral; that is, Wrightian fitness values of genotypes BB, Bb and bb are all equal to 1. Let AA, Aa and aa have the values of $1 + s$, 1, and $1 - t$, respectively. Assuming the multiplicative determination of genotypic fitness over the two loci,

$$
\begin{aligned}
W_{22} &= W_{21} = W_{20} = 1 + s \,, \\
W_{12} &= W_{11} = W_{10} = 1 \,, \\
W_{02} &= W_{01} = W_{00} = 1 - t \,.
\end{aligned}
$$

Substituting the above into the equations in Eq. (12), the change in the frequency of allele B, q, is obtained.

$$\Delta q = \Delta g_{AB} + \Delta g_{aB} = \frac{\Delta p}{p(1 - p)} D \tag{13}$$

where Δp is the change in the frequency of allele A. Since Δp divided by $p(1 - p)$ is proportional to the average effect of substituting allele a

by allele A, Eq. (13) may be expressed as

$$\Delta q = \alpha D/\bar{W} \qquad (13')$$

where α is the average effect of allele substitution. Δq in Eq. (13') is the product of the average effect at locus A adjusted for $\bar{W}$ and the covariance between the A and B loci. Eq. (13) and (13') say that q will change until D becomes zero or locus A comes to the point of equilibrium. When it is not zero, the sign of Δq depends upon the signs of α and D.

The expression of Δq in Eq. (13) or (13') does not contain R explicitly. However, the change in D over generations is a function of R as given in Eq. (3). Thus, the value of Δq over generations is a function of recombination fraction.

Using Eq. (13), Sved (1968) has shown that there will be an apparent selection at the neutral locus, whose magnitude depends upon D. Defining W'_{bb}, W'_{Bb}, and W'_{BB} as the apparent fitnesses of the three genotypes at the neutral locus. Then

$$W'_{Bb} - W'_{BB} = \frac{D^2(t-s)}{p_B^2 p_b}$$

and

$$W'_{Bb} - W'_{bb} = \frac{D^2(t-s)}{p_b^2 p_B}.$$

In case the A locus shows heterosis, s is negative and t positive so the two apparent fitness differences are positive. Thus the neutral locus will *appear* to show heterosis because of its association with the heterotic locus A. Moreover, since

$$\frac{\hat{p}_b}{\hat{p}_B} = \frac{W'_{Bb} - W'_{BB}}{W'_{Bb} - W'_{bb}} = \frac{p_b}{p_B}$$

the apparent selective values in any generation appear to predict an equilibrium gene frequency at the B locus equal to the observed frequency. This phenomenon is of special importance in interpreting the results of experiments on natural selection for arbitrary marker genes.

The preceding development clearly points out a possible mode by which a new mutation with no selective advantage or even with slight selective disadvantage may increase in frequency. A deterministic treatment of conditions for the increase of new alleles under a more general situation was given by Bodmer and Felsenstein (1967), while a stochastic treatment of joint evolution of linked pair of alleles was reported by Kojima and Schaffer (1964 and 1967). The genetic situation in this section corresponds to the case of "hitch-hiking" models by the latter authors.

Changes in Linkage Disequilibrium

As shown in Eq. (12) and (13), the degree of linkage disequilibrium affects the evolutionary course and evolutionary speed of linked genes. Consequently, the rate of change in D under selection has been one of the problems investigated in several theoretical papers in population genetics (e.g. Nei, 1963; Felsenstein, 1965).

Considering only the first order change,

$$\Delta D = \Delta(g_{AB}g_{ab} - g_{Ab}g_{aB})$$
$$= (\Delta g_{AB}) g_{ab} + (\Delta g_{ab}) g_{AB} - (\Delta g_{Ab}) g_{aB} - (\Delta g_{aB}) g_{Ab}.$$

Substituting Δg's by the equations in Eq. (12),

$$\Delta D = (1/\bar{W}) \{ g_{AB}g_{ab}(\bar{W}_{AB} + \bar{W}_{ab} - 2\bar{W}) \tag{14}$$
$$- g_{Ab}g_{aB}(\bar{W}_{Ab} + \bar{W}_{aB} - 2\bar{W}) - R W_{11} D \}.$$

Writing $g_{AB}g_{ab} + g_{Ab}g_{aB} = H$, which is the total frequency of double heterozygotes at the time of gametic union, Eq. (14) can be rewritten as

$$\Delta D = (1/\bar{W}) \{ \tfrac{1}{2} H(\bar{W}_{AB} - \bar{W}_{Ab} - \bar{W}_{aB} + \bar{W}_{ab}) \tag{15}$$
$$+ \tfrac{1}{2} D(\bar{W}_{AB} + \bar{W}_{Ab} + \bar{W}_{aB} + \bar{W}_{ab} - 4\bar{W}) - R W_{11} D \}.$$

The first difference, $(\bar{W}_{AB} - \bar{W}_{Ab} - \bar{W}_{aB} + \bar{W}_{ab})$, represents the additive $\times$ additive epistatic deviation (Kojima and Kelleher, 1961) and will be denoted by α_{AB}. The second difference is the sum of differences of individual gametic marginal means from the population mean and may be denoted as $\Delta(\bar{W}_{ij} - \bar{W})$ where the summation is applied to all the gametic types. Then,

$$\Delta D = (1/W) \{ \tfrac{1}{2} H\alpha_{AB} + \tfrac{1}{2} D \sum(W_{ij} - \bar{W}) - R W_{11} D \}. \tag{15'}$$

The first term in the bracket is independent of the value of D itself and proportional to the additive $\times$ additive interaction value. The order of magnitude and the sign of this term's contribution to ΔD are those of $\tfrac{1}{2} H\alpha_{AB}$. ΔD will not be zero at $D = 0$ as long as an additive $\times$ additive interaction exists.

The magnitude and sign of the second term in the bracket are more difficult to determine, but, in general, the magnitude is expected to be small, since a greater portion of deviations such as $(\bar{W}_{ij} - \bar{W})$ cancels out in the summation (note that $\sum g_{ij}(W_{ij} - W) = 0$). The contribution of the last term is to reduce the magnitude of D in a similar manner as in the case of Eq. (3). For the detailed discussion of the changes in D, Felsenstein's paper (1965) should be consulted.

One of the important observations made by Felsenstein was that the sign of α_{AB} and that of D will ultimately become the same. This point was proved by Felsenstein for haploid populations under selection, and indicated for diploid populations. The same indication is seen in Eq. (15′). However, that this relation for diploid holds true, can be shown from Kojima and Kelleher's equation (1961) on the change in population fitness. Their equation is

$$\Delta \bar{W} = (1/\bar{W}) \left[\sigma_A^2 + \sigma_B^2 + \sigma_{AB}^2 - 2\alpha_{AB} W_{11} RD \right] \qquad (16)$$

where $\sigma_A^2 + \sigma_B^2$ are the additive genetic variance in W's at loci A and B, and σ_{AB}^2 is a part of the additive $\times$ additive variance accountable among $\bar{W}_{AB}$, $\bar{W}_{Ab}$, $\bar{W}_{aB}$, and $\bar{W}_{ab}$. Kojima and Kelleher gave explicit expressions of these three variances. When the four equations in (12) are simultaneously equal to zero, $\Delta \bar{W}$ is also equal to zero. The examination of the expressions of σ_A^2 and σ_B^2 given shows that the additive genetic variance is also zero under this situation. Thus, at equilibrium Eq. (16) reduces to

$$\sigma_{AB}^2 = 2\alpha_{AB} W_{11} RD \qquad (17)$$

which states that α_{AB} and D must be of the same sign at the point of equilibrium. When $\sigma_A^2 + \sigma_B^2$ is sufficiently small, but there is sufficient amount of σ_{AB}^2, α_{AB} and D must also be of the same sign.

It is important to notice that expression (16) is not necessarily positive. If D should be rather large in magnitude and on the same side of zero as the additive by additive epistatic interaction, and if the additive genetic variance is not too large, fitness may actually *decrease*. This is in contradiction to the general rule for the evolution of single locus systems. The result can be understood in the following way. If there will be an excess of coupling gametes, say, at equilibrium, then an even greater excess of such gametes would result in yet a higher fitness. This greater excess cannot be achieved because of recombination. If the coupling gametes are initially above their equilibrium excess, fitness will be higher than at equilibrium and $\bar{W}$ will be reduced by recombination. Fitness at the same time will be increased proportional to the additive gametic variance. Which of these two phenomena is of greater magnitude depends upon the particular circumstances.

Another important aspect of the change in D in a slowly evolving population is seen in the case of Kimura's "quasi linkage equilibrium" (1965). He found that the change in linkage disequilibrium measured by the change in ratio $(g_{AB}g_{ab}/g_{Ab}g_{aB})$ becomes almost nil *very quickly* when linkage is loose and epistasis is relatively weak. Under this condition, he argued, the rate of change in the mean fitness of a population is equal to the additive genetic variance in fitness as stated in Fisher's fundamental theorem of natural selection (Fisher, 1930), and the stable equilibrium of gene frequencies corresponds to the local maximum of the mean fitness as pictorialized by Wright's adaptive surface. More recently, Wright (1967) made an analysis to clarify the adaptive surface concept under quasi linkage equilibrium using an optimum model of selection. Wright's calculations showed that the actual surface, where there was some deviation from the surface of random combination of alleles, differed little from that of random combination if selective differences were small and linkages were loose.

Gametic Frequency Equilibria

In 1930, R. A. Fisher described a simple epistatic model for the evolution of closer linkage between two genes governing a balanced polymorphism. Despite this work of Fisher's, population genetics assumed in its formulation that, except for special cases, the equilibrium frequency of gametic types in a population would simply be the product of the equilibrium frequencies of the genes at the separate loci. That is, it was assumed that at gene frequency equilibrium, $D = 0$. Wright in 1952 and Kimura in 1956 dealt with special cases of gene interaction in which this assumption broke down, but again it was supposed that these were very peculiar, special cases. Beginning with the papers of Lewontin and Kojima (1960) and Bodmer and Parsons (1962), however, there was a reversal of emphasis. By examining the problem from first principles, these authors showed that the assumption of linkage equilibrium at gene frequency equilibrium was generally unwarranted and that in general two alternative solution sets to the Eq. (12) had to be considered when $\Delta g_{ij} = 0$: those for which $D = 0$ and those for which $D \neq 0$.

Table 2. *General symmetric viability model*

	AA	Aa	aa
BB	$1 - \delta$	$1 - \beta$	$1 - \alpha$
Bb	$1 - \gamma$	1	$1 - \gamma$
bb	$1 - \alpha$	$1 - \beta$	$1 - \delta$

Setting the Eq. (12) to zero gives four simultaneous quadratics in the four gametic frequencies so that no completely general analytic solution is possible. In order to investigate the solutions it is necessary to make various symmetry restrictions on the fitnesses given in Table 1. All explicit literal solutions to the gametic equilibrium problem have used some variation of a symmetric viability model shown in Table 2. Of the eight possible independent fitnesses only four arbitrary values are allowed. This was further restricted by various authors. Kimura (1956) set

$\delta = (t-s)/(1+t)$, $\alpha = (t+s)/(1+t)$, $\beta = t(1+t)$ and $\gamma = 0$. Lewontin and Kojima set $\alpha = \delta$ while Bodmer and Parsons let $\beta = \gamma$. Each special case assumed that at equilibrium $g_{AB} = g_{ab}$ and $g_{Ab} = g_{aB}$ with the result that gene frequency at each locus would be $\frac{1}{2}$. For these solutions it was shown that there will be permanent linkage disequilibrium at gene frequency equilibrium provided that the recombination fraction R_{AB} is smaller than a value given by a function of α, β, δ, and γ. If R_{AB} is larger than that value, there may be stable gene frequency equilibrium with $D \neq 0$ or there may be no stable gene frequency equilibrium at all (Lewontin and Kojima, 1960; Lewontin, 1964a; and Ewens, 1968).

The most general analysis of the symmetric viability model of Table 2 has recently been completed by Karlin and Feldman (1969). Their findings can be summarized as follows.

(1) For any positive values of α, β, γ, and δ there is a value of R_{AB} sufficiently small to guarantee a locally stable equilibrium that is symmetrical, i.e. $g_{AB} = g_{ab}$ and $g_{aB} = g_{Ab}$.

(2) Under some conditions of loose linkage there may be two complementary asymmetrical locally stable equilibria in which $g_{AB} \neq g_{ab}$ although $g_{Ab} = g_{aB}$. Specifically:
Let

$$\hat{Z} = \left[\beta - \frac{\alpha}{2} + \frac{\delta R_{AB}}{2(R_{AB} - \delta)} \right] \bigg/ \left[2\beta - \frac{\alpha}{2} - \delta + \frac{\delta R_{AB}}{2(R_{AB} - \delta)} \right],$$

$$\hat{R} = \hat{Z}^2 - R_{AB}(1 - \hat{Z})^2/(R_{AB} - \delta). \tag{18}$$

Then if $\quad\quad\quad\quad R_{AB} > \delta$

and $\quad\quad\quad\quad \beta = \gamma$

and $\quad\quad\quad\quad \dfrac{\alpha}{2} - \dfrac{\delta R_{AB}}{2(R_{AB} - \delta)} > \delta > \beta \quad\quad\quad\quad (19)$

and $\quad\quad\quad\quad \hat{R} > 0$

there will be two complementary stable equilibria of the form

$$g_{AB} = \frac{\hat{Z} + \sqrt{\hat{R}}}{2}$$

$$g_{Ab} = g_{Ab} = \frac{1 - \hat{Z}}{2}. \tag{20}$$

$$g_{ab} = \frac{\hat{Z} - \sqrt{R}}{2}$$

(3) Under some conditions there may be four locally unstable asymmetric equilibria.

It was already shown by Bodmer and Felsenstein (1967) that the symmetric viability model in Table 2 gives three symmetric equilibria of the form

$$g_{AB} = g_{ab} = \tfrac{1}{4} + \hat{D}$$

and

$$g_{aB} = g_{Ab} = \tfrac{1}{4} - \hat{D}$$

where $\hat{D}$ is the solution of the cubic

$$64e D^3 - 16m D^2 - 4(e - 8R_{AB}) D + m = 0 \tag{21}$$

with

$$e = 2(\beta + \gamma) - (\alpha + \delta),$$

$$m = (\delta - \alpha).$$

At most, two of these equilibria are stable, but all three could be unstable. Thus putting together the symmetrical equilibria given by Eq. (21) with the finding by Karlin and Feldman of as many as four unstable asymmetrical equilibria, we find that up to seven interior equilibria may exist simultaneously for the two locus system, not including the trivial equilibria in which gene frequencies are fixed at zero. Of these seven, at most, two are stable. For any fixed set of fitnesses the number and stability of the equilibria depends upon the tightness of the linkage. Generally for tight linkage there will be symmetrical stable equilibria with $D \neq 0$. As linkage loosens, these disappear and may be replaced by a stable equilibrium with $D = 0$. There may, however, be a region of intermediate R_{AB} for which no equilibrium is stable. As linkage loosens, further stable unsymmetrical equilibria may appear. Except for the special situations of fitness relations given by conditions (19), the most common result will be stable equilibria with $D \neq 0$ when R_{AB} is less than a critical value while for larger values of R_{AB}, there will be a stable equilibrium of gene frequency with $D = 0$.

Asymmetrical Models

When the kind of symmetry shown in Table 2 is relaxed, there is virtually nothing analytic that can be said about the equilibria. Solutions of the Eq. (12) are not forthcoming for the general case so that recourse must be had to numerical work. It is not the purpose of this chapter to deal exhaustively with all varieties of numerical examples, but rather to give the broad outlines of the results so far obtained. A perspective on general numerical results can be obtained by reference to Eq. (15). This shows that the change in D will be zero when D itself is zero if and only if

$$\bar{W}_{AB} - \bar{W}_{Ab} - \bar{W}_{aB} + \bar{W}_{ab} = 0.$$

That is, $D = 0$ is an equilibrium point only under the condition that the additive × additive epistatic deviation is zero. One way in which that will be true is if there are no epistatic interactions so that the fitness of each genotype is an additive function of the fitnesses for each locus separately. In terms of the fitnesses in Table 1, $W_{22} - W_{21} - W_{12} + W_{11} = 0$ and so on for all four rows.

Since some deviation from additivity of fitnesses is a necessary condition for $D \neq 0$ at equilibrium, one way to investigate asymmetrical models is in terms of the strength of their epistatic deviations. In fact, for the symmetrical model with $\alpha = \delta$, the equilibrium value of D is a function of the epistatic deviation (e in Eq. (21)).

Table 3a. *Relative fitnesses for an asymmetric partially heterotic model with epistasis*

	AA	Aa	aa
BB	0.5000	0.5000	0.3750
Bb	0.5625	1.0000	0.3125
bb	0.3750	0.4375	0.3750

Table 3b. *Relative fitnesses for a quadratic deviations model. Phenotypic contributions at each locus are* $-6, 3.6, 6$ *respectively for AA, Aa and aa genotype.* $\overline{0} = 6$

	AA	Aa	aa
BB	0.8960	0.9651	0.8960
Bb	0.9651	1.0000	0.7919
bb	0.8960	0.7919	0.0301

A number of models with strong epistatic deviations have been investigated by Lewontin (1964a, b), Kojima (1965), and Singh and Lewontin (1966). These models fall into three classes. The first, illustrated in Table 3a, investigates cases where heterosis at each locus is not unconditional, but depends upon the genotype at the other locus. The second model shown in Table 3b arises from a form of natural selection known as the "quadratic deviation model" (Wright, 1935). The fitness of a genotype is given by

$$W = K - (P - \overline{0})^2 \qquad (22)$$

where $\overline{0}$ is some fixed phenotypic optimum, K is an arbitrary scaling constant, and P is the phenotype determined by adding the affects of each locus. Thus there is no epistasis on the underlying scale of phenotype, but very strong epistasis on the secondary fitness scale, arising from the non-monotonic relation between fitness and gene dose. The third model, given in Table 3c, was used for various aspects of the asymmetric epistasis

Table 3c. *Relative fitnesses which generate varying degrees of epistasis. α, β and γ are set in such a way that all fitnesses are between 0 and 1*

	AB	Ab	aB	ab
AB	$0.600 + \alpha$	$0.700 + \beta$	$0.700 + \beta$	$0.871 + \gamma$
Ab	$0.700 + \beta$	0.650	$0.871 + \gamma$	0.671
aB	$0.700 + \beta$	$0.871 + \gamma$	0.650	0.671
ab	$0.871 + \gamma$	0.671	0.671	0.571

model. As for the model in 3a, there is conditional overdominance. Both Models 3a and 3b show the same characteristic results,

1. stable equilibria of gene frequencies whose values depend upon R_{AB},
2. no stable gametic equilibria at $D = 0$ for any value of R_{AB} including $R_{AB} = 0.5$,
3. all values of R_{AB} produce stable repulsion equilibria $(D < 0)$,
4. small increases in $\bar{W}$ for tighter linkage.

The first model gave stable equilibria with $D > 0$ for tighter linkage in addition to the repulsion equilibria present. Despite the superficial resemblance of the fitness patterns in Models 3a and 3b, these coupling equilibria do not appear in the quadratic deviations models.

In the third model, Kojima attempted to answer several questions concerning the interplay of linkage and epistasis. Some of the findings are,

1. linkage in epistatic systems affects the locations of equilibrium points,
2. the tighter the linkage, the greater is the amount of linkage disequilibrium,
3. in general, there is more than one point of equilibrium in a system with a given fitness set and R, and a multiple equilibrium situation is more likely to occur with tighter linkage,
4. the range of attraction to different equilibria was investigated.

Multiplicative Models

There is one form of asymmetric model that can be partly analyzed and that has a certain intrinsic biological interest. This is a model in which the fitnesses at separate loci are multiplicative rather than additive, as shown in Table 4. For this model, Bodmer and Felsenstein (1967)

Table 4. *Fitnesses in a multiplicative model*

	AA	Aa	aa
BB	$(1-s_1)(1-s_2)$	$(1-s_1)$	$(1-s_1)(1-t_2)$
Bb	$(1-s_2)$	1	$(1-t_2)$
bb	$(1-t_1)(1-s_2)$	$(1-t_1)$	$(1-t_1)(1-t_2)$

have proved that there is always a stable gene frequency equilibrium, and that at this equilibrium $D \neq 0$ if

$$R_{AB} < \left(\frac{s_1 t_1}{s_1 + t_1} \right) \left(\frac{s_2 t_2}{s_2 + t_2} \right) \tag{23}$$

otherwise, $D = 0$.

The multiplicative model is especially enlightening because the degree of epistatic interaction is a simple function of the selection intensity. The four epistatic deviations are simply $s_1 s_2, s_1 t_2, s_2 t_1$, and $t_1 t_2$. Denoting these deviations as $\varepsilon_1, \varepsilon_2, \varepsilon_3$, and ε_4 then, Eq. (23) becomes

$$R_{AB} < \frac{\sqrt{\varepsilon_1 \varepsilon_2 \varepsilon_3 \varepsilon_4}}{\varepsilon_1 + \varepsilon_2 + \varepsilon_3 + \varepsilon_4} \tag{24}$$

which is of order $\bar{\varepsilon} \sim \bar{s}^2$. Thus, as selection grows weaker, the epistatic deviation grows weaker more rapidly, and the *relative* epistatic deviation is of order $\bar{s}$. In general, then, the weaker the selection the less important the linkage, the critical value of R_{AB} decreasing as $\bar{s}^2$.

We will see that for multiple-locus models, higher order interactions enter and cause an important modification of this conclusion.

Multiple-Locus Models

No analytic results have been obtained for multiple-locus models, even of the simplest symmetrical form. Attempts to make analytic formulations for more than two loci, such as those of Hill and Robertson (1966) and Sved (1968), are statements about two-locus interactions averaged over all pairs of loci on a chromosome. It is a trivial matter to write the analogous equations to Eq. (12) for any numbers of loci. These are given formally in Lewontin (1964a), and explicit derivations for a three locus case are found in Kojima and Klekar (1969). Explicit solutions are a different matter, however, and so far none has been found for three or more loci. As for asymmetrical cases, numerical work has been necessary, but some important generalizations have emerged. These generalizations are based on a variety of selection models, but can be understood most easily from the multiplicative model described in the previous section. In what follows, we will assume a symmetrical multiplicative model identical at all loci so that an individual homozygous at k out of n loci has fitness $(1 - s)^k$, the completely homozygous genotype having a fitness $(1 - s)^n$. The relative epistatic deviation per pair of loci is then s. Further we will assume that the recombination fraction between *adjacent* loci, R, is the same for all adjacent pairs of genes.

The result of such a model with $s = 0.5$ and $n = 5$ is shown in Table 5 from Lewontin (1964a). The results illustrate several general points.

1. The critical value of R is greater than that predicted from two locus theory. Relation (23) reduces to

$$R < \frac{s^2}{4}$$

for completely symmetrical models. In the present numerical case this value is $R = 0.0625$, yet there is significant stable linkage disequilibrium up to $R \cong 0.065$. The effect on the critical region is obviously small.

2. For any given R, there is a large increase in D at equilibrium over that expected from two locus theory. Table 5 shows all the pairwise D values (rescaled to $D' = 4D$) for the various values of R. For a symmetrical two locus multiplicative model the expected value of D' is given

Table 5. *The results of a five locus cumulative heterosis model* *(symbols are explained in the text)*

Gametes	000	0.01	0.02	0.03	0.04	0.05	0.06	0.063	0.0645	0.065
					R between Adjacent Loci					
00000	0.50000	0.46199	0.42053	0.37444	0.32183	0.25904	0.17488	0.13627	0.09817	0.03125
00001	0	0.01083	0.02193	0.03316	0.04418	0.05411	0.05997	0.05874	0.05413	0.03125
00010	0	0.00016	0.00074	0.00201	0.00438	0.00863	0.01675	0.02119	0.02567	0.03125
00011	0	0.00775	0.01572	0.02384	0.03192	0.03947	0.04495	0.04515	0.04336	0.03125
00100	0	0.00010	0.00048	0.00133	0.00299	0.00611	0.01254	0.01642	0.02087	0.03125
00101	0	0.00000	0.00003	0.00013	0.00044	0.00135	0.00458	0.00754	0.01213	0.03125
00110	0	0.00013	0.00061	0.00166	0.00363	0.00723	0.01443	0.01869	0.02344	0.03125
00111	0	0.00775	0.01572	0.02384	0.03192	0.03947	0.04497	0.04515	0.04336	0.03125
01000	0	0.00016	0.00074	0.00201	0.00438	0.00863	0.01675	0.02119	0.02567	0.03125
01001	0	0.00000	0.00003	0.00015	0.00050	0.00155	0.00524	0.00859	0.01370	0.03125
01010	0	0.00000	0.00000	0.00001	0.00006	0.00029	0.00164	0.00341	0.00700	0.03125
01011	0	0.00000	0.00003	0.00013	0.00044	0.00135	0.00458	0.00754	0.01213	0.03125
01100	0	0.00013	0.00061	0.00166	0.00363	0.00723	0.01443	0.01869	0.02344	0.03125
01101	0	0.00000	0.00003	0.00015	0.00050	0.00155	0.00524	0.00859	0.01370	0.03125
01110	0	0.00019	0.00088	0.00234	0.00504	0.00985	0.01905	0.02410	0.02912	0.03125
01111	0	0.01083	0.02193	0.03316	0.04418	0.05411	0.05997	0.05874	0.05413	0.03125
D'_{12}	1.00000	0.95476	0.90300	0.84164	0.76508	0.66172	0.49236	0.39660	0.28448	0
D'_{13}	1.00000	0.92352	0.83888	0.74296	0.63072	0.49236	0.29912	0.20836	0.11928	0
D'_{14}	1.00000	0.89284	0.77752	0.65208	0.51380	0.35836	0.17452	0.10408	0.04720	0
D'_{15}	1.00000	0.85140	0.69840	0.54184	0.38376	0.22808	0.08192	0.03984	0.01344	0
D'_{23}	1.00000	0.96744	0.92944	0.88300	0.82260	0.73604	0.58104	0.48584	0.36680	0
D'_{24}	1.00000	0.93572	0.86316	0.77876	0.67656	0.54488	0.34880	0.25104	0.15068	0
D'_{25}	1.00000	0.89284	0.77752	0.65208	0.51380	0.35836	0.17452	0.10408	0.04720	0
D'_{34}	1.00000	0.96744	0.92944	0.88300	0.82260	0.73604	0.58104	0.48584	0.36680	0
D'_{35}	1.00000	0.92352	0.83888	0.74296	0.63072	0.49236	0.29912	0.20836	0.11928	0
D'_{45}	1.00000	0.95476	0.90300	0.84164	0.76508	0.66172	0.49236	0.39660	0.28448	0
$\bar{W}$	0.49500	0.45688	0.41927	0.38203	0.34491	0.30738	0.26720	0.25240	0.24021	0.22781

by Lewontin and Kojima (1960) as

$$D' = \left[1 - \frac{4}{s^2} R \right]^{\frac{1}{2}}. \qquad (25)$$

Taking D'_{23} from Table 5, the observed value for two adjacent loci in the middle of the chromosome, we get the following comparison:

R	D'_{23}	D' from two locus theory	$\dfrac{D'_{23}}{D' \exp}$
0.04	0.823	0.600	1.37
0.05	0.736	0.447	1.64
0.06	0.581	0.200	2.91

Thus, although higher order interactions do not increase the critical value of R very much, they increase the intensity of the effect very considerably for any given R.

3. Loci well beyond the critical linkage distance from each other are held in linkage disequilibrium by the interactions of loci between them on the chromosome. For example with R between adjacent loci equal to 0.03, the outside genes are 0.12 recombination units apart, or nearly twice the critical distance given by Eq. (23). Yet $D'_{15} = 0.54184$. Thus whole chromosomes may be kept out of linkage equilibrium by cumulative effects.

A significant difference between two locus and multi-locus models is that *both $D = 0$ and $D \neq 0$ may be stable equilibria for the same recombination value.* As we previously pointed out, Karlin and Feldman

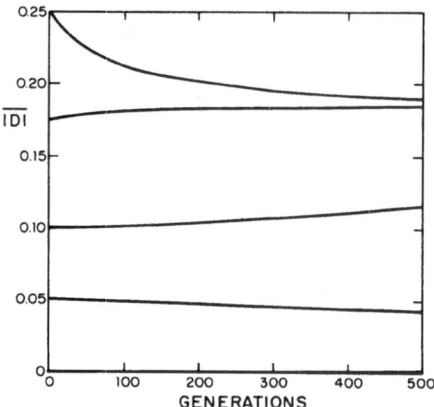

Fig. 1. Average absolute value of D (ordinate) in each generation (abscissa) of selection in a symmetrical multiplicative 5-locus model. Each line shows the history of a population beginning at a different initial condition, D_0

showed that for any given value of R_{AB} either $D=0$ or $D \neq 0$ were stable points but not both. For more than two loci this is no longer true. Fig. 1 shows the changes in $\overline{|D|}$, the average absolute value of D over all pairs of loci in a numerical solution of a five locus symmetrical multiplicative model with $s=0.1$ and $R=0.003$ between adjacent loci. According to (23) this value of R is above the critical level, so no linkage disequilibrium should be stable. As Fig. 1 shows if the initial value of $\overline{|D|}$ is greater than 0.1, $\overline{|D|}$ increases to an equilibrium value of ~ 0.1875, while if the initial value is below 0.05, it decreases to 0. We have not located the unstable point between these two attractions, but it is approximately at $\overline{|D|}=0.06$. It must be noted that $\overline{|D|}$ does not completely specify the genetic array except at $\overline{|D|}=0$ (complete randomness) and $\overline{|D|}=0.25$ (complete disequilibrium). Thus, in the multidimensional gametic space there may be points with the same value of $\overline{D}$ which lead to opposite equilibria. Since the entire 32 dimensional space would be impossible to explore, it might be that $\overline{|D|}$ and the variance of D among pairs of loci would be an appropriate space in which to explore the equilibrium problem. This remains to be investigated.

A Limiting Property

The existence of higher order interactions causing a much larger linkage disequilibrium than that predicted by two-locus theory raises an interesting possibility. Suppose we have a chromosome of fixed length and a fixed difference in fitness between a completely heterozygous genotype and a completely homozygous one. As we increase the number of loci in such a system, two things happen. The effect of substitution at each locus gets smaller and smaller, while the loci get closer and closer together. The first of these effects causes a decrease in $\overline{|D|}$ at equilibrium, while the second causes an increase. It is then conceivable that as the number of loci grows very large, the value of $\overline{|D|}$ at equilibrium will converge to a fixed value independent of the number of loci, but depending only on the total genetic map length and the total selection difference between the n-tuple homozygote and the n-triple heterozygote.

In the multiplicative model, if we fix the fitness of the n-tuple homozygote as k, then for each locus the ratio of the fitness of either homozygote to the heterozygote is

$$1 - s = k^{\frac{1}{n}} \tag{26}$$

where n is the number of loci. From Eq. (26) one can then assign the value of s for each locus for any number of loci.

Franklin and Lewontin (1969) have applied this model to several cases. Table 6 shows the values of s calculated from Eq. (26) for two of

Table 6. *The selective disadvantage s of a homozygote relative to a heterozygote at each locus when there are n loci acting multiplicatively and the n-tuple homozygote has fitness k*

	s	
n	k = 0.0225	k = 0.4832
2	0.8499	0.3049
5	0.5317	0.1354
18	0.1901	0.0396
36	0.1000	0.0200
360	0.0105	0.00213

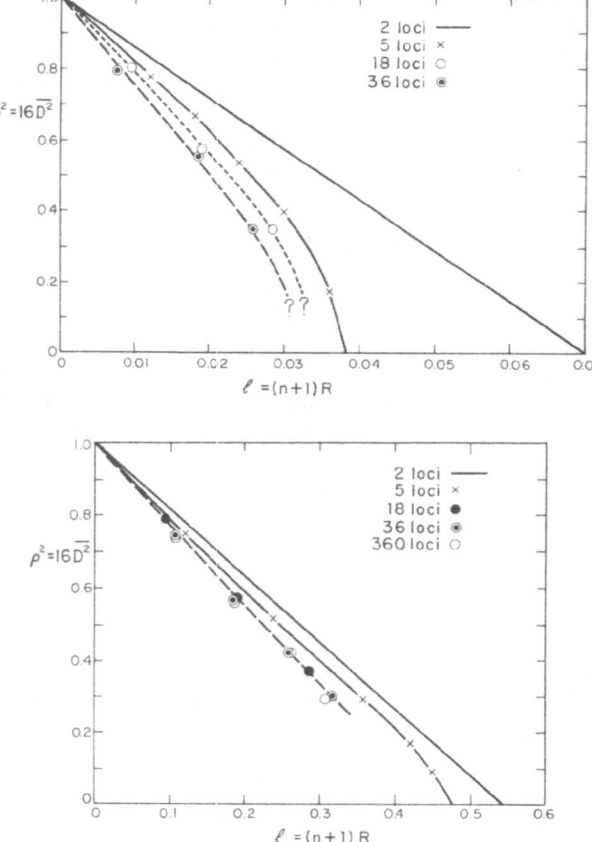

Fig. 2a and b. Relation between $\varrho^2 = 16\overline{D^2}$ at equilibrium and total map length of the chromosome for different numbers of loci. a $k = 0.0225$; b $k = 0.4832$

these. For large numbers of loci, the selection intensity per locus is very weak. The results of the calculations using these values of s are shown in Fig. 2a and 2b. On the abscissa is the total map length of the chromosome given as

$$l \doteq (n + 1) R$$

where R is small recombination function between adjacent genes. This metric is chosen because the average recombination between random loci on a chromosome of length l is $l/(n + 1)$. On the ordinate is the amount of linkage effect at gene frequency equilibrium, measured as $16\overline{D^2}$, which is the squared correlation between pairs of loci taken over all loci. We have chosen this measure of linkage effect because for the two locus case $\varrho^2 = 16D^2$ is related to l by the following linear relationship

$$\varrho^2 = 16D^2 = 1 - \frac{4}{3} l[1/(1 - \sqrt{k})]^2. \tag{27}$$

It is clearly impossible to calculate D over all pairs of loci for $n = 360$, and even for $n = 36$ this is very time consuming. For this purpose a relation devised by Sved (1968) can be utilized. If all gene frequencies are close to $\frac{1}{2}$, then the variance in the number of loci heterozygous among individuals has a relation to D as follows:

$$\bar{\varrho}^2 = 16\overline{D^2} = \frac{4 \operatorname{Var}(H) - n}{n(n - 1)} \tag{28}$$

where all terms are as previously defined.

The figures show that the suspected asymptotic property exists. For two loci the relation between $16\overline{D^2}$ at equilibrium and map length is a straight line as given by Eq. (27). For five loci the line is displaced and begins to curve downward at longer map distances. At 18 loci it is further displaced, but the line for 36 loci is indistinguishable from that for 18 loci in Fig. 2a and only slightly different from it in Fig. 2b. The results for 360 loci are absolutely indistinguishable from those for 36. Thus, somewhere between 18 and 36 loci for both models, the relation between linkage effect and total map length becomes independent of the number of loci.

We then have a limiting property of selection and linkage: *The equilibrium linkage structure of a chromosome is a function of the total map length and the slope of the inbreeding depression curve, independent of the number of loci involved.*

Of course this has been demonstrated only for the completely symmetrical multiplicative model and then only numerically. It suggests, however, that a continuous model of the chromosome might give analytic results of general validity, where the multi-dimensional discrete model is too cumbersome to handle.

Final Remarks

One of the important aspects in the evolution of multilocus systems is the rate of change in population fitness. Since the publication of Fisher's fundamental theorem of natural selection (1930), a considerable number of papers have been written on this subject (e.g. Kimura, 1958; Kojima and Kelleher, 1961; Moran, 1964; Kimura, 1965). The controversy in this connection, in most part, seems to stem from the question of what degree of departure from random combinations of alleles at different loci one is willing to accept in order to point our fundamental elements in the multilocus systems. Li (1957) and Turner (1967a) made the most recent discussions on this controversy, and Dr. Turner is expected to write on this subject in this volume.

The second remark is related to the question of "Why does the genotype not congeal?" (Turner, 1967b). As recognized by Fisher (1930) and Kimura (1956), an analysis of two locus system points out that linkage tends to be tightened under selectively favorable epistatic selection. Then, the question by Turner becomes the obvious one to be asked in view of very existence of successfully evolved organisms with many chromosomes. Upon a re-examination of the equation for population fitness increase in the three-locus system formulated by Kojima and Kelleher (1961), Turner (1967b) suggested "Selection tends on the whole to favor an optimum rather than minimum amount of recombination". This suggestion seems to be reasonable, but has not actually been proven due to the complexity existing in any multilocus system. For further details, the readers are referred to Kojima and Klekar (1969).

The final remark is concerned with the rate of changes in various parameters in multilocus systems. Even with relatively large differences among W's, gametic frequency changes become very slow as though they are in quasi stable equilibrium (Lewontin, 1964b). Kojima (1965) computed a measure of evolutionary rate by $r = (\sum \Delta g_{ij}^2)^{\frac{1}{2}}$ and has shown that the magnitude of this measure is strongly affected by the value of R. Moreover, the mode of change in r was not, in general, monotonic for epistatic models. The curve of r plotted against generation showed as many as three ups and downs before populations reached their equilibrium points. This indicates that a population may travel near many quasi stable or saddle points before it comes to rest after one hundred to several hundred generations. Thus, it seems likely that a population in actual environments may never come to the point of equilibrium due to various minor changes, including random changes, in the system of fitness values.

References

Bodmer, W. F., and P. A. Parsons: Linkage and recombination in evolution. Advan. Genet. **11**, 1–99 (1962).

—, and J. Felsenstein: Linkage and selections: theoretical analysis of the deterministic two-locus random mating model. Genetics **57**, 237–265 (1967).

Ewens, W. J.: A genetic model having complex linkage behavior. Theor. Appl. Genet. **38**, 140–144 (1968).

Felsenstein, J.: The effect of linkage on directional selection. Genetics **52**, 349–363 (1965).

Fisher, R. A.: The genetical theory of natural selection. Oxford: Clarendon Press 1930.

Franklin, I., and R. C. Lewontin: (In preparation) (1969).

Geiringer, H.: On the probability theory of linkage in Mendelian heredity. Ann. Math. Statist. **15**, 25–57 (1944).

Hill, W. G., and A. Robertson: The effects of linkage on limits to artificial selection. Genet. Res. **8**, 269–294 (1966).

Jain, S. K., and R. W. Allard: The effects of linkage, epistasis, and inbreeding on population changes under selection. Genetics **53**, 633–659 (1966).

Karlin, S., and M. W. Feldman: Linkage and selection: new equilibrium properties of the two-locus symmetric model. Proc. Natl. Acad. Sci. U.S. **62**, 70–74 (1969).

Kimura, M.: A model of a genetic system which leads to closer linkage by natural selection. Evolution **10**, 278–287 (1956).

— On the change of population fitness by natural selection. Heredity **12**, 145–167 (1958).

— Attainment of quasi linkage equilibrium when gene frequencies are changing by natural selection. Genetics **52**, 875–890 (1965).

Kojima, K.: Role of epistasis and overdominance in stability of equilibria with selection. Proc. Natl. Acad. Sci. U.S. **45**, 984–989 (1959 a).

— Stable equilibria for the optimum model. Proc. Natl. Acad. Sci. U.S. **45**, 989–993 (1959 b).

— The evolutionary dynamics of two-gene systems. In: R. W. Stacy and B. Waxman (eds.): Computers in biomedical research. New York: Academic Press 1965.

—, and T. M. Kelleher: Changes in mean fitness in a random mating population when epistasis and linkage are present. Genetics **46**, 527–540 (1961).

—, and Andrea Klekar: Deterministic simulation of evolutionary changes in three-locus genetic systems. In: Computer applications in genetics, pp. 147–159. Proc. Int. Conference on Computer Applications in Genetics. Ed. by N. E. Morton. Honolulu: University of Hawaii Press 1969.

—, and H. E. Schaffer: Accumulation of epistatic gene complexes. Evolution **18**, 127–129 (1964).

— — Survival process of linked mutant genes. Evolution **21**, 518–531 (1967).

Lewontin, R. C.: The interaction of selection and linkage. I. General considerations: heterotic models. Genetics **49**, 49–67 (1964a).

— The interaction of selection and linkage. II. Optimum models. Genetics **50**, 757–782 (1964 b).

—, and K. Kojima: The evolutionary dynamics of complex polymorphisms. Evolution **14**, 458–472 (1960).

Li, C. C.: Genetic equilibria under selection. Biometrics **23**, 397–484 (1957).

Lotka, A. J.: Elements of mathematical biology. New York: Dover Publ. 1956.

Moran, P. A. P.: On the nonexistence of adaptive topographies. Ann. Hum. Genet. **27**, 383–393 (1964).

Nei, M.: Effect of selection on the components of genetic variance. In: W. D. Hanson and H. F. Robinson (eds.) Statistical genetics and plant breeding. Natl. Acad. Sci. — Natl. Res. Council Publ. **982**, 501–515 (1963).

Singh, M., and R. C. Lewontin: Stable equilibria under optimizing selection. Proc. Natl. Acad. Sci. **56**, 1345–1348 (1966).

Sved, J.: The stability of linked systems of loci with a small population size. Genetics **59**, 543–565 (1968).

Turner, J. R. G.: Mean fitness and the equilibria in multilocus polymorphisms. Proc. Roy. Soc. B **169**, 31–58 (1967a).

— On supergenes. I. The evolution of supergenes. Amer. Natur. **101**, 195–221 (1967b).

Wright, S.: Evolution in populations in approximate equilibrium. J. Genet. **30**, 257–266 (1935).

— The genetics of quantitative variability. In: K. Mather (ed.): Quantitative inheritance, pp. 5–41. Her Majesty's Stationary Office 1952.

— Factor interaction and linkage in evolution. Proc. Roy. Soc. London B **162**, 80–104 (1965).

— "Surfaces" of selective value. Proc. Natl. Acad. Sci. U.S. **58**, 165–172 (1967).

Fitness and Optimization

R. Levins

Introduction

Fitness enters population biology as a vague heuristic notion, rich in metaphor but poor in precision. In the second edition of the origin of species, Darwin accepted Spencer's expression "the survival of the fittest" as equivalent to natural selection. It, therefore, refers to those properties of species or populations which can be invoked to account for the abundance, distribution, duration, and composition of species.

If we had precise information about the selection process for each situation, we could work directly with selection coefficients for the purpose of prediction, and only invoke fitness later as a summarizing explanation device. But the real value of the fitness concept for evolutionary biology is that it can be invoked to account for the increase of particular traits without knowledge of their genetics. Therefore, the central issue is the applicability of optimization arguments.

If we chose to remain at the level of tautological fitness, defined as that which is increased by selection, there is no need to demonstrate the legitimacy of an optimization argument, and all the difficulty of the problem is displaced onto identifying the fitness measure which is increased. On the other hand, as we make the fitness notion more precise and its measure less ambiguous, the optimization is less readily justified.

Wright and Fisher both demonstrated maximization theorems. Wright's theorem states that $\bar{W}$, the adaptive value of a population, will be maximized by selection. It is a strong statement, and correspondingly has a relatively narrow domain of applicability to situations in which there is no inbreeding, a constant environment, and genotypes whose relative fitnesses are independent of gene frequency. Wright himself (1948) showed how to modify the theorem for frequency-dependent selection. Li (1959) showed that for inbreeding a somewhat different function is modified. And Levene's (1953) theorem on genetic equilibrium in a spatially heterogeneous environment showed that the average $\log \bar{W}$ is maximized over the environments.

Fisher's fundamental theorem is a weaker statement, claiming only that fitness increases at a rate proportional to the additive genetic variance of fitness. It, therefore, allows selection to cease at nonoptimal gene frequencies if the additive genetic variance vanishes. This includes the cases of inbreeding and frequency-dependent selection.

One strategy for the extension of optimization principles to other situations is the search for the function which is optimized. In the case of inbreeding, Li invented a measure which is $\bar{W}$ for the whole population plus the fitness of the inbred population. More generally, Karlin (1967) argues that any process which converges can be shown to maximize something. The trouble is, the something which is maximized is not unique and depends on the particular process. But we want an optimization rule which can be applied without such knowledge.

Meanwhile, for ordinary situations of selection within populations in a stable environment, it is probably safe to claim that populations in nature are likely to differ in the same direction as their optima (Levins, 1968). Beyond these simple cases certain difficulties arise. The biological traits which maximize the rate of increase of a population, its size at equilibrium, the number of populations that are maintained, and the average duration of a population are not identical, so that it is not always clear what is being maximized. For instance, where the growth of a population is density-limited and obeys an equation of the type

$$\frac{dx}{dt} = r_0 x \left(1 - \frac{x}{K} \right)$$

where x is the population size, r_0 the intrinsic rate of increase, and K the carrying capacity, selection will act most strongly on r_0 if the population is subject to wide fluctuation and most strongly on K if the population is usually near the carrying capacity. And as we shall see below, the minimization of the extinction rate for local populations also introduces the variability of r and K.

Second, there is no general theory yet for second order selection.

Third, in a fluctuating environment, since the gene frequency is itself varying, there can be no optimization of the frequency itself but only of some statistic of gene frequency.

Second Order Selection

We define second order selection as selection for traits which do not appear directly in the expression for $\bar{W}$, but which do enter into the average $\bar{W}$ over time by virtue of entering into the expression for genetic change. Their effects enter the $\bar{W}$ expression indirectly, by their role in the determination of the gene frequencies. Thus, the mutation rate and recombination probabilities would be the results of second order selection. There is widespread belief that these traits are under genetic control, and that they evolve toward some optimum values. Kimura (1960) and Levins (1967) have attempted to show what determines the optimal values of mutation rates and offer some arguments indicating selection toward an optimum. However, there is as yet no general theory of second order selection.

The technical difficulties are such that only cases of lesser biological interest are amenable to analytic treatment. The interesting problems arise in the situations where the environment varies with certain autocorrelation. But there is no general solution available for these cases even with a single locus, and second order selection would require a bivariate diffusion process that usually results in incompatible partial differential equations.

Technical difficulties can perhaps be evaded by computer methods. A more serious problem arises in relation to the memory in a genetic system. In a variable environment, the optimum values for mutation rate or recombination rate depend on statistical properties of the environment which are stored in the form of gene frequencies. But we cannot invoke the law of large numbers here. The gene frequencies at any time depend on the recent past mostly, the rate of damping of effects of the remote past depending on the kind of selection and the mutation and recombination rates themselves. Only systems with very long memories can lead to even approximate convergence of gene frequencies, but such systems could only be established by genetic systems with even longer memories. Therefore, although the faith persists, the evidence is still lacking which would justify the invoking of optimization arguments for predicting the evolution of such population parameters.

Extinction and Colonization

When Thoday (1953) suggested the minimization of the extinction probability as a fitness component, he was thinking in terms of a geological time scale. However, recent work on islands (MacArthur and Wilson, 1967) demonstrates that the turnover rate of local populations is very high, that extinction and recolonization are recurrent events to which statistical parameters can be assigned. Further, the patchiness of most habitats makes the population structure of many species insular. Therefore, we will look more closely at extinction and recolonization as fitness criteria.

MacArthur and Wilson considered a model of population growth for which the birth and death rates are density independent below some limiting value K and then either the birth rate falls or the death rate increases so as to prevent increase above this value. They then calculate the expected time to extinction. The major results are the following:

1. If the birth rate is less than the death rate, a population starting with a single propagule will survive less than $1/\lambda$ in $(2K-1)$ years, where λ is the per capita reproduction per year. Even for very large K, such populations will not survive long.

2. If the birth rate exceeds the death rate, survival is very sensitive to K. The authors' figures show that with the death rate 1.82 per capita per year and the birth rate 2 per capita per year, a carrying capacity of 100 would allow survival for about 860 years while $K = 10,000$ allows for survival to 10^{41} years! Clearly then a maximization of the life expectancy for local populations depends on a large K as well as a high intrinsic rate of increase. But once K exceeds about 1,000–10,000 the extinction rate is too low to allow differences in K to be very important.

This model of MacArthur and Wilson is concerned with stochastic extinction in conditions of constant environment. We will modify the model in the following ways:

First, density dependence will be allowed to act at all densities. Therefore, the mean rate of increase will be

$$\frac{dx}{dt} = rx\left(1 - \frac{x}{K}\right).$$

Secondly, we will allow three kinds of random variation: sampling variance, fluctuations in r, and fluctuations in K. Finally, we will allow the extinction probability to be a function of the genetic makeup of the local populations.

Case 1. Random variation due to adult mortality

Then the mean is given by the equation above and the variance of change is proportional to the mean rate of increase. The methods of diffusion processes can be applied (see Kimura's monograph, 1964) to give a limit distribution

$$F(x) = \frac{C}{x\left(1 - \dfrac{x}{K}\right)} e^{\frac{r}{a}x}$$

where a is a constant of proportionality for the variance as a multiple of the mean change and c is

$$\int_0^K e^{\frac{r}{a}x} \left/ \frac{1}{x\left(1 - \dfrac{x}{K}\right)} \right. dx.$$

The distribution is U-shaped. Thus, the populations which have not disappeared are mostly near K or very small, a result which is also given by MacArthur's and Wilson's model. The exponential term increases with x, so that the proportion of the populations which are near K goes up with r/a.

Case 2. Random variation in r

The mean change is $\bar{r}x\left(1 - \dfrac{x}{K}\right)$ and the variance is $\sigma_r^2 x^2 \left(1 - \dfrac{x}{K}\right)^2$ so that the limit distribution is

$$F(x) = Cx^{\frac{\bar{r}}{\sigma^2} - 2} \left(1 - \frac{x}{K}\right)^{-\frac{2\bar{r}}{K^2} - 2}.$$

Thus there is an accumulation of populations near K. If $2\sigma_r^2 > \bar{r}$, there is also an accumulation of populations near zero, as in Case 1. Therefore "fitness" is reduced by variance of r for a fixed mean r, and the appropriate fitness measure would be $\dfrac{\bar{r}}{\sigma^2}$.

Case 3. Random variation in K

For convenience we define $\theta = 1/K$. Then the mean change is

$$E\left(\frac{dx}{dt}\right) = rx(1 - \bar{\theta}x),$$

and the variance is $r^2 x^2 \sigma_\theta^2$. The limit distribution is now

$$F = Cx^{\frac{2}{r\sigma^2} - 2} e^{-\frac{2\bar{\theta}x}{r\sigma^2}}$$

Now the distribution decreases exponentially for large values of x. At the low end, there will be an accumulation of probability near zero if $r\sigma^2 > 1$, and an appropriate fitness measure would seem to be $\dfrac{1}{r\sigma_\theta^2}$.

Thus fitness is increased by reducing the random variation in the carrying capacity, and this may be achieved by a broad tolerance for diverse kinds of habitat or food.

A direct solution of the differential equation is also possible. The substitution

$$\frac{1}{y} = x$$

gives

$$\frac{-dy}{y^2 dt} = \frac{r}{y}\left(1 - \frac{\theta}{y}\right),$$

or

$$\frac{dy}{dt} = -ry + r\theta.$$

Thus

$$y = y_0 + \int e^{-r(t-\tau)}\theta(\tau)\,d\tau$$

and

$$x = \frac{x_0}{1 + x_0 \int e^{-r(t-\tau)}\frac{1}{K(\tau)}\,d\tau}.$$

This tells us that in a fluctuating environment the effective K is a weighted harmonic mean of the K's of the recent past, that a low variance in K is an important fitness component, and that rapid damping makes the population more sensitive to fluctuations in K. The first results are consistent with the conclusions of other models while the last may be an artifact of the model.

But an extinction-migration model must also consider migration. If we assume that the number of new populations founded depends on the number of migrants sent out, then the appropriate fitness measure is the product of average duration and average population size. In fact, if p is the proportion of all potential population sites which are occupied, and if the migration and extinction rates are m and e respectively, then

$$\frac{dp}{dt} = mp(1-p) - ep$$

and at equilibrium

$$\hat{p} = 1 - \frac{e}{m}.$$

In all the models considered, m increases and e decreases with K, and m decreases while e increases with the variance of K. The role of r is not robust. However, we believe that the last model is least realistic in this regard (population decreases toward K from above at rate r). Thus our final conclusion is that in extinction-migration population selection models fitness is a monotonic increasing function of r and K, and a decreasing function of the variances of r and K. But the exact form of the expression remains unknown.

However, selection within populations of given K will act to increase F without reference to the var (r). Therefore if there are genotypes which increase r at the price of increasing its variance then selection at the intrapopulational and interpopulation levels may be antagonistic.

Unit of Selection

Selection takes place at several levels of organization. Phenomena such as differential fertilization by sperm bearing different alleles at the t-locus in mice (Dunn, 1953) can be regarded either as an anomaly of segregation or as a form of selection among gametes in which the appropriate fitness measure would be the relative probabilities of successful fertilization. At the individual level we have the familiar Mendelian selection. Where close relatives also live together, familial or nest selection can occur, and in situations where a local population strongly modifies its environment population selection may be important.

Although the details of the mathematical expression for change will differ, the process of selection can be described unambiguously at each level, a measure of fitness can be prescribed, and the consequences of selection can be determined in terms of fitness at that level. Further, if selection at various levels is in the same direction the outcome is still readily describable in fitness terms at any level. The interesting situations are those in which selection at different levels are in opposing directions.

We now consider two cases of the interaction of selection on the mendelian and population levels. In each case, mendelian selection is pushing the gene frequency x toward zero and population selection is moving x toward one. The problem is, what is the outcome of these antagonistic processes at different levels?

Consider first a series of populations such that the proportion with gene frequency x is $F(x)$. Suppose that populations with gene frequency x go extinct at the rate of $e(x)\,dt$. Then in the next time interval

$$F(x, \tau + d\tau) = \frac{F(x, \tau)\,[1 - e(x)\,d\tau]}{1 - d\tau \int F(x)\,e(x)\,dx}$$

and the change in F is

$$\frac{F(x, \tau + d\tau) - F(x, \tau)}{d\tau} = \frac{F(x, \tau) \left[\int F(x) \, e(x) \, dx - e(x) \right]}{1 - d\tau \int F(x) \, e(x) \, dx}$$

But $\int F(x) \, e(x) \, dx$ is the average extinction rate $\bar{e}$. Thus

$$\frac{dF}{dt} = -\left(e(x) - \bar{e} \right) F(x).$$

This is an unstable situation. Those gene frequencies for which the extinction rate is greater than average will decrease in abundance while those for which the rate is lower will increase. Meanwhile the average rate will decrease since

$$\frac{d\bar{e}}{d\tau} = \frac{d}{d\tau} \int e(x) \, F(x) \, dx$$

which is

$$\frac{d\bar{e}}{d\tau} = -e(x) \left[e(x) - \bar{e} \right] F(x) \, dx.$$

This gives

$$\frac{d\bar{e}}{d\tau} = -\sigma^2(e)$$

and since $\bar{e}$ is an inverse measure of fitness we have the familiar result for group selection that fitness increases at a rate proportional to its variance in the population. Further, since the extinction probability lies between zero and one, its variance is less than $\bar{e}(1 - \bar{e})$ and therefore

$$\frac{d\bar{e}}{d\tau} > -\bar{e}$$

so that

$$\bar{e}(\tau) > \bar{e}(o) \exp(-\tau).$$

From the equation for $\dfrac{dF}{dt}$ we can derive equations for all the moments of $F(x)$:

$$\frac{d\bar{x}}{dt} = \frac{d}{dt} \int x F(x) \, dx$$

or

$$\frac{d\bar{x}}{dt} = \int x \frac{dF}{dt} dx$$

and for the higher moments

$$\frac{d\mu k}{dt} = \int (x - \bar{x})^k \frac{dF}{dt} dx - k \frac{d\bar{x}}{dt} \int (x - \bar{x})^{k-1} F dx.$$

For convenience we will assume that differential extinction increases the gene frequency $\bar{x}$ and that the counter selection pressures alone would drive $\bar{x}$ to 0. Then we can ask, is $F(x)$ a stable solution to the simultaneous equations for its moments near $\bar{x} = 1$, $k = 0$ and for $\bar{x} - 0$, $k = 0$. In the former case population level selection can overwhelm the other pressures, in the latter case the counter pressures swamp the effects of differential extinction. If both are true outcome depends on the initial conditions, while if the solutions are unstable at both $\bar{x} = 0$ and $\bar{x} = 1$ a polymorphism will result.

A detailed analysis of this system appears in Levins (1969). Here we will limit the discussion to some results of the stability analysis based on the characteristic roots of the Lyapunov matrix for the equations of the moments.

Case 1. Migrant selection vs. differential extinction

The extinction rate is a decreasing function of gene frequency x, and the new populations formed by migrants have a binomial distribution with mean

$$x^* = \bar{x} - s\bar{x}(1 - \bar{x})$$

and variance

$$\bar{x}(1 - \bar{x})/n$$

for populations founded by n migrants. Here s is the migrant selection coefficient. Then

$$\frac{dF}{dt} = -e(x) F(x) + \bar{e}B(x, x^*).$$

the Lyapunov matrix near $\bar{x} = 0$ or 1 is

$$L = \begin{vmatrix} -s\bar{e}(1 - 2\bar{x}) & -e_1 & 0 & 0 & 0 \\ \dfrac{\bar{e}(1 - 2\bar{x})}{n} & -e_0 & 0 & & \\ 0 & & -e_0 & & \end{vmatrix}$$

where

$$e_1 = \frac{de}{dx} x = \bar{x}.$$

If the extinction rate is a linear function of gene frequencies,

$$e(x) = a(1 - bx),$$

we can show that for

$$b > \frac{ns}{1 + ns},$$

migrant selection swamps the effects of differential extinction. When

$$ns > b > \frac{ns}{1 + ns},$$

the gene frequency will go to 0 or 1 depending on initial conditions, and for

$$b > ns$$

group selection will prevail over migrant selection. With a non-linear $e(x)$, polymorphism may be possible.

Case 2. Intrapopulation mendelian selection vs. differential extinction

Here

$$\frac{dF}{dt} = -e(x) F(x) + \frac{d}{dx} (MF) + \bar{e}B(x, \bar{x})$$

where M is the mean change under mendelian selection. Now the matrix is

$$L = \begin{vmatrix} -m_1 & -e_1 - m_2 & 0 & 0 & 0 \\ \dfrac{\bar{e}(1-2\bar{x})}{n} & -e_0 - 2M_1 & 0 & 0 & 0 \\ 0 & 0 & -e_0 - 3M_1 & & \\ 0 & & & & \end{vmatrix}$$

where

$$M_i = \frac{dM}{dx} \quad \text{at} \quad \overline{x}.$$

Thus, the system is unstable at $\overline{x} = 1$. The only alternatives are permanent polymorphism if $-e'(o) > (n-1)S + \frac{2ns^2}{e(o)}$ or the prevalence of mendelian over population selection if the inequality is reversed.

Other particular models give different results, but in general the selection at a lower level (intrapopulation or migrant) has an advantage over selection at a higher level. A stronger differential extinction is required to offset the other kinds of selection, so that we cannot assert the general proposition of maximization of population fitness unless lower levels of selection operate in the same direction or are very weak.

Conclusions

1. The notion of fitness may be used in the cases of simple Mendelian selection to invoke an optimization principle "populations will differ in nature in the same direction as their optima".

2. In a fluctuating environment, when selection occurs at several levels and in relation to second order selection, the optimization principle is not generally justified.

3. In all cases examined, the appropriate measure of fitness depends on the ecological parameters r_0 and the variances of these parameters over environments.

References

Dunn, L. C.: Variations in the segregation ratio as causes of variations of gene frequency. Acta. Genet, et statist, Med. **4**, 139–147 (1953).

Karlin, Sam: Quoted by P.A.P. Moran in 5th Berkeley Symposium on probability and statistics (1967).

Kimura, M.: Optimum mutation rate and degree of dominance as determined by the principle of minimum genetic load. J. Genet. **59**, 21–34 (1960).

— Diffusion models in population genetics. Methven's Review Series in applied probability Vol. 2 (1964).

Levene, Howard: Genetic equilibrium when more than one niche is available. Amer. Natur. **87**, 331–333 (1953).

Levins, R.: Theory of fitness in a heterogeneous environment, V. Optimum genetic systems. Genetics **52**, 891–904 (1965).

— Theory of fitness in a heterogeneous environment, VI. The adaptive significance of mutation. Genetics **56**, 163–178 (1967).

— Evolution in changing environments. Princeton University Press 1968.

— (in press) On extinction. Amer. Math. Assoc. symposium on mathematical problems in biology.

Li, C. C.: Population genetics. University of Chicago Press (1959).

MacArthur, R., and E. O. Wilson: The theory of island biogeography. Princeton University Press 1967.

Thoday, J.: Components of fitness. S.E.B. symposium (1953).

Wright, S.: Adaptation and selection. In Genetics, Paleontology and Evolution. (Ed.) Jepsen *et al.* Princeton University Press 1947.